CSR, Sustainability, Ethics & Governance

Series editors

Samuel O. Idowu, London Metropolitan University, London, United Kingdom
René Schmidpeter, Cologne Business School, Germany

More information about this series at http://www.springer.com/series/11565

Samuel O. Idowu • Stephen Vertigans •
Adriana Schiopoiu Burlea
Editors

Corporate Social Responsibility in Times of Crisis

Practices and Cases from Europe, Africa and the World

Springer

Editors
Samuel O. Idowu
Guildhall School of Business and Law
London Metropolitan University
London, United Kingdom

Stephen Vertigans
School of Applied Social Studies
Robert Gordon University
Aberdeen, United Kingdom

Adriana Schiopoiu Burlea
Economics and Business Administration
University of Craiova
Craiova, Romania

ISSN 2196-7075 ISSN 2196-7083 (electronic)
CSR, Sustainability, Ethics & Governance
ISBN 978-3-319-52838-0 ISBN 978-3-319-52839-7 (eBook)
DOI 10.1007/978-3-319-52839-7

Library of Congress Control Number: 2017936477

Printed on acid-free paper

This Springer imprint is published by Springer Nature
The registered company is Springer International Publishing AG
The registered company address is: Gewerbestrasse 11, 6330 Cham, Switzerland

Foreword

Corporate Social Responsibility (CSR) has become a compelling business argument that is becoming universally accepted and articulates how companies are expected to operate in a world of global scrutiny. What this actually means is, perhaps, a question that anyone new to the field of CSR might wish to ask. Both individual and corporate citizens of the world are now expected to behave responsibly and ethically in all regards! This is because CSR has over approximately 50 years continued to reflect and reshape how we operate and behave, either as individuals or as corporate entities. Consumers and clients occasionally still see some lapses in the way businesses operate, for example the global financial meltdown in 2008 that shook the capitalised world to its roots, and in a perfect CSR economy, it would not have taken place. The devastating effects of the global financial meltdown are still as fresh in the minds of policymakers, politicians and wider societies as if it occurred yesterday. There are numerous lessons that have come into consumer consciousness as a result of the 2008 crisis; for example, we are now aware that we are globally economically more dependent on each other than we were before (or perhaps realised), and a mistake or corrupt individual in one part of the world may bring the rest of the world to its financial knees! What an extraordinary lesson indeed!

It's a great honour for me to be asked to write the *foreword* to the book by these three editors who explore how corporate entities cope with their CSR activities in competitive economic times. The 14 chapters of the book draw on examples from different parts of the world and the global economy and have assembled what I believe is both a scholarly and interesting read on the issue. I will take this opportunity to congratulate the editors and writers for putting the book together, knowing it positively contributes to that which is understood about CSR.

Guildhall School of Business and Law
London Metropolitan University
London, UK
Summer, 2016

Simon Jones

Preface

Corporate Social Responsibility (CSR) has become a topic of increasing importance in all areas of human existence—in business, politics, academia and the civil society in general. Issues relating to CSR are discussed, researched and propagated in all economies around the globe. There is practically nowhere in the world today where issues relating to CSR, Sustainability, Ethics and Governance are not debated, researched, encouraged, practised and perceived as being desirable. In fact, it is now a core part of corporate strategies of many companies in most economies.

CSR activities and the financial resources corporate entities devote to them, some scholars have argued, are just like an insurance policy premium on an insurable interest: when times are hard one either cancels the policy or allows the current policy to expire unrenewed. How has CSR fared during the recent global financial crisis? Have companies abandoned or cancelled planned CSR activities because times are hard? Or do companies perceive hard times as good time to even be more socially responsible by devoting more resources to CSR activities? Chapters in this book provide answers to these and many other questions people would like to have answers to in times when things are not so great for companies either on a micro or macro level.

The objective of this book is to add to knowledge by exploring and providing an overview of how corporate entities from around the world cope with financial resources challenges in terms of their CSR activities in periods when things are difficult. These issues are explored from the standpoint of corporate entities' quest to help find solutions to our social, economic and environmentalchallenges. Understanding how corporations cope with life during a period of uncertainty like the one which besieged our world in 2008 is the motivation for wanting to consolidate in a single book information on how corporate entities in different countries around the world dealt with their planned and pressing CSR activities.

It is believed that the way different dimensions of CSR, Sustainability, Ethics and Governance are propagated and practised when times are hard would be of interest to CSR scholars, practitioners and stakeholders in general worldwide. Thus, these three editors provide original contributions from a dozen countries in

14 chapters that explore different versions of CSR practices in difficult times from around the world as at this point in time in the twenty-first century.

London, UK
Aberdeen, UK
Craiova, Romania
Summer, 2016

Samuel O. Idowu
Stephen Vertigans
Adriana Schiopoiu Burlea

Acknowledgements

We are eternally grateful to a number of people who have made the publication of this book on CSR when times are hard possible. We express our thank you to all our contributors who have worked tirelessly despite their heavy professional commitments to put together their chapters and were also asked to anonymously peer review all the chapters that made their way into the book for quality assurance purposes. Thank you to everyone. We are also grateful to Dr Simon Jones, the lead editor's Dean, for writing a befitting foreword to the book.

We would like to thank our publishing team at Springer headed by the Senior Editor, Christian Rauscher, Barbara Bethke and other members of the publishing team who have supported this project and all our other projects.

We are also grateful to our respective family for bearing with us during those periods we should be with them but chose to spend time meeting our obligations towards the book.

Finally, we apologise for any errors or omissions that may appear anywhere in this book; please be assured that no harm was intended to anybody. Causing harm or discomfort to others is simply not the spirit of corporate social responsibility.

Corporate Social Responsibility in Times of Crisis: An Introduction

Following the 2008 global crash, the optimism surrounding CSR potentials changed overnight to a period of challenge. Prior to the economic meltdown, there had been considerable investment in CSR-related activities with leading transnational corporations (TNCs) spending large sums of money which they duly reported. Small and medium enterprises (SMEs) also became more active, with some realising that their local commitments were now considered to be part of something new and exciting whilst other organisations realising the significance of CSR approaches within the TNC-dominated supply and value chains. Projects incorporated local and global ambitions, employees were being encouraged to participate and beneath the public relations veneer committed substance could be witnessed. The crash rapidly changed expectations as we plunged into challenging times. Popular CSR expectations became located within the scepticism that during the financial halcyon days had surrounded the underlying intentions of TNCs in particular. This suspicion focused upon the 'real' reasons why companies were investing in projects that were not immediately connected to the profit bottom line. When reviewing the public relation campaigns of TNCs, the level of cynicism could be understood. Corporate promotions often orientated around short-term projects that were accompanied by photographs of the opening of medical centres and schools, the digging of new wells and the health care provided to sick people which emphasised the problems communities face that the corporate benefactor solves. In some respects, the financial downturn was expected to support the critics who considered CSR to be glorified PR campaigns. In particular when resources are restricted, PR would be reduced and CSR activities will diminish accordingly. Hence, TNCs were considered to be fair-weather CSR advocates whose commitment would end when things turned stormy.

With significantly reduced budgets, TNCs did indeed reduce their financial contributions to CSR. Nevertheless, many continued to be committed to CSR-related practices and do so after in-depth evaluations of expenditure and the value of returns. That they continue to be committed provides a retrospective insight into why they initially became involved. As Toker (2015: 433) explains, owing to the increasing attention to corporate practices from consumer and other stakeholder groups, marketing and management scholars have also shown an intense interest in the field of CSR and therefore public relations scholars and practitioners are paying a lot of attention to CSR as an integral part of public relations practices.

Consequently rather than viewing CSR as an approach to be adopted as a corporate whim, it is important not to lose sight of the factors which were instrumental in the late twentieth-century resurgence of the concept. For instance, globalisation and international implementation of neo-liberalism during the 1980s weakened levels of social welfare and governance at a time when TNCs were gaining in prominence and wealth. What had hitherto been considered to be government responsibilities were shifted onto companies. These shifting perceptions also coincided with mounting evidence into environmental damage and climate change, whilst enhanced levels of communication resulted in much greater interest and scrutiny of corporate actions by civil actors. With respect to the high-profile marketing of the environment, Banerjee (2008: 66) argues that this stems from 'public perceptions of environmental problems along with increased environmental legislation are two key reasons why the environment became an important issue for corporations resulting in the need for companies to "sell environmentalism" in order to be perceived as green'. For critics such as Beder (2000), environmental emphasis within CSR-related policies can be referred to as greenwashing, a process whereby companies manipulate an image of environmental, social and cultural commitments. And within 'greenwashing' image makers connect into contemporary buzzwords such as sustainable development and environmentally friendly and social investment.

Clearly there are examples of corporate manipulation of green messages. Similarly there are also instances where TNCs and SMEs have developed environmental programmes far beyond compliance and which they do not promote. Whilst recognising the extremes of intention, our position is one that advocates 'win-win' situations. Companies that are both able to reduce environmental damage and provide sustainable welfare programmes while enhancing their reputation do not strike us as the basis for criticism. Nor should they be the source for unadulterated praise. Following on from our earlier discussion, we should not lose sight that in some regards TNCs and SMEs have become more responsible not through altruism but through expectations, and even demands, that they should do so. Moreover, companies also became more understanding about the value of CSR for the corporate or local brands. Guzman and Becker-Olsen (2010: 203) observe how many companies have responded with more visible investor relations material focused on CSR initiatives and employee/communityengagement programmes. Strategic CSR programmes can serve a marketing purpose. In these instances, companies may choose to engage in CSR programmes as public relations opportunities, reputational insurance that will help in times of crisis or brand building.

Furthermore, the global economic risks have meant that when facing international and regional uncertainties, economic competition and rising public awareness and distrust of global corporations, company reputation is an integral part of the evaluation carried out by institutional and individual investors. Individual and corporate investors such as pension schemes are less inclined to buy shares, banks are not willing to loan money for expansion or consolidation and governments refuse to allocate concessions to companies with higher reputational risk. In other words, a bad reputation can mean the end of TNCs and SMEs, a fact which may have sharpened the enthusiasm of some executives and local owners for CSR.

Pragmatically there is an argument that approaches to CSR can provide companies with competitive advantages. When we look at strategic approaches, policies that connect with core business and stakeholder requirements can be productive. Irrespective of their motivations, companies have to be serious about their approaches to CSR if they are to secure new business in emergent markets. For instance, changing legislation within many African countries now emphasises the centrality of local content within concessions. Higher percentages of indigenous populations must be employed by foreign companies. This change is causing difficulties and is requiring the acceleration of education and training programmes to enable sufficient qualified nationals and SMEs to be involved and participate in the supply chain.

These changes emphasise the need to contribute towards a social and business environment that is conducive to and supportive of CSR approaches and which becomes embedded within surrounding relationships. To be successful, CSR objectives need to be shared across stakeholders. Frankental (2001: 23) asserts that CSR 'can only have real substance if it embraces all the stakeholders of a company' with social and environmental sustainable goals 'embedded across the organization horizontally and vertically'. As Maon et al. (2010) have stressed, approaches must consolidate different stakeholders and integrate different expectations and demands. Local engagement with communities and the incorporation of their ideas and practices will strengthen support for, and durability of, initiatives. In so doing, CSR can be adapted to local needs and demands making the outcomes more meaningful for the people most affected.

In summary, CSR programmes are unquestionably being implemented in challenging times. Companies face challenges of development, consultation, implementation, integration, measurement and consolidation that they are perhaps not best placed to address, certainly not alone. Within the uncertainties over corporate and financial resources, global financial stability and strengthening evidence for climate change, we could expect that academics to be at the forefront of solutions that can help position CSR within a sustainable framework to robustly work towards improvements. Yet as Maon et al. (2010) observe until recently academics have rarely considered organisational and practical aspects of CSR implementation. There is now a much needed emergent emphasis on what triggers engagement in initiatives, internal and external factors behind social change within organisations, the design and structure of strategies and policies and how organisations integrate into business models and processes. And it is for this point that our book has been written to contribute towards ways in which CSR is, and is not, contributing to the challenges faced across different parts of the world. Some challenges are shared whilst others differ, heavily influenced by the nature of industries and the conditions and expectations of governments, local communities and markets. Therefore, contributions are included from different parts of the world and from different types of the economy with different stakeholders and initiatives. The diversity provides important lessons to be learnt from original fieldwork about what has worked and may be replicated. Attention is also placed upon enhancing understanding about what has worked less well. If CSR is to be better positioned in helping to address the challenging times, then there needs to be much greater sharing of knowledge and understanding about successes and failures which can provide the basis for international collaborations that the magnitude of the problems demands. We hope that this book makes at least a small contribution towards this end. Certainly, the imaginative ways in which the challenges are being approached and in turn challenged provide us with optimism.

Split by continent, the first part of this book explores European case studies and perspectives before progressing onto North America and concluding with Africa. Switzerland provides the starting point as Looser, Clark and Wehrmeyer examine the impact of financial crisis on SMEs' CSRs attitudes. Emphasis is placed upon SMEs' culture with a business model that incorporates traditionalvalues of craftsmanship, stewardship and the concern with societal contributions and stakeholder relationships including communities. Swiss idiosyncrasies surrounding civic rights and strong sense of identity and community which evolve from common history help to explain the approaches that the SMEs adopt and reactions to the 2008/2009 financial crisis which was less severe in Switzerland than most neighbouring countries. Although not using the technical term CSR, Looser et al. show through primary research how SMEs continued to implement ethical values throughout 2008/2009 and then across subsequent local crises. With informal CSR adopted by Swiss SMEs embedded within culture and part of the basis for stakeholders trust, these commitments remain integral within relationships and activities during boom and crisis periods.

Del Baldo develops a theoretical and normative perspective to the financial crisis in chapter "Authenticity: Is Corporate Social Responsibility the key to overcoming crisis?". Commencing with a review of Pope Francis Encyclical Letter, she outlines how the care of the planet would require collective contributions to the common good. Through content analysis of a range of sized Italian companies, Del Baldo focuses upon behaviours which contribute to responsible, durable decision-making processes. The case studies highlight how company ethical leadership contributes to the socio-economic terrain with values that are integral to corporate development. These companies enable employees to prosper, commit to local and global communities and in so doing adhere to the Encyclical Letter principles.

Fourth-generationsustainability-orientated business models are the focal point of chapter "Developing fourth generationsustainability-oriented business models: Towards naïve, native, and narrative intelligence" by Melissen and Moratis. Acknowledging failure within CSR initiatives in meeting sustainability-related indicators, they propose a 'fourth-generation' business model. Drawing on insights from evolutionary psychology, three types of 'sustainability intelligence' are identified to help enhance understanding human behaviour, partnerships and leadership.

Integrating these elements within the model will provide the basis for 'sustainability intelligence' and gain the support of consumers that extends beyond the short term. This is because greater sustainability requires consumer commitment to sustainable products and services which enhanced intelligence can enable.

Romanian CSR activities during crisis are reviewed in chapter "Corporate Social Responsibility in times of crisis: Evidence from Romania" by Sitnikov and Bocean. Against a backdrop of the global financial crisis, which first hit Romania in 2008, they note the polarity which ranges from the threat to CSR commitments to the opportunities. Examining what they describe as the microeconomic and macroeconomic levels, Sitnikov and Bocean study CSR in OMV Petrov, the oil and gas group, and the Romanian Government's National Strategy for Corporate Social Responsibility. Although there has been progress with the implementation of CSR programmes, hitherto this had tended to concentrate on sponsorship and charity projects. However, OMV Petrom has strengthened CSR integration into corporate strategy as a sustainability concept. At the macro level, the national strategy has brought together corporate commitments and stakeholder expectations. However, the formality of the document has contributed to limited impact.

The Mediterranean region, in particular Cyprus, is the focus in chapter "The practice of internal Corporate Social Responsibility in SMEs in Cyprus". Papasolomou brings in analysis of SMEs building upon arguments for greater internal integrated and employee-driven CSR practice and the benefits which ensue. An empirical study of 180 employees in 12 SMEs across four regions provides the application as the author establishes perceptions of training and development, organisational justice, work–life balance and health and safety. The data showed companies prioritise employee training, health and safety, organisational justice and work–life balance. For Papasolomou, the imbalance is a source of unhappiness and dissatisfaction as workers' requirements are only partially met. Reviewing and revising the spectrum of CSR activities would better enable more SMEs to address employee concerns.

The Romanian public sector is the subject matter for Ogarca and Puiu in Chapter "Corporate Social Responsibility in the Romanian public sector". In so doing, they extend the scope of CSR study which tends to concentrate on the private sector and also incorporate the National Strategy mentioned in chapter "Corporate Social Responsibility in times of crisis: Evidence from Romania". Data is collated from a total of 94 public sector employees and managers who represented different domains such as army, education, health care, justice and public administration. Insights are gained into levels of knowledge and understanding. The results enable Ogarca and Puiu to suggest solutions to barriers such as lack of financial resources and awareness alongside the insufficient promotion of the National Strategy and lack of political will.

Chapter "Corporate Social Responsibility Policy in the United States of America" sees a change of continent through Camilleri's examination of policy in the USA, arguably the nation with the least sustainable consumption patterns. He examines government policies in the American context of low levels of state welfare provision. Areas such as education, health care and community investment which are provided by many European governments are often left to corporate initiatives and programmes. Without that commitment investment does not happen. The establishment of corporate citizenship procedures and socially responsible practices are becoming more noticeably embedded into corporate business functions and strategic decisions. Nevertheless, national institutional structures face considerable challenges in addressing hitherto limited corporate and consumer willingness to address the imbalance between consumption and sustainable development.

Africa provides the geographical focal point of Part Three commencing in chapter "Corporate Social Responsibility in Challenging Times: A consideration of how Small and Medium Scale Enterprises attempt to deal with CSR Challenges in Nigeria" with the most populous state in Nigeria. SMEs are the focal point as Lincoln explores the CSR challenges facing smaller organisations. Drawing upon original fieldwork, Lincoln identifies the need for enhanced socially responsible behaviour across businesses. The potential for CSR is currently being restricted by the lack of relevant guidance and need for greater efforts across government, NGOs and all sizes of business to position the triple bottom line as integral within strategic frameworks. Government institutions in particular could play a more prominent role in promoting the significance of CSR and wide ranging benefits. Lincoln identifies gaps that can be filled through greater training and skills development to build capacity within SMEs that would be significant in meeting the challenges.

Shifting across to East Africa, Kenya is the case study for Cheruiyot and Tarus in chapter "Corporate Social Responsibility in Kenya: Blessing, Curse or a Necessary evil?". Examining the Kenyan context leads the authors to refer to the country at an 'adolescent age' of development facing extensive challenges at a time of considerable economic growth. Across political, legal, social, economic and environmental spheres, there are challenges and opportunities for CSR commitments. Highlighting the historical roots of CSR, regulatory framework, market access and a range of stakeholders allied to particular catastrophic events provide motivations for CSR development. Facing these challenges and opportunities leads Cheruiyot and Tarus to propose three scenarios, considering CSR to be a blessing, curse or a necessary evil.

Sarpong reviews in chapter "Corporate Social Responsibility in Ghana: Issues and Concerns" issues and concerns that have emerged during recent years in Ghana, one of the most rapidly developing African countries. Following civil society scrutiny, not least surrounding negative impacts such as pollution, waste and soil erosion from mining activities, public demands for greater social responsibility have led to enhanced corporate management of social, environmental and economic impacts. Social justice has become a more integral component as companies incorporate local demands within international guiding frameworks. Tackling social problems has been spearheaded by large firms restricting the space for SMEs. Moreover, the expansion of CSR-related rhetoric is often not matched by corporate practices which can mean that the potential for CSR to assist conflict prevention, resolution and management is underachieved.

Mugova, Mudenda and Sachs consider difficulties in the implementation of CSR in African countries, in particular Zimbabwe in chapter "Corporate Social Responsibility in challenging times in developing countries". CSR within Zimbabwe has received relatively little academic attention which Mugova et al. help to address. They point out how Zimbabwe presents a number of challenges for the development, implementation and maintenance of CSR programmes, including the weak institutionalenvironment and poor governance structure with conditions exacerbated by political, economic and social tensions. Indeed, challenging times are the norm rather than the exception. By reviewing several Zimbabwean companies, the authors consider the key factors which support or hinder CSR implementations. Leadership, SME predominance, lack of indigenous businessownership, industry variations and government operations are shown to be instrumental. Together Mugova et al. argue that the study provides important lessons for other developing countries.

The mining sector is discussed in the second chapter on Ghana when Amponsah-Tawiah and Tuokuu consider the impact of the decreasing gold prices in chapter "Effects of Dwindling Gold Prices on Corporate Social Responsibility (CSR) performance in Ghana's Mining Sector". Five large mining companies are studied to examine the extent of CSR with a representative in a relevant position from each organisation interviewed. The authors discovered that dwindling gold prices had led to the shape and nature of CSR activities being redefined. However, unlike some of the examples discussed in earlier chapters, Amponsah-Tawiah and Tuokuu argue that the activities are driven by profitability and not necessarily strategy. In Ghana, this has led to CSR activities being operationalised, shifted from the mainstream to the margins. To address this marginalisation requires a broader strategic plan which incorporates more extensive stakeholderengagement and board commitment.

In the final case study, Agbeko, Blok, Omta and Van der Velde explore microfinancial institution strategies in chapter "Corporate Social Responsibility strategies adopted by Micro Finance Institutions: A case study". Like microfinance institutions (MFIs) in other locations, within Ghana MFIs provide socially responsible enterprises easier access to loans alongside social empowerment investment programmes. Returns on the latter have tended to be modest, and as a consequence, the credit crunch from 2008 onwards led to MFIs changing emphasis from social responsibility more generally and empowerment in particular to economicsustainability and profitability. This shift in business model led to mergers between MFIs in the search for greater profits and priority placed on loans for larger companies in the search for larger profits. In applying academic literature to empirical research into an MFI branch, the authors are able to identify best practice and make a number of recommendations to enhance levels of CSR investments.

Robert Gordon University, Aberdeen, UK — Stephen Vertigans
Guildhall School of Business and Law, London, UK — Samuel O. Idowu
University of Craiova, Craiova, Romania — Adriana Schiopoiu Burlea

References

Banerjee, S. (2008). Corporate social responsibility: The good, the bad and the ugly. *Critical Sociology, 34*(1), 51–79.

Beder, S. (2000). *Global spin: The corporate assault on environmentalism*. Foxhole: Green Books.

Frankental, P. (2001). Corporate social responsibility: A PR invention? *Corporate Communications, 6*(1), 18–23.

Guzman, F., & Becker-Olsen, K. (2010). Strategic corporate social responsibility: A brand building tool. In C. Louche et al. (Eds.), *Innovative CSR: From risk management to value creation*. Sheffield: Greenleaf Publishing.

Maon, F., Lindgreen, A., & Swaen, V. (2010). Organizational stages and cultural phases: A critical review and a consolidation model of corporate social responsibility development. *International Journal of Management Reviews, 12*(1), 20–38.

Toker, H. (2015). Public relations and finance. In S. O. Idowu et al. (Eds.), *Dictionary of corporate social responsibility*. Heidelberg: Springer.

Contents

Part III Corporate Social Responsibility in Times of Crisis: Africa

Part IV Corporate Social Responsibility in Times of Crisis: Summary

Editors and Contributors

About the Editors

Adriana Schiopoiu Burlea is a professor of Management at the University of Craiova, Romania; professor teaching and research in Human Resources Management, Project Management, Social Audit, Corporate Social Responsibility and Negotiation Techniques at the Faculty of Economics and Business Administration, University of Craiova, Romania; visiting professor teaching and research in Cross-Cultural Management BEM Bordeaux Management School, 680, Cours de la Libération, Talence, France; visiting professor teaching and research in Human Resources Management at the University of Corsica "Pasquale Paoli", IAE Corte, Corte Corsica, France; and visiting professor teaching and research in Human Resources Management, Intercultural Management at Saint-Etienne School of Management (SESOM). Professor Adriana Schiopoiu Burlea is Romanian ambassador for ARGH (Association Francophone de Gestion des Ressources Humaines—http://www.agrh.fr/les-ambassadeurs/) and member of the Académie des Sciences du Management de Paris—http://www.asmparis.org/csdr/view/30. She is member in many organising committees for national and international conferences. She stimulated and encouraged the cooperation between her university and the other universities around the world (e.g. cooperation with staff and teachers from other international universities in various scientific projects—for example, universities from France, the USA, the UK, Switzerland, Belgium, Italy, Portugal, Norway and Holland).

Samuel O. Idowu is a Senior Lecturer in Accounting and Corporate Social Responsibility at Guildhall School of Business and Law, London Metropolitan University, UK. He researches in the fields of Corporate Social Responsibility (CSR), Corporate Governance, Business Ethics and Accounting and has published in both professional and academic journals since 1989. He is a freeman of the City of London and a Liveryman of the Worshipful Company of Chartered Secretaries and Administrators. Samuel is the Deputy CEO and First Vice President of the

Global Corporate Governance Institute. He has led several edited books in CSR, is the Editor-in-Chief of two Springer's reference books—the Encyclopedia of Corporate Social Responsibility and the Dictionary of Corporate Social Responsibility, and is an Editor-in-Chief of the International Journal of Corporate Social Responsibility. He is also a Series Editor for Springer's books on CSR, Sustainability, Ethics and Governance. One of his edited books won the most Outstanding Business Reference Book Award of the American Library Association (ALA) in 2016 and another was ranked 18th in the 2010 Top 40 Sustainability Books by *Cambridge University Sustainability Leadership Programme*. Samuel is a member of the Committee of the Corporate Governance Special Interest Group of the British Academy of Management (BAM). He is on the Editorial Boards of the International Journal of Business Administration, Canada, and Amfiteatru Economic Journal, Romania. Samuel has delivered a number of Keynote Speeches at national and international conferences and workshops on CSR and has on two occasions 2008 and 2014 won Emerald's Highly Commended Literati Network Awards for Excellence. To date, Samuel has edited several books in the field of CSR, Sustainability and Governance and has written four forewords to books. Samuel has served as an external examiner to the following UK Universities—Sunderland, Ulster, Anglia Ruskin and Plymouth. He is currently an external examiner at Robert Gordon University, Aberdeen; Teesside University, Middlesbrough; and Sheffield Hallam University, UK.

Stephen Vertigans is currently Head of School of Applied Social Studies at Robert Gordon University, Aberdeen, UK. His research interests include corporate social responsibility and political violence with particular interest at present in energy-related development in East and West Africa. In his publications, Stephen applies a sociological approach in order to try to enhance levels of understanding about the wider social, cultural and political processes which are instrumental in the success or failure of CSR policies.

Contributors

Daniel Agbeko has 18 years of practical Banking experience and technical knowledge in rural finance, entrepreneurship and microfinance with strong skills in credit analysis for micro-, small- and medium-scale enterprises. Between 2008 and 2015, Daniel worked with uniCredit as a branch manager, between 1998 and 2004, was the credit manager of Wamfie Rural Bank and, later in 2001, became the supervising manager. In 2005, he was the chief executive officer of Women and development project (a financial NGO). Currently, he is an international partner of Quente NL. and the founder and CEO of Quente Africa. Daniel is also involved in microfinance and entrepreneurship training programmes for the rural banks and other private microfinance institutions across Ghana and other developing countries. He has also presented technical papers on microfinance, entrepreneurship and rural banking at several seminars in Ghana and microfinance and entrepreneurship conferences in India, the Netherlands, Italy, the UK and the USA. He is currently in the final year of his PhD in microfinance and entrepreneurship in the University of Wageningen, the Netherlands. He completed his MBA in Finance from the University of Leicester, UK, in 2007. He had his BSc. in Agriculture concurrently with his Diploma in Education in 1997 from the University of Cape Coast Ghana. He has also obtained numerous certificates in microfinance both locally and abroad.

Kwesi Amponsah-Tawiah is a Senior Lecturer at the Department of Organization and Human Resource Management (OHRM), University of Ghana Business School. He is an Applied Psychologist and an Occupational Health and Safety specialist, with considerable research and consultancy experience in Organisational Development and Change Management, Occupational Health and Safety Management, Organisational Behaviour Management and Psychosocial issues at the workplace. He has consulted for a number of local and international organisations including the ILO and the Ministry of Employment and Social Welfare. Kwesi is currently a fellow of the COIL Institute for Global Networked Learning and member of the British Psychological Society, the Ghana Psychological Association, the International Commission on Occupational Health and the Ghana Institute of Health Safety and Environment Professionals. As a young academic and a researcher, Kwesi has published extensively in some International Journals in the area of Occupational Health and Safety, Corporate Social Responsibility and Human Capital Management. He currently coordinates Executive Development at the University of Ghana Business School. Kwesi Amponsah Tawiah holds a PhD in Applied Psychology from the Institute of Work Health & Organisations (I-WHO), University of Nottingham, a BA (Hons) and an MPhil degree in Psychology and Industrial and Organisational Psychology, respectively, from the University of Ghana.

Vincent Blok is Assistant Professor in sustainable entrepreneurship, business ethics and responsible innovation at the Management Studies and Philosophy Chair Group, Wageningen University. From 2002 to 2006, Blok held various management functions in the health care sector. In 2006, he became director of the Louis Blok Institute, an international research institute in the field of organic and sustainable agriculture, nutrition and health care. In 2005, he received his PhD degree in philosophy at Leiden University with a specialisation in philosophy of technology. Together with nine PhDs and two postdocs, Blok pursues three lines of research—Business Ethical Issues in Sustainable Entrepreneurship, Business Models for Sustainability and Responsible Innovation in the private sector—in several (European) research projects. Together with the philosophy group of Wageningen University, he works in the field of environmental philosophy, hermeneutics of technology and innovation and philosophy of management. Blok's work appeared amongst others in *Journal of Business Ethics, Journal of Cleaner*

Production, Journal of Agricultural and Environmental Ethics, Journal of the British Society for Phenomenology, Studia Phasenomenologica and Journal of Responsible Innovation.

Mark Anthony Camilleri is a resident academic lecturer in the Department of Corporate Communication within the Faculty of Media and Knowledge Sciences at the University of Malta. He lectures marketing-related subjects in a joint/dual Master's programme run by the University of Malta in collaboration with King's College, University of London. Dr Camilleri is a PhD graduate from the University of Edinburgh in Scotland. He taught business subjects at graduate and postgraduate levels in Hong Kong, Malta and the UK. His current research interests include corporate social responsibility, stakeholder engagement, sustainable tourism and creating shared value.

Dr Camilleri is a frequent speaker and reviewer at the American Marketing Association's (AMA) Marketing and Public Policy annual conference. He has published papers in *Sustainability Accounting, Management and Policy Journal, Corporate Reputation Review, Tourism Planning and Development, Tourism and Hospitality Research and the International Journal of Leadership in Education* as well as in the proceedings of AMA Conferences. His short contributions are often featured in popular media outlets such as the Times of Malta, Business2Community, Social Media Today, Triple Pundit, CSRwire and the Shared Value Initiative.

Thomas Kimeli Cheruiyot is Professor of Corporate Social Responsibility and Dean at Moi University School of Business and Economics. He obtained his doctorate in business management from Moi University, MSc from Wageningen University and bachelor of agribusiness management from Egerton University. He has over 20 years' experience in teaching and research at the university. He has published in *African Journal of Economic and Management Studies, TQM Journal, International Journal of Law and Management, International Journal of Emerging Markets, Sage, Information Age, Springer International Publishing*, amongst others. Current research interests include amongst others corporate social responsibility, stakeholder theory, corporate governance and organisation studies.

Philip Evans Clark is a DBA candidate at the Business School Lausanne in business sustainability with a special interest in collective learning and collective action. He is the founder and president of Projet Socrate, a not-for-profit organisation looking at the intersection of philosophy and business. For many years, he led the innovation team of Orange Business Services Global Services, a major international telecom carrier. Prior to this, he was a journalist and an entrepreneur.

Mara Del Baldo is Associate Professor of Small Business Management, Financial Accounting and Economics of Sustainability at 'Carlo Bo' University of Urbino, Italy. Her main research interests include entrepreneurship and small businesses management; corporate Social Responsibility; sustainability and business ethics; SMEs and networking strategies; accountability; financial and integrated reporting; and social and environmental accounting (SEAR). She is a member of the European Council for Small Business, the Centre for Social and Environmental Accounting Research (CSEAR), the SPES Forum, the Global Corporate Governance Institute (GCI), and the European Business Ethics Network (EBEN) Italian Chapter. She is editorial board member and reviewer of several international scientific journals. She is also author of numerous scientific publications, including articles in Italian and foreign journals, chapter books, conference proceedings and books. She has given numerous lectures and didactic seminars on invitation in various Italian and foreign universities.

Claudiu Bocean is an Associate Prof. Bocean is a PhD supervisor at the Faculty of Economics and Business Administration, the University of Craiova, Romania. In 2000, he graduated with a Bachelor's Degree, major in Accountancy and Informatics, from the Faculty of Economics, University of Craiova, Romania. In 2004, he graduated with a Master's degree in Business Administration (MBA) from the Faculty of Economics, University of Craiova, Romania. In 2007, he was awarded a PhD degree in Economics from the same Faculty. In 2015, he was awarded his Habilitation title in Management by the Academy of Economic Sciences Bucharest, Romania. Since 2002 to present date, he has taught and researched at the Faculty of Economics and Business Administration, University of Craiova, on topics such as Human Resource Management, Corporate Social Responsibility, Organisation Theory and Business Economics whilst cooperating with scholars from other national and international universities and organisations on projects.

Adebimpe Lincoln holds an LLB and an LLM in Commercial Law from Cardiff University and an MBA in International Business and PhD in Entrepreneurship and Business Development from the University of South Wales. She holds a Postgraduate Diploma in Legal Practice and also holds various teaching qualifications in Higher Education. She is a Fellow of the Higher Education academy and an Associate Member of the Chartered Institute of Personnel Development. She has over 12 years' experience in Higher Education. She has held positions in Cardiff University and the University of South Wales and was a Senior Lecturer in Law at Cardiff Metropolitan University before taking up a position as an Assistant Professor in Saudi Arabia. She has extensive experience working with the University of Liverpool online Master's in law programmes. She is the External Examiner for Undergraduate and Postgraduate Law and Professional Practice Programmes at Manchester Metropolitan University.

Dr Lincoln has vast experience in supervising research students. She has supervised a wide range of Master's level research including MBA and LLM dissertations. She has also supervised students at Doctoral Level in the area of Entrepreneurship and Corporate Governance. Her research interests lie in the area of Female Entrepreneurship, Small and Medium-Sized Enterprise (SME) sector, Leadership practices of Nigerian Entrepreneurs, Corporate Governance and Board Diversity and Corporate Social Responsibility amongst Nigerian SMEs. She has published and presented a number of articles on SMEs and female entrepreneurs in Nigeria.

Stéphanie Looser after gaining work experience in the accounting/consulting, auditing, public and nonprofit sectors is currently working towards a PhD in the field of Environmental Strategy at the University of Surrey, focusing on Small and Medium-Sized Enterprises (SMEs) and their unconventional but sophisticated approach to Corporate Social Responsibility (CSR). She is evaluating these heuristic and/or ethics-driven business models for comparison with standard systems in order to hasten them towards policy.

Frans Melisen is a professor of sustainable business models at NHTV Breda University of Applied Sciences. He holds a PhD in industrial engineering and management science. His research focuses on the link between sustainable development and human behaviour, with special emphasis on mitigating the social dilemma by means of sustainable business models. Frans has authored and co-authored a number of papers in refereed journals, has presented at various conferences and has (co-)edited/ (co-)authored and contributed chapters to various (text) books. Frans is also the Scientific Director of the EU research project Smart City Hospitality (SCITHOS).

Lars Moratis is a professor of sustainable business at NHTV Breda University of Applied Sciences and academic director of the Competence Center Corporate Responsibility at Antwerp Management School. He holds a PhD in management science and has worked in the field of corporate social responsibility in both the academic and the business context. His main research interests are the credibility of corporate CSR claims, CSR standards, CSR strategy development and implementation, sustainability-oriented business models and responsible management education. He has published several books, book chapters and articles in academic and practitioner-oriented journals on these topics.

Moono Mudenda received a Bachelor of Business Administration degree from the Copperbelt University in Kitwe, Zambia, in 2006. She received a Master of Business Administration Degree from Solusi University in Bulawayo, Zimbabwe, in 2012. Moono Mudenda was also a recipient of an entrance fellowship at the East West Centre in Honolulu, Hawaii, for the Asia Pacific Leadership Program 2014–2015 cohort. She has worked in various positions within the financial sector. Ms. Mudenda is a business development consultant. Currently, Moono Mudenda sits on the Board of Directors at the Alliance Française de Lusaka for a mandate of 1 year.

Shame Mugova is a financial analyst, currently working as a financial planner/ business planner in South Africa, and is a PhD Candidate with Durban University of Technology (South Africa). He received a Master's Degree (MBA) from Solusi University (Zimbabwe) and a Bachelor of Commerce Honours Degree in Finance from National University of Science and Technology (Zimbabwe) in 2012 and 2007, respectively. His research interests are in corporate social responsibility, corporate governance and trade credit. He has 8 years of work experience in Zimbabwe's tax, medical, private voluntary organisation, engineering, clothing and textile industry sectors. His work experience is in the field of finance and accounting and has worked for 3 years as an independent business consultant for Small-to-Medium Enterprises. Shame Mugova lives in Durban, South Africa.

Radu Ogarca is an Associate Professor at the Department of Management, Marketing and Business Administration within the Faculty of Economics and Business Administration at the University of Craiova, Romania. He has a master's degree in business administration at the University of Craiova and a PhD in management at Bucharest University of Economics Studies. Since 2002, he has been teaching bachelor's and master's students. He teaches Management, Small Business Management, Entrepreneurship and Business Ethics. His research is focused on business ethics, corporate social responsibility, entrepreneurship and social entrepreneurship. In December 2013, he completed postdoctoral studies after a 36-month period in which he did research on *small business ethics in Romania*. In 2014, he won a grant at the University of Craiova and carried out research on *Ethics Management in Higher Education System* of Romania together with PhD Lecturer Silvia Puiu.

Onno Omta is Chaired professor in Business Administration at Wageningen University and Research Centre. In 1978, he graduated in Biology (BSc in Biochemistry, MSc major in tropical horticulture/plant physiology and minor in Microbiology) and defended his PhD thesis on 'The management control of biomedical research and pharmaceutical innovation' in 1995, both at the University of Groningen. After two 1-year teaching appointments, he became Head of the Department Office of the Faculty of Biology of the Radboud University in Nijmegen. In the mid-1980s, he led a multidisciplinary team of content, education and media experts as author and editor of the Biology I course of the Dutch Open University. From 1988 till 1993, he was Managing Director of the Institute for Research in Extramural Medicine (EMGO institute) at the Vrije Universiteit in Amsterdam. Then he was appointed as Associate professor in Innovation Management at the University of Groningen. In April 2000, he was appointed as Head of the Department of Business Administration/Management Studies, currently consisting of an international Faculty of about 12 FTE and 26 PhD students from Europe, Africa, Asia and South America, 7 external PhD students-lecturers from Universities of Applied Research and 1.5 FTE Administrative support staff. In the last 5 years, about 3 PhD students and 60 MSc students graduated each year, and on average, 32 articles per year were published in ISI and/or Scopus rated scientific journals. In July 2015, an international peer review committee concluded, based on an international comparison, that the research quality of the Department, as well as its viability, was excellent (both 1: highest rating), whereas its relevance to society was very good (2), and congratulated him for his 'academic and strategic leadership'.

Ioanna Papasolomou is a Professor and the Head of the Department of Marketing, School of Business Administration, at the University of Nicosia in Cyprus. She has teaching and research experience in the UK. She holds a PhD in Marketing from Keele University as well as an MBA, an MPhil in Marketing Public Relations and a Postgraduate Certificate in Teaching and Learning in Higher Education. Dr Papasolomou has contributed papers to a number of international conferences and journals such as the Journal of Marketing Management, Marketing Intelligence and Planning, Journal of Brand Management, Journal of Product and Brand Management and the Journal of Marketing Communications. She has also contributed chapters and case studies to several academic books. She was the Chair for the 10th and 14th International Conference on Corporate and Marketing Communications hosted in 2005 and 2009, respectively, at the University of Nicosia. She is a reviewer for a number of academic journals such as the International Journal of Bank Marketing, Journal of Marketing Communications, International Journal of Corporate Communications, International Journal of Social Responsibility and Euro Mediterranean Journal in Business.

Silvia Puiu is a PhD Lecturer at the Department of Management, Marketing and Business Administration within the Faculty of Economics and Business Administration, Craiova, Romania. She has a doctorate degree in Management since March 2012, and between 2008 and 2015, she taught courses and seminars in Management, Business Ethics, Marketing, Public Marketing and Political Marketing. In 2014, she won a grant at the University of Craiova and did research on *Ethics Management in Higher Education System of Romania* together with PhD Associate Professor Radu Ogarca. In December 2015, Silvia Puiu completed postdoctoral studies after she has completed her research on *Ethics Management in the Public Sector of Romania.* During the scholarship, she spent 2 months in Italy at the University of Milano-Bicocca conducting a comparative research between Romania and Italy regarding Corruption Perception Index, ethics management in the public sector and regulations on whistle-blowing. During the last years, Silvia Puiu published more than 40 articles in national and international journals or in the proceedings of international conferences. Her research covers topics from strategic management, ethics management, public marketing and management. She is also a reviewer for two journals indexed in DOAJ, Cabell's, REPEC, EBSCO and Copernicus (Management & Marketing, The Young Economists Journal). Since 2014, Silvia Puiu is also a member of Eurasia Business and Economics Society (EBES).

Paul R. Sachs is Executive Director of NHS Human Services—Philadelphia, a multistate nonprofit organisation in the USA serving people with mental health, addictions and intellectual disability challenges He received a PhD in Clinical Psychology from Vanderbilt University in Nashville, TN, and an MBA from Saint Joseph's University in Philadelphia, PA. He is a licensed psychologist in Pennsylvania, USA, and resides in the Philadelphia, PA, USA, area.

Sam Sarpong is a Lecturer at the Centre for Communication and Entrepreneurial Skills (CENCES) at the University of Mines and Technology at Tarkwa, Ghana. He received his PhD from Cardiff University. He also holds an MBA and an MA from the University of South Wales and Cardiff University, respectively. He has taught at Cardiff University, University of London, Birkbeck College and Swansea Metropolitan University, all in the UK. His research is mainly on management issues with a focus on Corporate Social Responsibility and Business Ethics. He has also published widely in these fields.

Catalina Sitnikov is Professor at the University of Craiova (Romania), Faculty of Economics and Business Administration. She received PhD title in Management in 2000 and Habilitation title in Management in 2014, and since February 2015, she is PhD supervisor in Management.

For 3 years she was a Visiting Lecturer at the Helsinki University of Technology, Lahti Center (Finland). Since 1995, she has taught undergraduate, master's and PhD students. She teaches Quality Management, Total Quality Management and Management. Her main research areas include management, strategic management and mostly quality management, instruments and models specific to the stages of quality planning, control and improvement, quality management strategies, ISO standards, CSR from the perspective of quality management and specific standards and instruments.

Daniel Kipkirong Tarus is a Professor in the Department of Accounting and Finance and Acting Deputy Vice Chancellor, Finance, Moi University. He obtained a PhD in Finance from Moi University (Kenya) with research focus on corporate governance and corporate social responsibility. He has over 10 years' experience in university teaching and research in the areas of corporate governance, corporate leadership and social responsibility. He has published widely in peer-reviewed journals such as *Journal of African Business, Management Research Review, Corporate Governance, TQM Journal, International Journal of Emerging Markets, International Journal of Commerce and Management, African Journal of Business and Economics, Springer International Publishing*, amongst others. Research interests include areas of corporate governance and social responsibility.

Francis Xavier Dery Tuokuu is a Doctoral Researcher in Environmental Studies at Antioch University New England, USA. He is a 2015 Finalist of the Mandela Washington Fellowship for African Young Leaders (an initiative of President Obama). He is also a Sustainability Fellow at Franklin Pierce University, where he coordinates their climate change and energy conservation programme. Francis also serves as the Editor-In-Chief of the PhD Newsletter at Antioch New England Graduate School. He is the Founder and Executive Director of Centre for Responsible Business-Ghana and works as Project Officer with CSR Training Institute, Canada. He has over 50 published articles and over three hundred daily news/reports and stories in various newspapers, websites and magazines across the world, including ihe Daily Graphic (Ghana's leading newspaper), Business and Financial Times, the Finder newspaper, the Chronicle newspaper, The Development Analyst Magazine, tekedia.com, myjoyonline.com,ghanaweb, etc. He has two published peer-reviewed articles, two peer-reviewed articles under review and one published book on Corporate Social Responsibility (CSR). He holds a BA (Hons) Degree in Geography and Resource Development with Philosophy from the University of Ghana, Legon, and an MSc Degree in Corporate Social Responsibility and Energy from the Robert Gordon University, Aberdeen, in the UK.

Gerben Van der Velde holds a PhD in Economics from Delft University of Technology. From 2004 to 2013, he was Assistant Professor in Economics, Delft University of Technology. He supervises Accountancy research projects at Windesheim University of Applied Sciences and Wageningen University. At Nyenrode Business University he lectures Economics.

Walter Wehrmeyer joined CES in 1998 as the BG Surrey Scholar, focusing on contaminated land and ways in which current processes towards recycling land can be improved, in terms of both the speed of recycling and the quality of the site remediation. Before that he was at the University of Kent, jointly appointed between the School of Biological Sciences and the (now) Kent Business School, working on environmental performance measures and management systems.

Part I
Corporate Social Responsibility in Times of Crisis: Europe

Switzerland
Italy
The Netherlands
Romania 1
Cyprus
Romania 2

To (Crafts) Man Up: How Swiss SMEs Cope with CSR in Harsh Times

Stéphanie Looser, Philip Evans Clark, and Walter Wehrmeyer

1 Introduction

Corporate Social Responsibility (CSR) is usually associated with the integration of ecological and social issues into corporate activities (Mayr, 2015). The question of why firms should be interested in CSR—based on stakeholder, social, shareholder, or philanthropic reasons—is a matter of ongoing debate (e.g., McWilliams & Siegel, 2001). Whereas for Godfrey (2005) philanthropy is the marriage between CSR as strategy and CSR in terms of legitimacy, so that companies declaring strategic philanthropy have CSR in their managing portfolio (Vveinhardt & Zygmantaite, 2015).

This study defines CSR as: "In general, corporate sustainability and CSR refer to company activities—voluntary by stakeholder definition—demonstrating the inclusion of social and environmental concerns in business operations and in interactions with stakeholders" (Van Marrewijk, 2003). This definition can be traced back to Carroll's (1979) three-dimensional conceptual model of CSR (see Fig. 1).

Carroll's model captures both social responsiveness to CSR (reaction, defence, accommodation, and proaction and social issues (i.e., Stakeholder). Social responsibilities categories include economic, legal, ethical and discretionary responsibility. Van Marrewijk adopts several elements from Carroll's framework such as the inclusion of various stakeholders (e.g., consumers, shareholders) where "different business operations and interactions" take place among them (Van Marrewijk, 2003) and he refers directly to Carroll's (1979) social responsibility categories, deriving his concept of "voluntarines" (Van Marrewijk, 2003) from an expression of "the philosophy of social responsiveness" (Carroll, 1979).

S. Looser (✉) • W. Wehrmeyer
University of Surrey, Surrey, UK
e-mail: s.looser@surrey.ac.uk

P.E. Clark
Business School Lausanne, Lausanne, Switzerland

S.O. Idowu et al. (eds.), *Corporate Social Responsibility in Times of Crisis*,
CSR, Sustainability, Ethics & Governance, DOI 10.1007/978-3-319-52839-7_1

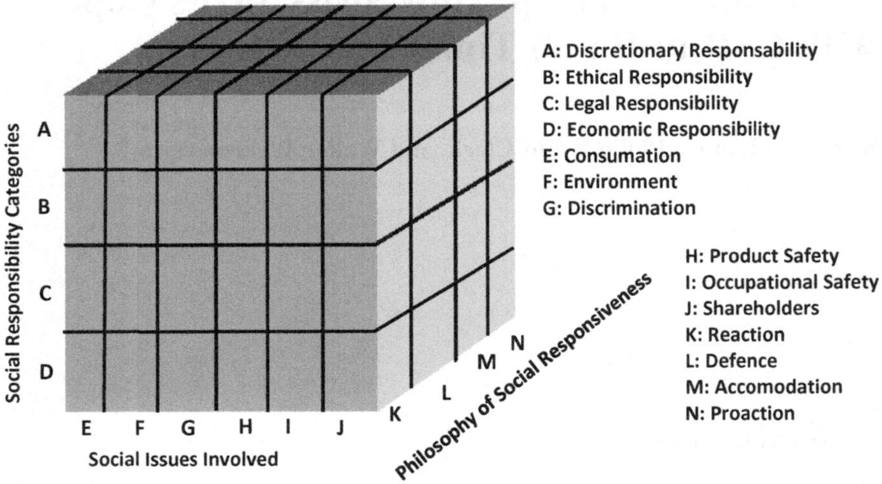

Fig. 1 The three-dimensional conceptual model of CSR - adapted from Carroll (1979)

Carroll's model has been criticised for inconsistency failing to explain why and how CSR should be hierarchical, and also for attempting to combine various heterogeneous concepts such as business ethics, corporate citizenship, and stakeholder management (Nkiko, 2013).

CSR is also often discussed within the triple bottom line framework—in the tradition of Friedman's (1962) neoclassical theory—with the aim to add social and environmental issues to the bottom line of profit-maximisation (Elkington, 1998).

As L'EPOQuE is a business model derived from companies deeply embedded in their communions and communities, to combine van Marrewijk's (2003) with Carrols's (Carroll, 1979) quite broad definition provides an adequate foundation from which to test the consistency of a business model during crises.

Social responsibility as a concept is not easily assessed as it pits individual against community (Orlitzki, Schmidt, & Rynes, 2003) business against society (Richter, 2010), and economic goals versus ethical inclination (Lorenzo-Molo & Siloran Udani, 2013; Wang & Juslin, 2009). This study pays a closer look at a business model that advocates a more applied, habitual, inherent, heuristic, and practical approach to CSR as it goes back to the essence of social responsibility, which occurs at tight interfaces between business, employees, and communities.

2 CSR for Its Own Sake

Academics assume that companies would be generally more willing to do some CSR if there exists a win-win outcome between "good business" (Epstein & Roy, 2003; McWilliams & Siegel, 2000) and enhanced financial performance (Margolis

& Walsh, 2003)—evolving "the business case for CSR" (Filbeck & Gorman, 2004; Nelling & Webb, 2009).

However, extrinsic motives for improving the bottom-line by the "business case for CSR" are not the only possible reasons to engage in responsibility issues (Wagner, 2005; Wagner, Schaltegger, & Wehrmeyer, 2001). On the contrary, many companies and their leaders developed a corporate culture driven primarily by a social and intrinsic motive where CSR is a moral duty (Graafland & Mazereeuw-Van der Duijn Schouten, 2012; Polonsky & Wood, 2001). Mintzberg (1983) calls this strain "the noble way for corporations to behave", thus as the purest form of CSR. In this case, an ethical approach drives CSR "for its own sake" (Mintzberg, 1983).

There is evidence of a significant number of small firms trailblazing this path of intrinsically motivated and ethics driven CSR (Jenkins, 2006; Fassin et al., 2010). These companies are often led by their owner or the owner family, operating his/her business according to personal, family values (Del Baldo, 2010a; El Baz, Laguir, Marais, & Staglianò, 2014; Looser, 2015; Looser & Wehrmeyer, 2015b). Evidently, this has the power to evolve CSR as "social good" (Osuji, 2011).

Such idiosyncrasies were also found in Switzerland where Small and Medium-sized Enterprises (SMEs) have highly sophisticated CSR agendas embedded in corporate cultures that nurture an implicit, intrinsic "raison d'être" beyond external pressure often resulting in extreme formalisation (Looser & Wehrmeyer, 2015a). Previous research culminated in the characterisation of this core logic as L'EPOQuE, the overarching Swiss small business model, making Switzerland, arguably, a hidden champion in CSR (Looser & Wehrmeyer, 2015b). This model is determined by several features that will be outlined in more detail in Sect. 7. It underlines the particularly important roles, for instance, the owner, the concept of "Swissness" as a general orientation towards quality, niche products, and networks play.

This paper further explores the consistency and steadiness of this modus operandi in times of crises.

In particular it asks: How did the global 2008/2009 financial crisis influence Swiss SMEs in terms of their CSR business model? What was in comparison the effect of the local abolition of the minimum price for the Euro by the Swiss National Bank in early 2015? Did SMEs change any key or sub features of their business model? Did they sacrifice any of their values? Or to what extent helped their model and approach to CSR in surviving the downturns? And if so, did it help to have strong ties with employees, customers, suppliers, and other SMEs, not only in Switzerland but on a worldwide scale? Did it lead to innovative solutions? How did it affect their actions? Did they reduce their investments? Or, if they had to lay-off people, did they do it in a responsible manner? What were the impacts, in the short-, mid- and/or long-term? Has commitment to CSR declined or increased since 2008/2009? Are there differences between Swiss German-, French-, Italian-, or Romansh-speaking regions?

To answer these questions, the remainder of this paper is structured in a theoretical section (Sect. 3) outlining the concepts applied, followed by the data gathering methodology by the presentation of results and by a discussion and according conclusions in the final sections.

3 Theoretical Framework

Porter and Kramer (2006) argued that "successful corporations need a healthy society and at the same time a healthy society needs successful companies". Empirical findings indicate evolutionary changes in companies as they move from superficial CSR to culturally embedded CSR (i.e., in a business model) and the application of relative business practices (Høgevold, Svensson, Wagner, Petzer, Klopper, et al. 2014). Other researchers found different companies having different motives for CSR based on their sensitivity to outside pressure ranging from passive conformity, active resistance (Zheng, Luo, & Maksimov, 2014) to voluntary, unintended deployment (Looser, 2015).

3.1 The Convergence of CSR and Business Models

Given the growing recognition of the critical role companies play in the overall health and functioning of societies, the practices of CSR principles are key indicators for measuring society's expectations (Asif, Searcy, Zutshi, & Fisscher, 2013). However, what became evident is that to shift the focus from theory to practice requires paying attention to SMEs, their business practices (Brown & King, 1982), their social capital (Russo & Perrini, 2010; Spence, Schmidpeter, & Habisch, 2003) and their linking of CSR to innovation (Ferauge, 2013; MacGregor & Fontrodona, 2008; Telle, 2006) through craftsman—and stewardship.

Currently the conviction is growing that if CSR is to have a meaningful impact it should be a matter of corporate culture and core logic (Du & Vieira, 2012; Osuji, 2011) not of formalised processes or regulatory strategy (Buhmann, 2015) etc. To help address this demand a closer look should be paid at intrinsic, morally motivated, and informal processes and relationships fostering organisational cultures that embed and drive CSR (Lorenzo-Molo & Siloran Udani, 2013).

Though academics and practitioners have recognised this need (Gond, Grubnic, Herzig, & Moon, 2012; Jamali, 2008; Margolis & Walsh, 2003; Zadek, 2004) there is still a dearth of deeper knowledge about CSR implemented in culture or in business models as its manifestation (Asif et al., 2013). While there exist research on top-down processes, starting at the strategic level and implementing extrinsic CSR through various management systems (Pondeville, Swae, & De Rongé, 2013) there is a knowledge gap on the evolutionary, traditional, historically grown business models with little or no formalisation resulting in unintentional but state-of-the-art CSR agendas. This research analyses the stability of such informal business models that are built on CSR values (and not on a strategic decision or commercial pursuit) under external (financial) pressure. This study examines the importance the businesses' core logic, and/or motives have on the perception, and thus, the design and effectiveness of CSR frameworks (Looser & Wehrmeyer, 2015a).

A recent shift in the CSR debate emphasises a move from strategic processes to organisations that maintain CSR because it is a part of their culture (Matten & Moon, 2008; Asif et al., 2013; Looser & Wehrmeyer, 2015b, 2015d; Lorenzo-Molo & Siloran Udani, 2013). In the case of Switzerland, most of these companies are small ones with informal and unsystematic CSR agendas (Del Baldo, 2010b; Morsing & Perrini, 2009; Nkiko, 2013) but with business models based on the traditional values of craftsmanship, stewardship, and its concern to contribute to society, build and maintain a "proper" business involving the community, and including close relationships with employees, customers, suppliers, and other small firms (Il Park & Ghauri, 2015; Looser & Wehrmeyer, 2014).

In return, these business practices are highly valued by communities and meaningful stakeholders. This recalls the bottom-up approach proposed by Asif et al. (2013). Furthermore, it compromises institutional dynamics of CSR as "soft law" (e.g., Nolan, 2013; Zeyen, Beckmann, & Wolters, 2014) aligned with concepts of legitimacy (Zheng et al., 2014), virtue ethics, and morale.

At the same time, there is an overall attempt and trend in the business world to conceptualise day-to-day operations in models (Arjaliès & Mundy, 2013). Such models are called "business models", which in turn are regarded as tool kits to improve performance and increase revenues (Kalakou & Macário, 2013). This study defines a business model as "the rationale of how an organization creates, delivers and captures value" (Osterwalder & Pigneur, 2010).

Above all, how business is organised mirrors what values, processes, or systems drive a company and which bricks it is built on (Del Baldo, 2010a; Jenkins, 2006). This intersection, in turn, is closely linked to the motivation and set-up of CSR for a given company (Newman & Sheikh, 2012; Schultz, 2013).

In times of crises, the business model plays an important role in how companies cope with external challenges (Cordeiro & Sarkis, 1997; Jones & Wicks, 1999; Neron & Norman, 2008); how able it is to react to unpredictable, inexplicable, and/or uncontrollable events (Hassel, Nilsson, & Nyquist, 2005; McWilliams & Siegel, 2000). Yet, there is little information how or if business models and their key features sustain rough situations. It is not yet clearly how consistent they remain when faced with external pressure in terms of values and attitudes. It is reported from other contexts, (Høgevold et al., 2014; Wagner, 2005; Wagner et al., 2001) that such developments led in many cases—especially in Multinational Enterprises (MNEs)—to significantly less responsibility, which is in the long-run somewhat myopic (Fassin, 2008).

Crises are not homogenous and can be classified. The next section describes some well-known typologies in order to assess whether different types of crises have different consequences on the Swiss SMEs' business model for CSR.

4 The Nature of Crises

With regard to crises, much attention is devoted to the management of accidental events that perturb ongoing activities and puts a company's brand or organisation at risk (Bowen, 1953; Martinet & Reynaud, 2001; Wartick & Cochran, 1985). And indeed, CSR has been at times used or rather misused in these instances to explain or diffuse their effects. Critical voices underline the fact that "strategic giving may be used to the detriment of consumers and society" (Polonsky & Wood, 2001) and relative CSR or sustainability reports, largely applied by MNEs, may serve as "veils hiding activities" (Deegan, 2002) with the primary purpose to reconstruct an eroded legitimacy (Banerjee, 2008; Gond et al., 2012).

For example Lindblom (1994) mentions four possible strategies that companies have adopted in the context of crises: (1) inform stakeholders about performance improvement planned by the organisation to manage the event; (2) modify stakeholders' perception on the facts themselves; (3) distract stakeholders from the issue at hand; or (4) change external expectations vis a vis the performance of the organisation. These are well-honed techniques that competent public relations offices can master and apply in the event of accidents or untimely revelations (i.e., information leeks). On deeper and more ethical grounds, protecting the environment, managing human resources, health and security in the workplace, relations with local communities and relations with suppliers are complex issues that CSR can and does address.

There is evidence that responses to crisis will vary dependent on the commitment of organisations to CSR. For example, Martinet and Payaud (2008) describe four levels of CSR strategies, which will predict firm behaviour in times of turmoil.

Wartick and Cochran (1985) also detail four categories based on the reactivity of firms, pointing to their level of social performance. Enterprises can either be: (1) reactive; (2) defensive; (3) resigned; or (4) proactive. Reactive firms are in denial of their responsibilities. Defensive enterprises recognise them but do not accept them. Resigned enterprises acknowledge their accountability and address the concerns of their stakeholders. Finally, proactive enterprises not only accept their responsibilities but forecast them, thus going beyond stakeholders' expectations. But these frameworks do not tell us much about the type of crisis. Thus, there must be a difference between the 2008 financial meltdown, the hurricane Katrina, and variations in exchange rates (two of them apart from the hurricane Katrina were looked at here).

Before exploring the impact of crises on CSR of Swiss SMEs, it might help to anchor the term in a typology. Jönsson, Lundin, and Sjöberg (1978) for example classified crisis in relation to the decision process they trigger. Westphalen (1992) arranged them in terms of their character (objective or subjective), their nature (technical or political) and their origin (endogenous or exogenous). Same insights were found by Lagadec (1993), Lagadec (1995), Rosenthal and Koumin (1993) and more recently by Roux-Dufort (2003) who classified crisis along two dimensions: their origin which can be either technical/economic (such as boycotts or industrial accident) or social/human (such as terrorist actions or social conflicts) leading to either external or internal effects on organisations. Sartre (2003) added legal crisis (i.e., frauds, white collar crimes, etc.) and informational crisis (e.g., confidential leaks).

Gundel (2005) on the other hand proposed a typology based on the crises' predictability and the capacity of organisations to influence them (see Fig. 2). Predictability is straightforward (Gundel, 2005). Influence is the ability of a firm to manage and reduce the crises' effect. As a consequence, Gundel (2005) identifies four types of crisis: (1) conventional crisis are predictable and influenceable; (2) unexpected crisis are unpredictable but influenceable; (3) intractable crisis are predictable but uncontrollable; (4) fundamental crisis are unpredictable and uncontrollable.

The advantage of this typology is its basement not only on the level of responsiveness but at the same time on the efficiency of the response as well. Predictability is crucial for the capacity to plan and run routines, which are core to production. When planning is not possible, then it is important to assess resilience. As outlined in the following, the quality of resilience that defines Swiss SMEs CSR is crucial. Of these types, two are of special interest herein: fundamentally global crisis like the financial collapse of 2008 and intractable crisis, such as the free floating of exchange rates decided by the central bank of Switzerland—or even climate change—because they represent the most acute kind of crisis in times of extreme volatility. Being one of the severest kind of crisis, it is assumed that it will affect CSR strategies especially considered in the "business case" approach expandable.

In order to analyse to what extent and how crises can effect countries and their firms' business models, it is crucial to understand the social, political, economic, and environmental situation. Thus, the next section sheds light on these circumstances in Switzerland.

Fig. 2 Crisis matrix (adapted from Gundel, 2005)

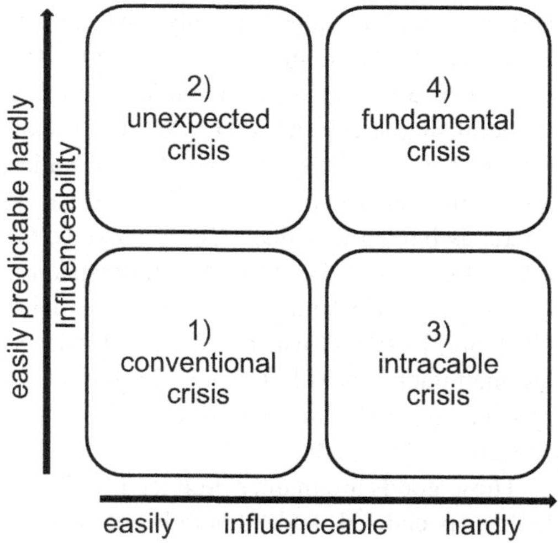

5 Swiss Idiosyncrasies

Switzerland, officially known as the Swiss Confederation, is a federal parliamentary republic. Federalism is one of the most important features of the Swiss state. The central government must pay respect to 26 individual cantons, with regard of their budgets, languages, unique geographies, social practices, etc. (Linder, 2005). Switzerland is characterised by a form of direct democracy where society retains the civic rights to challenge any law passed by parliament (by referendum) and introduce supplements to the federal constitution (by initiatives) (Kriesi, 1980).

Although predominantly German-speaking (the other linguistic and cultural regions are French-, Italian-, and Romansh-speaking), Switzerland is not bound by a sense of common ethnic or linguistic identity. Instead, the strong sense of identity and community evolves from a common historical background, traditionally common values, its direct democracy and federalist system, and from "Alpine symbolism" (Federal Administration, 2008).

Principles of subsidiarity and liberalism determine the relationship between state, the economy, and the third sector (mostly professional associations). As the state is traditionally weak, there is a strong incentive for the other sectors to organise themselves.

Switzerland has a strong third sector and a deep philosophical, economic, political, and social sense of independencies (Helmig, Lichtsteiner, & Gmür, 2010). The Red Cross, founded in 1863 in Geneva illustrates this long tradition in third sector movements (Helmig et al., 2010). Switzerland also has a long history of "armed neutrality". Having avoided two World Wars, it was able to rely on traditional businesses based on family capitalism and "Mittelstand"/small business culture (Linder, 2005). As an independent and neutral state, Switzerland is not a member of the European Union (EU) or the European Economic Area, and business is mainly embedded in local value chains (Federal Administration, 2008).

Switzerland has a stable and prosperous economy based on high-tech products and services. Its wealth is mainly borne by SMEs as the economic backbone (State Secretariat for Economic Affairs (SECO), 2012a, 2012b). Indeed, the majority (i.e., 99.8%) of all companies are SMEs, which employ about 70% of Swiss labour (Federal Statistical Office (FSO), 2013), contribute 20% of export value (Credit Suisse, 2014), and 60% of Swiss Gross Domestic Product (GDP) (Schweizer Radio und Fernsehen (SRF), 2013). The Federal Statistics Office (FSO, 2003) defines SMEs as having less than 250 employees, their turnover or balance sheet total is below Euro 50 Mio or Euro 43 Mio respectively.

Tough Switzerland is home to several large multinational corporations (e.g., Glencore, Nestlé, Novartis, ABB, and Adecco), its predominant economic sectors are manufacturing (FSO, 2013), largely of specialist chemicals, health and pharmaceutical goods, and scientific and precision instruments (SECO, 2005; Federal Administration, 2008).

These goods are mainly produced by Swiss SMEs which are often niche actors, highly specialised, and, although not widely known, in many cases world, or at least, Swiss market leaders (SECO, 2011, 2012c, 2012d, 2012e). Small companies may have international business operations but their location, strategic and organisational orientation remains regionally focused (Gabler, 2015). However, size does not fully define companies; it is only one way to categorise them.

Hence, Switzerland is regarded as the land of the "hidden champions", not only in an economic sense, as it relates to its highly specialised and quality-oriented SMEs (e.g., Brûlé, 2005), but also with respect to CSR, as shown by the Swiss Corporate Sustainability Report (Berger, Winistörfer, Weissert, Heim, & Schüz, 2012) or Enderle (2004).

According to Berger et al. (2012) responsible values are widespread, anchored in tradition, and sometimes also in "mission statements". And yet CSR activities are still not often openly communicated, not connected with markets, and not used as differentiation strategy. This might be related to the cultural context of direct democracy and federalism, where the community counts and not the individual.

Due to a (so far) balanced Federal State, which implies a certain capacity to absorb financial volatility, the Swiss economy—largely made of SMEs—was not as badly affected than most other Western countries by the financial crises in 2008/2009. There was in particular not a real credit crunch. Admittedly, the country's growth rate diminished in 2009 but, as the next section will show the character of Swiss SMEs as stakeholders, does not change much under such external impacts. Swiss business leaders are both stubborn and reluctant to abandon their beliefs and habits. They trust the quality of their products and uphold the way they do business and lead their life.

Although decency characterises Swiss business, Berger et al. (2012) found some trends towards more reporting (e.g., Code of Conducts, CSR or sustainability reports), which could be traced back to the need to increase stakeholder engagement and dialogue in order to sustain competitive advantage. Yet Gentile (2012), Gentile and Lorenz (2012), Lorenz and Spescha (2012), and Samuel, Schilling, and Wehner (2012) showed in a set of comparative studies, that Swiss CSR seems to be mainly driven by traditionalvalues and ethics.

Considering the significant role of SMEs on Switzerland's economicdevelopment, their collective "grandness", and cumulative impact, an in-depth understanding of SME CSR practices seems necessary (SECO, 2012e, 2012f, 2012g, 2012h). In particular, it would help to trace the effect of crises on local society as well as on SMEs themselves. Given the importance of lead actors especially in small companies (as elaborated in the previous sections) this cannot be achieved without looking at the values, beliefs, and ethics of Swiss small firm owner-managers.

6 SMEs as Stakeholder in Swiss CSR

SMEs, unlike large companies and Multinational Enterprises (MNEs), are often unfamiliar with the scientific or technical term "CSR"; consequently, they do not usually pursue formalised programmes to manage CSR, nor issue external reports (Looser & Wehrmeyer, 2015a). Nevertheless, SMEs are very active in the field of CSR in ways somewhat subtle and different from MNEs. This is confirmed by primary research with owner-managers of SMEs. For more details on the quoted owner-managers see Table 1.

> [. . .] A report does not prove the embeddedness of CSR in corporate culture or values [. . .] (SME 5)
>
> [. . .] We all can issue nice reports but what matters is the substance [. . .] (SME 1)

Thus, SMEs do not engage in CSR for strategic reasons or in a systematic way; instead they want to "give something back to society"—the "mantra" of their family tradition and/or the owner-manager's personal values and community-embeddedness.

> [. . .] It is just standard to help poor or jobless people in the village [. . .] (SME 2)

This is unconventional because, although a clear-cut system or strategy towards CSR might be missing; there is clear evidence that SMEs are very active and engaged in CSR. Examples of this strong CSR practice beyond formal systems include handshake instead of formal contracts with customers, employees, suppliers, etc., democracy and transparency, commitment to quality, altruistic and philanthropicvalues, and visionary instead of instrumental goals. By this core logic or "raison d'être" they save on transaction costs (e.g., for lawyers setting-up contracts), establish a certain ground level of trustworthiness among entities, and induce incentives to imitate such practices.

These peculiarities are strongly supported for instance by SME 6's and 7's owner-managers who stated:

> [. . .] I work physically among my employees on our various construction sites in order to personally take care of my staff's health and safety [. . .] we provide apples and water and in the summer sun cream for free [. . .] (SME 7)
>
> [. . .] we try to teach them how to integrate into Swiss society, that's not easy since most of our employees are coming from countries that are currently or were recently in war against each other [. . .] thus, we try to establish peace between them [. . .] (SME 6)

Looser and Wehrmeyer (2015a) found many SME offices literally without an executive floor and CEOs sitting among their employees, clearly indicating that owner-managers see themselves more as co-workers than as bosses.

> [. . .] Ultimately, our performance depends on people. And that is the reason why CSR is a good thing to do [. . .] (SME 3)
>
> [. . .] Our employees are working for my company and we have the responsibility for their safety. That's my aim and goal since they are my success factor. [. . .] (SME 4)

Table 1 Sample description

Attribute		Companies	
		Count	Code ID
Gender	Male	5	1, 3, 4, 5, 7
	Female	2	2, 6
Number of years in business	Less than 3	–	–
	3–5	–	–
	6–10	–	–
	More than 10	7	1–7
Position or title in company	Owner and manager	7	1–7
	Employed manager	–	–
Level of education	National level certificate	6	2–7
	Bachelor's degree	1	1
	Master's degree or higher	–	–
Legal form of enterprise	Limited company	1	4
	Incorporated company	6	1–3, 5–7
	Cooperative	–	–
Sector	Finance	–	–
	Manufacturing	2	2, 5
	Service	1	1
	Agriculture	1	3
	Construction	2	4, 6
	Food	1	7
Number of employees	Less than 10	1	2
	10–50	1	3
	51–100	1	5
	101–200	3	1, 4, 7
	201–250	1	6
	251–300	–	–
	301–500	–	–
	501–1000	–	–
	More than 1000	–	–
Language region	German-speaking	3	2, 6, 7
	French-speaking	2	1, 4
	Italian-speaking	1	3
	Romansh-speaking	–	–
Annual turnover in Euro Mio	Less than 10	1	2
	10–20	3	1, 3, 5
	21–50	1	6
	51–100	–	–
	101–500	–	–
	501–1000	–	–
	More than 1000	–	–
	Missing system	2	4, 7

SME 1, for example, has publicly communicated core elements of their CSR that include equal treatment of all employees, transparency in all respects (especially performance indicators, salaries, and bonuses), and majority votes where every employee has the option to veto management decisions. For example, the staff jointly decided against entering global markets—despite perfectly valid economic arguments—because they feared that they would no longer enjoy the kind of transparency and democratic procedures if exposed to world-wide competition (Looser & Wehrmeyer, 2014, 2015a).

The existence of such idiosyncrasies is confirmed by the statements of many stakeholders, first and foremost by the Swiss government (SECO, 2012b, 2012g, 2012h) which recognises that SMEs are led by an "Unternehmer" (in contrast to MNEs' managers). To hire disabled people is an ordinary practice of Swiss SMEs as Baumgartner, Greiwe, and Schwarb (2004), CDI (2011), Schur, Kruse, and Blanck (2005) confirm. This orientation of taking responsibility for others goes definitely beyond CSR as "façade" (Castelló & Lozano, 2011) and the "business case" approach particularly targeting at shareholder value.

> [...] since I remember we always employed handicapped or socially weak and infortune people [...] (SME 4)

Swiss SMEs support the Swiss apprenticeship system by hiring about 200,000 apprentices per year (FSO, 2012a, 2012b). By doing so, responsible business practices are rooted at the beginning of careers and expressions of personal ethics at work deeply embedded during the process of work socialisation.

> [...] We have always educated apprentices; by this we secure our own staff and the level of education that is needed to fill this very complicate production process. Furthermore, I try to implement in these young people the seeds of trust, quality, ethics and a sense of responsibility for themselves but also for society and environment [...] (SME 2)

The development of an ethical groundwork by experience points to the guidance by virtues, or moral. Publicly available reports, e.g., Swiss Corporate Sustainability Report (Berger et al., 2012) and research e.g., Baumann-Pauly, Wickert, Spence, and Scherer (2013) or Looser and Wehrmeyer (2015a, 2015c, 2015d) suggest that running a SME is more a lifestyle decision to chase innovation based on visions rather than primarily a wealth-maximising strategy. In many cases, SME owner-managers are envied for this libertarian lifestyle, even though they sacrifice security to pursue their ideal. Responsibility and altruism are often important elements of their upbringing and ethical behaviour a habit of doing business.

> [...] Since I am aware, my parents taught me to take care of others because we as business owners have a life-long responsibility for our region [...] (SME 4)

Many SMEs' children go to private schools that focus on fostering the pupils' strengths and not trying to fix their weaknesses.

These schools teach social responsibility for instance by letting the pupils clean the school house, organise and cook lunch, and look after the younger pupils during school breaks. Later, children of family business owner-managers are encourage pursuing trainings in other companies. They often travel and work in companies that belong to their network (e.g., in China, India, USA) in order to be learn about cultural differences that may serve their later role as leader of the family business.

> [...] I established an education because there is no adequate education supplied by the public system [...] (SME 6)

This explains the relative ease with which SME owner-managers set up often highly developed and far-reaching CSR programmes disengaged a the "bottom line" perspective, often informal and not publicly shared. Swiss SMEs also experience extensive social support due to their philanthropic profile, their function as societal pillars, and their long-term continuity passing the business down to their children.

Overall, SMEs show high investment in social capital, backed by strong regional networks of customers, who mutually rely on the SMEs' traditional image. Due to regional supply chains, SMEs are often more sustainable for the same product as a result of shorter transportation distances. Further, their business strongly relies on reciprocal, trust-based deals.

> [...] I personally take care of all our suppliers and customers in the market hall in order to foster relationships; however this is very informal mostly leading to a coffee break or a beer after closing time [...] (SME 3)

Many SMEs emphasised that they only do business within the family business network and would rather sacrifice a deal, contract or opportunity to increase their profit if business partners ask for costly contractual formalisation or do not rely on the handshake between them. International trade partners (SMEs e.g., from China, India, or Thailand) often rely on this trust-based business practice. In other words, traditional, SME specific values determine their global business operations as well.

> [...] All our international trade partners from China, India, or Thailand rely on business sealed by handshake [...] (SME 7)

Such corporate cultures built on trust and ethics result in very low staff turnover rates (3% or less) and in lost lasting employee relationships that can span several generations.

Moreover, due to their limited size, SMEs are very innovative and capable of adapting new trends very quickly, accounting for the Swiss marketleadership of many SMEs.

> [...] We are world market leader and solely developed this production process [...] (SME 1)

To conclude, SMEs are important actors of Swiss CSR and their approach, or well, business model, and their idiosyncratic features should be looked at further. Therefore, the next section analysis this model in more detail.

7 An Emerging Template for CSR in Switzerland

The Swiss model is the manifestation of informal, horizontal, and trust-based business approaches; their proximity to and involvement in local environments, by e.g., recruiting from the local community, and their willingness to pace their growth while avoiding atomic markets. It often mirrors the commitment to quality and efficiency borne by the strict and, in some cases, almost conservative value-set of their leaders (Looser & Wehrmeyer, 2015a; Looser, 2015).

These aspects can be aggregated and visualised as in the L'EPOQuE model (see Fig. 3), which consists in six key features (Looser & Wehrmeyer, 2015b):

L A liberalistic, visionary leader—and ownership—driven approach, where the Leader "is" the business and vice versa,

E A focus on tight, long-term, trust- and democracy-based relations to Employees,

P A preference for niche Products (specialised, small scale production with high entry barriers),

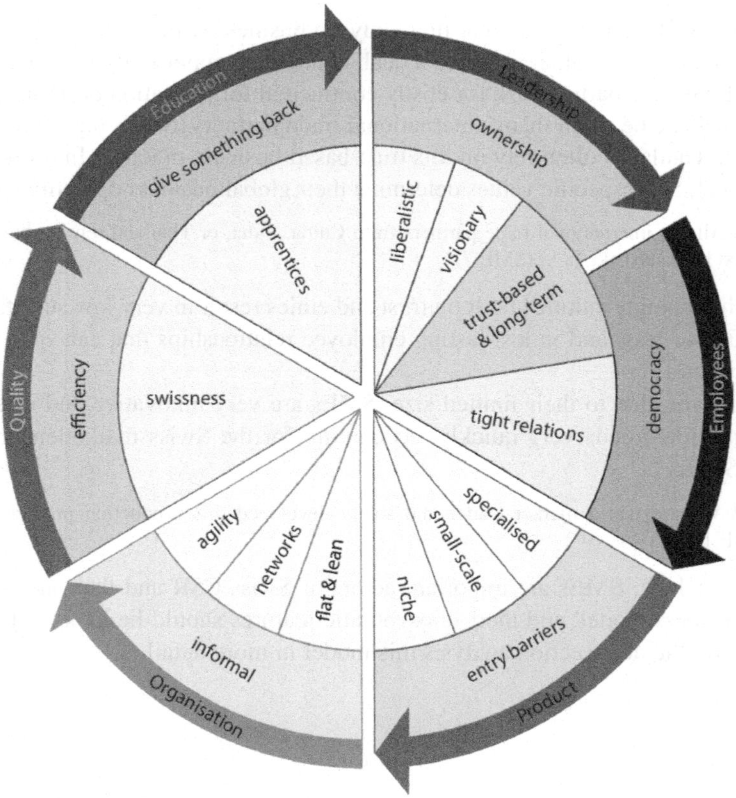

Fig. 3 L'EPOQuE—an emerging template of CSR in Switzerland

O Driven by networks and informal, flat Organisations (therefore agility),
Qu Relying on "Swiss" Quality (i.e., Swissness and efficiency).
E Giving priority to Education, e.g., the apprenticeship system, as part of a commitment to "give something back" and establish their ethics during the process of work socialisation.

It is worth noting that concrete environmental actions like recycling and circular economy initiatives, buying organic and fair-trade food/textiles, or at least local products, are standard practices in Switzerland. Organic products, for instance, make 7.7% of total food sales with a turnover in 2015 of CHF 2323 Mia. 12.8% of total available surface is used for organic production. Within the 6'031 organic food producers all are Swiss SMEs (Bio Suisse, 2016). They are driven by voluntarism, not by regulation (Baumann-Pauly et al., 2013). Thus, environmental key drivers can remain understated since they are deeply embedded in the DNA of Swiss society.

L'EPOQuE represents a valuable business model gained through extensive research (interviews and Delphi process) (Looser & Wehrmeyer, 2015b).

At this stage, it is not clear how or if the business model and its key features could be affected by crises. This research explores this question by looking at two crises: (1) the 2008/2009 global financial crisis and (2) the local crisis caused by the abolition of the minimum price for the Euro by the Swiss National Bank.

This might serve as a test of the Swiss business model "L'EPOQuE", eventually strengthening its overall model and, more importantly, examining the resistance as template for CSR-driven businesses—see also Looser and Wehrmeyer (2015d).

8 Methodology

The aim of the empirical part is to identify possible influences financial crises might have on the key features of the Swiss SMEs business model. To look at the global 2008/2009 crisis and at the influence of the local Swiss Franc-Euro crisis the method of focus group discussion in combination with the method of case study (see Sect. 10) were chosen. The former is specifically useful for the analysis of inherent concepts, heterogeneity, and consistency for CSR among companies belonging to different sectors, language regions, etc.

Combining information from focus groups with case studies is especially useful since information from different sources is triangulated, which increases data reliability and internal validity; whereas the use of a set of different theories and research topics enhances external validity.

9 Focus Group Discussion

In general, the method of focus group discussion is applied in the context of different institutional logics and foundations (Lounsbury, 2008). It is heuristic since it examines how ideas develop and operate within a given cultural background (Kitzinger, 1994). The data gathered is interpersonal, negotiated, challenged, and to some extent validated (Kitzinger, 1994) since this method "investigates what participants think, but it excels at uncovering why participants think they do" (Morgan, 1988).

This study particularly seeks to learn how the group's attitudes towards CSR and especially their business model for CSR is affected by external pressure in the form of global as well as local financial crises. Although the analysis is based on individual statements, the unit of analysis remains the group, so as to provide a shared understanding of the participants' views (Morgan, 1996). This approach allows a qualitative exploration emerging from intensive, personal discussions about the stability of CSR patterns in Switzerland. Arguably, the cross-sectorial approach limits the degree to which the results can be generalised.

An initial sample of two companies (SMEs 1 and 2) was randomly selected from the online telephone book tel.search.ch (2015) that allows for sectorial and/or corporate search, which secured that only companies were reached. The businesses were contacted by phone to seek their participation and to confirm their expertise in the topic of CSR. By looking up Zefix (2014), the commercial register, as well as the company websites the size was verified.

The initial contact by phone was further applied for snowballing, i.e., to gather contacts to other, somehow aligned, companies, and to have a viral sample distribution.

Some companies asked for anonymity (by an ID as stated in the fourth column) thus, it was decided to anonymise the whole sample.

The group discussion questions underwent a pre-test. The meeting, with an average duration of one hour, took place in October 2015. It was recorded digitally, subsequently transcribed and qualitatively analysed. Some authors (Gephard, 2004) strongly recommend the use of software for the qualitative analysis of text since this allows a systematic, category based examination of individual statements. The decision here fell in favour of MAXQDA. Firstly, the transcript was coded in-vivo; the second coding looked more at the steadiness and consistency of theoretical constructs of CSR according to L'EPOQuE. By doing so, each code represents "a particular system of beliefs, values, and images of the ideal" (Philipsen, 1987).

10 Case Study

A case study often refers to published reports about companies while they were studied over a period of time (Creswell, 2007). This is the approach followed here. In general, the "case" examined can be either an individual, organisation, an event, action, existing in a specific time and place (Atteslander, 2010).

Based on that, certain patterns of behaviour and attitudes could be identified. Often informal research methods lead to such case studies that are unlikely to appear in formal research venues, journals and professional conferences, rather in popular works, such as the news, newspapers etc. (Creswell, 2007).

11 Findings

The focus group discussion and case studies showed surprisingly strong business attitudes and confirmed the existence and application of business features as predicted by L'EPOQuE (see Fig. 3) even during serious global (Sect. 11.1.) and local (Sect. 11.2.) crises.

11.1 Effects of the 2008/2009 Global Financial Crisis on Business Model

There are many reasons that explaining why confronted with a fundamental crisis like the one of 2008 Swiss SMEs did not recant on their CSR commitments. One reason is that Switzerland, unlike other countries, has not outsourced manufacturing and agriculture. Safeguarding both the first and second sector might explain why the Swiss economy managed to circumvent the disastrous effects of the financial crisis. Instead of lowering quality, switching to mass production or delocalising, Swiss SMEs have streamlined their production processes. This included the integration of services into solutions.

> [. . .] smaller companies have in the long run no other choice than to rely on high-quality products, thus, our survival is determined by our ability to be as efficient as possible with the world's best products [. . .] (SME 7)

> [. . .] On a worldwide scale, we are the only one having this knowledge and thus we are the only supplier of such a combination of product and service [. . .] (SME 3)

Another reason was found in the simple, direct way of doing business that Swiss SMEs adopt, which saves expensive transaction costs.

> [. . .] this is about our wish to be as efficient as possible [. . .] (SME 1)

Swiss SME's are often world leaders in niche products. To survive they have developed strategies of resilience, reinforced their agility, in a word adapted to strong market demands.

> [...] Since 30 years we are Swiss market leaders, but in the next year I have to put all my effort in finding new niches [...] (SME 5)

Their life is a permanent struggle for survival, which have made them somewhat immune to strong swings in market conditions, so crises are "usual" to them, lowering the effect of surprise of negative externalities

> [...] My life is dedicated to risk, insecurity, abandonment, and loneliness, however, you cannot imagine how much envy I daily experience for my libertarian lifestyle [...] (SME 6)

Swiss SMEs, some of them quite "old" and well-established, have survived all type of crises: conventional, unexpected, intractable, and fundamental (compare with Sect. 6). In general they favour a pragmatic approach, and view crisis like a storm that comes and goes away with time. Many of these companies are more than 100 years old, they survived two world wars, black Tuesdays, the rise of oil price and all kind of other financial hardships.

> [...] I once decided to run this company though I was aware of sacrificing a lot of comfort, money and adopt a lot of stress, sweat, and blood running this company [...] (SME 4)

This is not to say, that they are generally just used to crises, but that they see them from a more neutral point of view and with some "ease". Likewise, they have their product, their customers, and as long as this system works, a crisis is [...] much ado about nothing [...]—as the owner-manager of SME 2 stated.

Furthermore, as most of them are owners and managers of their company, they are self-funded, self-financed and only partially dependent on financial markets. This might be the most important reason for the Swiss SMEs' sustainable business approach as found in this study. A characteristic of Swiss SMEs is that they are often clients of small regional and in many cases cooperatively organised banks with little ties to international finance. The consequences are that in downturns such as the 2008 meltdown, the Swiss market was not hit as badly as others. Above all, this explains why Swiss SMEs have little need to measure and or exhibit their CSR performance.

As demonstrated in their stakeholder analysis (Looser & Wehrmeyer, 2015a), the response of NGOs, consumers, trade unions, and the Swiss government's to SMEs and their business model during the 2008/2009 financial crises validated their high levels of legitimacy, power, and urgency emphasising solidarity, their approach of stewardship, and their "friendship in the case of hardship".

Nevertheless, given the unbroken popularity of formal CSR, the Swiss SME culture may experience pressure to move to a more utilitarian model if they grow to a significant size or if they are included in global market systems where major trade partners (B2B) or customers (B2C) require CSR standards, certifications, and reports in respect to their formal CSR systems. Yet it is important to point out that growth patterns are probably as diverse as business models and that we must beware of generalisation.

As most SMEs in this study stated, it is possible to do business on a worldwide scale without sacrificing informality, provided trade partners belong to the SME or family business network and share, within their own tradition, the same moral idiosyncrasies and ethical standards (e.g., handshake quality). In other words, the SME culture can be somewhat stronger than formal processes as long as they exhibit similar patterns across cultural boundaries and paragons the traditional values fostered by their owner-managers.

To conclude the 2008/2009 crisis, although unpredictable and uncontrollable, had only little effect on the Swiss SMEs' business model and/or values as stated by the SMEs that participated in the focus group discussion. This conclusion is supported by a very significant example of sustaining values in the case of Bobst.

Bobst (2015) is a leading packaging company based in the canton de Vaud with a deeply embedded culture of ethical CSR that found a surprising solution during and in the aftermath of the 2008/2009 financial crisis.

11.2 The Case of Bobst

In 2009, the business of Bobst (2015) was in a rut. Orders had dropped significantly while little could be done in terms of process optimisation or productivity boosts. Bobst was looking at the dire prospect of having to fire 150 employees. Working together with trade unions, which see themselves as social partners rather than enemies of management—resulting in teams on both sides sat down to find an alternative.

While it was obvious that Bobst had severe problems, they found that other companies, like Scheuchzer (2015), which specialises in train construction and the maintenance of railways lines, were flooded by orders, mostly international, that they had problem fulfilling. In other words, the reality was that while some companies were losing business, others were striving. This led to a meeting between Bobst and Scheuchzer employees committees (not only management). It did not take long for them to realise that they shared the same kind competences in machine assembling and tooling.

Thus, it was decided that for a certain period Scheuchzer take over Bobst's employees while remunerating Bobst for this "loan". Bobst employees went to work for Scheuchzer for a period of 6 months after which they would be reintegrated into their "original" company. The whole "exchange-action" saved 150 jobs, which would not have happened without a deep commitment to the well-being of employees, excellent relations between social partners, and deeply embedded regional networks (Fondation IPT, 2015). This solution was so successful that it became a model for the entire canton of Vaud, which, in sum, helped to save hundreds of jobs (Pereira Pündrich & Triki, 2014; RTS, 2015).

In a study made by the Eidgenössische Technische Hochschule (ETH) in 2008–, 2009 on 2000 companies 95% of respondents stated that they would maintain or increase their commitment to CSR. This study confirms that most CSR activities are the result of personal engagement without any economic counterpart. Moreover, it underlines the modesty and indifference to publicising accomplishments in that field, as 25% of companies admitted not advertising their social engagement.

11.3 Effects of the 2015 Local Crisis on Business Model

The shock caused by the rise of Swiss franc following the Swiss National Bank decision (from 15 January 2015) to abandon a fixed rate between the Euro and Swiss franc was both predictable and controllable crisis (see Sect. 4) In other words, Swiss SMEs were aware the possibility and to some extent prepared, but only to a certain extent.

In the aftermath, for a short time, there was parity between the Euro and Swiss franc, which then changed to a disadvantageous exchange rate for the Swiss economy. In one single swoop, the Swiss franc went from 0.83 Euro to 1.02 Euro (Travail Suisse, 2015) losing 20% of its value. This was a sharp measure especially for exporting companies: margins fell by 15–20%, competiveness dropped, and so did foreign investments.

According to the interviewees, the genuine problem was not that this happened, since it was to some extent predictable as the Federal Bank's decision of 2011 to fix the exchange rate to 1.20 Swiss francs for 1 Euro was per definition a temporary measure. However, what the SMEs resented most was the manner in which the Federal Bank acted, namely, without involving SMEs though they count for 99% of all Swiss companies and despite the fact that Switzerland was not hit as badly as others by the 2008/2009 crisis thanks to Swiss SMEs outstanding effort during that rough period (SECO, 2010; KMU Portal, 2012).

Beyond the obvious stress this decision imposed on the Swiss economy, what upset Swiss SMEs the most was the rupture of their culture of dialogue and consensus, where alternatives are examined and risk minimised. In this instance, the Federal Bank had acted univocally and undemocratically, arbitrarily by-passing the opinion of the country's main stakeholders. It was a break of trust.

Swiss people are disciplined and hard working with a strong sense of continuity. Values such as dialogue, discussion, collaborative decision making (i.e., social capital) are the key to Switzerland's success (as outlined in Sects. 5 and 7). Any infringement of these values are deeply felt and rejected. As a consequence the break of and the non-consulting was from the SMEs' point of view less predictable and more shocking than the 2008/2009 crisis.

Whether this in total will induce an abolishment of any of the SMEs' key values or a change in their business model is not clear yet. What is confirmed is that responsibility must be felt and acted as something mutual that works on the basis of trust. It is yet to be analysed in forthcoming studies how Swiss SMEs adjust their culturally embedded CSR practices to this "break of trust" and if they have the strength, will, and ability to cope with the situation to the same extent as they did in the 2008/2009 crisis. According to several experts (e.g., KMU Portal, 2015) by the beginning of 2016 Swiss economy seems to recover thanks to innovative solutions of Swiss SMEs particularly in the second sector as described in Sect. 2 and in the "Bobst case".

12 Discussion and Conclusions

Qualitative research methods (literature research, focus group discussions, and case studies) have clear limitations in respect to objectivity, reliability and validity, because they involve interpretation and selection. The reliance on experts' statements in this study is substantial, which may be a source of partiality if they were made insincerely and/or incorrectly. In addition, the vast majority of interviewees were male, which, again, may have created a systematic bias towards male leadership attributes (word given, handshakes, etc.).

Other limits to reliability and internal validity may stem from the relatively small size of the sample used in these case studies. However, qualitative methods abide to different rules regarding sample generation and size. Utmost care was given to secure correct data.

Thus, samples were randomly picked from the Swiss online telephone book, and the conduct of interviews and the research methodologies grounded on the rules for sample generation within qualitative research were strictly followed (Atteslander, 2010).

It is doubtful that more meaningful results would have been achieved by questioning larger samples (e.g., by a quantitative survey) because the intent of this research was not geared toward quantitative results, but rather driven by a qualitative exploration of CSR (e.g., open ended questions) in the context of Switzerland and Swiss SMEs, as a first step toward further investigations. The responses gathered with regard to the six research questions never implied nor suggested that more people would have given a substantially different perspective.

The broad definition of CSR adopted in this study could possibly be a drawback especially with regard to internal validity: the use of more specific terms might have yielded different outcomes. However, Carroll's and van Marrewijk's definitions enabled the identification of patterns and key drivers, the various motives and manifestations at play in CSR within SME's. This research does not intend to provide a new definition for CSR since this would have added to "the jungle" (Crane et al., 2013, p. 15).

Data was gathered in all three linguistics regions (i.e., German, Italian, and French speaking parts) of Switzerland and no difference in values and attitudes of SME owner-managers were found albeit different languages and cultural contexts. More insights into overarching SME attributes were brought about by the comparative analysis.

This book chapter is based primarily on a quantitative definition of SMEs violating qualitative aspects, such as property situation, personal management by the owner etc., as suggested by Curran and Blackburn (1994), although such data was gathered and examined at each step in our research. The organisations and companies under review were randomly selected, however, their willingness to participate may have created a self-selection bias (Atteslander, 2010). Their consent to participate in this study may in part rely on their awareness of their somewhat sophisticated approach to CSR thus tilting the results towards best-in-class examples. However, none of the participating organisations used this study to promote their company, nor advertise their participation in their marketing material.

To summarise, the validity of this research rests on chronological and easily traceable data collection, the hermeneutical integration of new data, rival theories, and research findings from multiple sources, transcribed and coded using the MAXQDA process.

To avoid falsification of data or biases resulting from single-expert answers various methods—i.e., focus groups discussion, secondary data analysis, and case studies—and different means of analysis were used.

Finally, this field of research is both relatively new and admittedly complex. However, the insights gained in this research allow for a unique opportunity to peer at the various modalities used by SME's in their approach to CSR (as proposed in the next section) and can contribute to practitioners as well as public institutions, the scientific community, and to society at large.

SMEs in Switzerland often have highly sophisticated CSR agendas. In many cases they are unintended, informal as they are coming from corporate cultures that nurture a "raison d'être" beyond profit-maximisation. Previous research aggregated this core logic to an overarching SME business model making Swiss SMEs, arguably, hidden CSR champions with regard to social, economic, and environmental responsibility. This model is borne by a set of key features: i.e., the process of work socialisation, soft assets, proximity and informality, agility, the nexus of company ownership and government, a particular focus on education, and long-range planning.

By the methods of focus group discussion with seven SME owner-managers and of case studies this research looked at the influence a financially rough situation might have on such a CSR business model. The results show that there is no substantial leverage. This research also found no significant differences in the stability and sustainability of business values during crises between three of the four language regions (currently neglecting Romansh-speaking SMEs).

The same can be said with regard to investments: measured in absolute figures, cuts were made, since earnings dropped during both crises. However, the SMEs considered in this research paper, did not diminish their relative investments (measured in percentage of earnings)—on the contrary, in the first and second sector, SMEs invested in streamlining of their production processes, in product innovation, and in the customisation of their products from pure "hardware" to systems including services and solutions. This helped Swiss SMEs to build up high entry barriers to their specific markets making it very difficult for competitors to enter those markets. Currently, many of the very successful companies have become suppliers of complete business solutions. These comprehensive packages are the result of constant innovation driven not by basic research but coming from their philosophy of stewardship and their aspiration and ambition to provide excellent craftsmanship that surprisingly work best under pressure caused by crises.

Since the Swiss model is culturally embedded and driven by intrinsic and implicit motives, it is not exposed to often expensive CSR activities and therefore not influenced by harsh economic situations. On the contrary, its features and values—especially the focus on high quality in niche products and the resulting avoidance of atomic markets—are responsible for the survival of their moral commitments. The psychological and sociological tradition of stewardship coupled with an ambition toward excellent craftsmanship is a method that sustains morale and ethical behaviours despite economic downturns.

Evolving from a moral standpoint, it prevents the potential sacrifice of ethics in the face of financial scarcity. At the same time, it mirrors a highly competitive business approach that has the potential to be generalised to other unconventional, "non-standard" milieus where money might be short (e.g., in start-up companies). Thus, it fills what is needed when asking for new templates that are working best in such a convergence of mission, ethics, culture, economic rationale, innovation and, most importantly, under external pressure.

The way SMEs combine the psychological and economic features of their trust-based business model, L'EPOQuE, provides a couple of advantages not only to them but also to societies they have "a social contract with" (Reidenbach & Robin, 1993). It establishes a certain ground level of trustworthiness among entities and induces incentives to imitate those practices (Bhattacharya, Devinney, & Pilluta, 1998).

Seeing trust as an integrated part of business makes the concept of trust more tangible for standard economic analysis and approaches (e.g., the business case for CSR) without losing any of its facets or without making unrealistic or too idealistic assumptions (Bhattacharya et al., 1998).

Formal CSR, on the contrary, focuses on the satisfaction of diffuse stakeholder interests and on profit-maximisation, consequently an arduous task in financially difficult times. Considering the economic, political, and social competitiveness of Swiss SMEs and their relevance to CSR, this study demonstrates the power of the small business approach as it works detached from economic situations. This questions the primacy of the profit motive and formalisation to support CSR.

> [. . .] this report (as expression of formal CSR, note of the authors) have the sole aim "to cover my ass" and, honestly, after one year the only thing we actually did related to CSR was writing this superficial piece of paper [. . .] (SME 5)

The results imply that policy makers, public institutions, the scientific community etc. should be careful when establishing systems that favour financial returns from CSR engagement (Looser & Wehrmeyer, 2015d). Furthermore, not consulting, or well consciously neglecting SMEs in decision processes is inappropriate in light of the social and economic role they play especially in Switzerland. According to several SME owner-managers, in a worst case scenario, such a stress field could result in demotivation and corrosion of the values as found and aggregated to L'EPOQuE.

What may be missing at this particular stage of research is a detailed analysis and validation of the motives of owner-managers and their inherent role as drivers for Swiss CSR. The L'EPOQuE business model aims at understanding the nexus of CSR concerns and business models showing how the former is embedded in the latter. Arguably, this model represents a regional interpretation of a specific set of circumstances likely to reveal differences in emphasis. Other contexts may have led to different overarching factors. Even within this research some heterogeneity related to sector, region, or size was observed.

There is, for instance, a tendency towards more democracy but less commitment to life-long training in service companies. Manufacturing companies, often located in less populated regions, have in general tighter links to their employees (evolving a "family feeling") than tertiary one as evidenced by their strictly regional hiring policies. Their owner-managers, often very active in local associations and/or politics, play a bigger role in their community than those in urban settings.

One question remains: how did NGOs', local citizens and working people experienced and assessed L'EPOQuE? To answer this question and to rekindle the dialogue between all groups a roundtable, as part of an action research, was organised. The media, NPOs, the government, and 15 SMEs participated in this exercise to stimulate more attention within the wider public but also to exchange their expertise on CSR issues. The participants were either involved in the stakeholder analysis or in the later stages that looked at SME patterns.

The contribution of NGOs in particular explored the boundaries between their "social enterprise approach" and SMEs and confirmed that there is not much difference. Both parties agreed that the roundtable had initiated a mutual learning process wherefrom societal benefit arose. The exchange with the media and government helped clarify and give answers on the issue of how SME practices should be more strongly respected and involved in future CSR developments, as well as how Swiss society could profit (and vice versa for SMEs) from unconventional CSR practices without sacrificing its altruistic or philanthropic inclinations. This roundtable also enabled further discussions and confirmation of L'EPOQuE 2.0 business model by gaining insights into shared understandings of the SME business from other sectors.

As a result, the level of confidence in the model was increased and allowed its results to be generalizable. More information with regard to the acceptance of the model can be found in the extensive stakeholder analysis (Looser & Wehrmeyer, 2014), in the Delphi process conducted during the model evolution (Looser & Wehrmeyer, 2015b), in the consistency analysis of the model with conventional business models (Looser & Wehrmeyer, 2016), and in a comparative analysis with 15 different other countries (Looser & Wehrmeyer, 2015c) etc.

This is not to say that Switzerland will not evolve CSR and build new forms of constituencies extending beyond discreteness and tradition. An example is given by Swiss Valais, an association of 160 SMEs, paired up to create a "sustainable" label. The main drivers here are visibility and differentiation, not bureaucracy. Swiss culture is driven by largely protestant ethical values that underline hard work, business acumen, wealth creation but also modesty. However, CSR is felt in some part of the country like an added value, which should be promoted to a wider audience to serve and to reinforce local ties. Closeness, trust, a shared sense of responsibility can develop through public commitment to CSR.

> [. . .] the way we do our business deserves wider recognition [. . .] (SME 6)

This is a conclusion strongly supported by others, e.g., Del Baldo (2010a, 2010b); Delley (2013); Looser and Wehrmeyer (2015d). Evidently, this experience can be shared more widely—the way CSR is practised by SMEs is good in general, but particularly for the firm, communities, and society. And when advertised, it can reinforce a regional sense of belonging. Swiss firms' scores on cultural dimensions show that social capital and institutional pressure are indeed powerful drivers for CSR. This is similar to findings from Italy (e.g., Calace, 2014), Germany (Schwalbach & Klink, 2012), but also from developing countries, for instance, Bangladesh (Mahmudur Rahim & Wisuttisak, 2013). But most importantly, what the "Valais case" showed is that CSR is not monolithic. Thus, the problems formalisation based on e.g., public regulation ending up in certifications and reports show, is that it tends to impose a "one size fits all" mentality and focuses only on ex-post accountability and not on, what might be useful, ex-ante prevention of adverse impacts (Buhmann, 2015). Notably, the latter is exactly what the modus operandi of Swiss SMEs achieves.

Above and beyond the disadvantages of being cumbersome and expensive, formalisation is unrealistic and a very "un-Swiss" proposition. The lesson learnt from the "Valais case" is that CSR is linked to the territory and the size of companies. Companies with local imprints should develop local CSR practices that are different from other SMEs with regional or national reach.

It serves no purpose for SMEs to emulate global models. Thus, it is useless to squeeze companies into sets of best practices—see also Looser and Wehrmeyer (2015c). What matters is accountability and this varies with size, industry, and scope of business. Getting together to demonstrate commitment to CSR and constituency can foster a revived sense of belonging, develop innovative solutions, and give CSR more consistency.

13 Further Research Steps

The relationship between leadership, good business, religiosity, and spirituality might be an area needing further investigations. For instance, the relation between L'EPOQuE and the EoC project (Baldarelli, Del Baldo, & Ferrone, 2015). Since L'EPOQuE is built on reciprocity, it may chime with the principles of the over 1000 companies participating in the EoC project such as "love one another as I loved you" (John 13: 34) or "give and it will be given to you" (Luke, 6: 38) (Bruni & Uelmen, 2006). The shift from weak to strong CSR reported by companies following the EoC principles and the wide spread of these values not only as productive ends in themselves but as new forms of humanistic management (Del Baldo &

Baldarelli, 2015) snugly fits with the L'EPOQuE's insistence on subsidiarity in decision making processes.

Other approaches could look at employees' opinion—thus probably requiring a large quantitative survey. Such studies could help refine L'EPOQuE model built on site visits and a small number of interviews. It could serve as means of crosschecking the results obtained here with trade unions (as some sort of employees' representatives). It could also address issue about diversity in culturally driven contexts. Given that culture is a process of normalisation, it could trace the socialisation trends of employees working within L'EPOQuE's model. Questions like are employees training easier under such circumstances would open up large research opportunities that should not be ignored.

Finally, the relationship between social enterprises and socially driven SMEs should be investigated. Are they similar or not? If so, would it lead to problems or rather to social benefits? And if they weren't, what dimensions of dissent would be concerned? How would they relate to each other? Would they be co-dependent, mutually exclusive, incompatible, or just substantially different? Which one is better and by what yardstick? These are questions very well worth further research.

References

Arjaliès, D.-L., & Mundy, J. (2013). The use of management control systems to manage CSR: A levers of control perspective. *Management Accounting Research, 24*, 284–300.

Asif, M., Searcy, C., Zutshi, A., & Fisscher, O. A. M. (2013). An integrated management system approach to corporate social responsibility. *Journal of Cleaner Production, 56*, 7–17.

Atteslander, P. (2010). *Methoden der empirischen Sozialforschung* (13., neu bearbeitete und erweiterte Auflage). Berlin: Erich Schmidt Verlag.

Baldarelli, M.-G., Del Baldo, M., & Ferrone, C. (2015). The relationship between CSR, good governance, and accountability in the economy of communion (EoC) enterprises. In S. O. Idowu, C. S. Frederiksen, A. Y. Mermod, & M. E. J. Nielsen (Eds.), *Corporate social responsibility and governance—Theory and practice*. Zürich: Springer.

Banerjee, S. B. (2008). Corporate social responsibility: The good, the bad and the ugly. *Critical Sociology, 34*(1), 51–79.

Baumann-Pauly, D., Wickert, C., Spence, L. J., & Scherer, A. G. (2013). Organising corporate social responsibility in small and large firms: Size matters. *Journal of Business Ethics, 115*(4), 693–705.

Baumgartner, E., Greiwe, S., & Schwarb, T. (2004). Die berufliche integration von behinderten Personen in der Schweiz. Studie zur Beschäftigungssituation und zu Eingliederungsbemühungen. *Beiträge zur Sozialen Sicherheit, 5*(4), 1–36.

Berger, V., Winistörfer, H., Weissert, S., Heim, E., & Schüz, M. (2012). *Swiss corporate sustainability survey 2012: Nachhaltigkeit in Schweizer Unternehmen*. Winterthur: ZHAW.

Bhattacharya, R., Devinney, T. M., & Pilluta, M. M. (1998). A formal model of trust based on outcomes. *Academy of Management Review, 23*(3), 450–472.

Bio Suisse. (2016). *Bio in Zahlen*. Retrieved April 27, 2016, from http://www.bio-suisse.ch/de/bioinzahlen.php

Bobst. (2015). *Annual profile 2015*. Retrieved January 17, 2016, from http://www.bobst.com/fileadmin/user_upload/Investors/Reports/Annual_Profile_2015_EN.pdf

Bowen, H. P. (1953). *Social responsibility of the businessmen*. New York: Harper.

Brown, D. J., & King, J. B. (1982). Small business ethics: Influences and perceptions. *Journal of Small Business Management, 20*(1), 11–18.
Brûlé, T. (2005, November 19). Frustration and the great European office furniture hunt. *Financial Times*. Retrieved May 31, 2015, from http://www.ft.com/intl/cms/s/b680f346-58a0-11da-90dd-0000779e2340,Authorised=false.html?_i_location=http%3A%2F%2Fwww.ft.com%2Fcms%2Fs%2F0%2Fb680f346-58a0-11da-90dd-0000779e2340.html%3Fsiteedition%3Dintl&siteedition=intl&_i_referer=
Bruni, L., & Uelmen, A. J. (2006). Religious values and corporate decision making: The economy of communion project. *Fordham University Journal of Corporate & Financial Law, 9*, 645–680.
Buhmann, K. (2015). Juridifying corporate social responsibility through public law: Assessing coherence and inconsistencies against UN guidance on business and human rights. *International and Comparative Corporate Law Journal, 11*(3), 1–24.
Calace, D. (2014). Non-financial reporting in Italian SMEs: An exploratory study on strategic and cultural motivations. *International Journal of Business Administration, 5*(3), 1–35.
Carroll, A. B. (1979). A three-dimensional conceptual model of corporate performance. *Academy of Management Review, 4*, 497–505.
Castelló, I., & Lozano, J. M. (2011). Searching for new forms of legitimacy through corporate responsibility rhetoric. *Journal of Business Ethics, 100*(1), 11–29.
CDI. (2011). Supported education. Forschungsbericht Schweiz. Retrieved April 24, 2014, from http://www.cdi.unisg.ch/de/volkswirtschaft/forschung/supportededucation
Cordeiro, J. J., & Sarkis, J. (1997). Environmental proactivism and firm performance: Evidence from security analyst earnings forecasts. *Business Strategy and the Environment, 6*(2), 104–114.
Crane, A., Matten, D., & Spence, L. J. (2013). Corporate social responsibility in a global context. In A. Crane, D. Matten, & L. J. Spence (Eds.), *Corporate social responsibility: Readings and cases in a global context* (2nd ed., pp. 3–26). Abington: Routledge.
Creswell, J. W. (2007). *Qualitative inquiry & research design: Choosing among five approaches*. Thousand Oaks: Sage.
Credit Suisse. (2014, June). Erfolgsfaktoren für Schweizer KMUs—Perspektiven und Herausforderungenim Export. *Swiss Issue Branches*. Retrieved August 6, 2015, from https://www.creditsuisse.com/meda/production/pd/docs/unternehmen/kmugrossunternehmen/cs-kmu-studie-de.pdf
Curran, J., & Blackburn, R. (1994). *Small firms and local economic networks: The death of the local economy?* London: Paul Chapman.
Deegan, C. (2002). The legitimising effect of social and environmental disclosures—A theoretical foundation. *Accounting, Auditing & Accountability Journal, 15*(3), 282–311.
Del Baldo, M. (2010a). Corporate social responsibility and corporate governance in Italian SMEs: The experience of some "spirited businesses". *Journal of Management Governance, 16*, 1–36.
Del Baldo, M. (2010b). Corporate social responsibility and corporate governance in Italian SMEs: Towards a "territorial" model based on small "champions" of CSR? *International Journal of Sustainable Society, 2*(3), 215–247.
Del Baldo, M., & Baldarelli, M. G. (2015). From weak to strong CSR: The experience of the EoC (Economy of Communion) industrial parks in Germany and Italy. *uwf UmweltWirtschaftsforum, 23*, 213–226.
Delley, R. (2013). La responsabilité sociale d'entreprise. *Repères*. Fondation IPT (p. 6).
Du, S., & Vieira, E. T. (2012). Striving for legitimacy through corporate social responsibility: Insights from oil companies. *Journal of Business Ethics, 110*(4), 413–427.
Eidgenössisch Technische Hochschule (ETH). (2009). *Grand Challenges in Environmental and Sustainability Science and Technology*. CCES Conference. Retrieved January 28, 2016, from http://www.cces.ethz.ch/index
El Baz, J., Laguir, I., Marais, M., & Staglianò, R. (2014). Influence of national institutions on the corporate social responsibility practices of small- and medium-sized enterprises in the food-

processing industry: Differences between France and Morocco. *Journal of Business Ethics, 134*(1), 117-133.
Elkington, J. (1998). Partnership from cannibals with forks: The triple bottom line of the 21st-centrury business. *Environmental Quality Management, 8*(1), 37–51.
Enderle, G. (2004). Global competition and corporate responsibilities of small and medium-sized enterprises. *Business Ethics: A European Review, 13*(1), 51–63.
Epstein, M., & Roy, M. (2003). Making the business case for sustainability. *Journal of Corporate Citizenship, 9*, 79–96.
Fassin, Y. (2008). SMEs and the fallacy of formalising CSR. *Business Ethics: A European Review, 17*(4), 364–378.
Federal Administration. (2008). *Information about Switzerland*. 7 July 2008. Retrieved August 3, 2015, from http://web.archive.org/web/20100123153543/http://www.eda.admin.ch/eda/en/home/reps/ocea/vaus/infoch.html
Ferauge, P. (2013). The complementarity of corporate social responsibility and innovation: Evidence from Belgian firms. *Global Journal of Business Research, 7*(5), 99–113.
Filbeck, G., & Gorman, R. F. (2004). The relationship between environmental and financial performance of public utilities. *Environmental and Resource Economics, 29*, 137–157.
Fondation IPT. (2015). *Die Brücke zur Arbeit*. Retrieved January 11, 2016, from www.fondation-ipt.ch/
Friedman, M. (1962). *Capitalism and freedom*. Chicago, IL: University of Chicago Press.
FSO Federal Statistical Office. (2003). *Definitionen*. 06 May 2003. Retrieved January 1, 2015, from http://www.bfs.admin.ch/bfs/portal/de/index/themen/06/11/def.html
FSO Federal Statistical Office. (2012a). *Bildungssystem Schweiz—Indikatoren: Umfeld des Bildungssystems—Ausbildende Betriebe*. Retrieved April 20, 2014, from http://www.bfs.admin.ch/bfs/portal/de/index/themen/15/17/blank/01.indicator.403302.4031.html
FSO Federal Statistical Office. (2012b). *Panorama: Industrie und Dienstleistungen*. Retrieved September 3, 2014, from http://www.bfs.admin.ch/bfs/portal/de/index/themen/06/01/pan.Document.118137.pdf
FSO Federal Statistical Office. (2013). *Statistik der Unternehmensstruktur 2011*. Retrieved March 6, 2014, from http://www.bfs.admin.ch/bfs/portal/de/index/themen/06/01/new/nip_detail.html?gnpID=2013-716
Gabler. (2015). *Gabler Wirtschaftslexikon*. Retrieved August 8, 2015, from http://wirtschaftslexikon.gabler.de/
Gentile, G. C. (2012). Gemeinschaft und Gesellschaft. In T. Wehner & G. C. Gentile (Eds.), *Corporate Volunteering. uniscope: Publikationen der SGO Stiftung* (pp. 173–180). Wiesbaden: Gabler Verlag Springer Fachmedien.
Gentile, G. C., & Lorenz, C. (2012). Schweizer Unternehmen als gute Bürger—Eine Tradition im Wandel der Zeit. In T. Wehner & G. C. Gentile (Eds.), *Corporate volunteering uniscope: Publikationen der SGO Stiftung* (pp. 79–89). Wiesbaden: Gabler Verlag Springer Fachmedien.
Gephard, R. P. (2004). From the editors: Qualitative research and the academy of management journal. *Academy of Management Journal, 47*(4), 454–462.
Godfrey, P. (2005). The relationship between corporate philanthropy and shareholder wealth: A risk management perspective. *Academy of Management Review, 30*, 777–798.
Gond, J.-P., Grubnic, S., Herzig, C., & Moon, J. (2012). Configuring management control systems: Theorizing the integration of strategy and sustainability. *Management Accounting Research, 23*, 205–223.
Graafland, J. J., & Mazereeuw-Van der Duijn Schouten, C. (2012). Motives for corporate social responsibility. *De Economist, 160*, 377–396.
Gundel, S. (2005). Towards a new typology of crises. *Journal of Contingencies and Crisis Management, 13*(3), 106–115.
Hassel, L., Nilsson, H., & Nyquist, S. (2005). The value relevance of environmental performance. *European Accounting Review, 14*, 41–62.

Helmig, B., Lichtsteiner, H., & Gmür, M. (2010). *Der dritte Sektor in der Schweiz. Landerstudie zum Johns Hopkins comparative non-profit sector project (CNP)*. Bern: Haupt Verlag.

Høgevold, N. M., Svensson, G., Wagner, B., Petzer, D. J., Klopper, H. B., Sosa Varela, J. C., et al. (2014). Sustainable business models—Corporate reasons, economic effects, social boundaries, environmental actions and organisational challenges in sustainable business practices. *Baltic Journal of Management, 9*(3), 357–380.

Il Park, B., & Ghauri, P. N. (2015). Determinants influencing CSR practices in small and medium sized MNE subsidiaries: A stakeholder perspective. *Journal of World Business, 50*, 192–204.

Jamali, D. (2008). A stakeholder approach to corporate social responsibility: A fresh perspective into theory and practice. *Journal of Business Ethics, 82*(2), 213–231.

Jenkins, H. (2006). Small business champions for corporate social responsibility. *Journal of Business Ethics, 67*(3), 241–256.

Jones, T. M., & Wicks, A. C. (1999). Convergent stakeholder theory. *Academy of Management Review, 24*, 206–221.

Jönsson, S. A., Lundin, R. A., & Sjöberg, L. (1978). Frustration in decision making processes: A tentative frame of reference. *International Studies of Management & Organization, 7*(3), 6–19.

Kalakou, S., & Macário, R. (2013). An innovative framework for the study and structure of airport business models. *Case Studies on Transport Policy, 1*, 2–17.

Kitzinger, J. (1994). The methodology of focus groups: The importance of interaction between research participants. *Sociology of Health and Illness, 16*(1), 103–121.

KMU Portal. (2012). Eurokrise: Die Schweiz halt sich, so gut es geht. Retrieved January 20, 2016, from http://www.kmu.admin.ch/aktuell/00305/02432/02687/index.html?lang=de

KMU Portal. (2015). *Moderates Wachstum im Jahre 2015*. Retrieved January 20, 2016, from http://www.kmu.admin.ch/aktuell/00305/02432/02687/index.html?lang=de

Kriesi, H. (1980). *Entscheidungsstrukturen und Entscheidungsprozesse in der Schweizer Politik*. Frankfurt: Campus Verlag.

Lagadec, P. (1993). *Preventing chaos in a crisis*. London: McGraw Hill.

Lagadec, P. (1995). *Cellules de Crise: Les conditions d'une conduite efficace*. Paris: Editions d'organisation.

Lindblom C. K. (1994). *The implications of organizational legitimacy for corporate social performance and disclosure*. Critical Perspectives on Accounting Conference, New York.

Linder, W. (2005). *Schweizerische Demokratie* (2., vollständig überarbeitete und aktualisierte Auflage). Bern: Haupt Verlag.

Looser, S., & Wehrmeyer W. (2014). *CSR mapping: Swiss stakeholder salience, concerns, and ethics*. Social Science Research Network. Retrieved August 27, 2015, from http://papers.ssrn.com/sol3/papers.cfm?abstract_id=2541199

Looser, S., & Wehrmeyer, W. (2015a). Stakeholder mapping of CSR in Switzerland. *Social Responsibility Journal, 11*(1), 780–830.

Looser, S., & Wehrmeyer, W. (2015b). An emerging template of CSR in Switzerland. *Corporate Ownership and Control Journal, 12*(3), 541–560.

Looser, S., & Wehrmeyer, W. (2015c). Varieties of capitalism and small business CSR: A comparative overview. *International Journal of Social, Behavioral, Educational, Economic and Management Engineering, 9*(7), 1953–1962.

Looser, S. and Wehrmeyer, W. (2015d). Doing well or doing good? *UmweltWirtschaftsForum, 4*. (published online).

Looser, S. and Wehrmeyer, W. (2016). Swiss CSR-driven business models: Extending the mainstream or the need for new templates? *Corporate Ownership and Control Journal, 13*(4), 130–141.

Lorenz, C., & Spescha, G. (2012). Engagement bei Profit- und Non-Profit-Unternehmen. In T. Wehner & G. C. Gentile (Eds.), *Corporate Volunteering. uniscope: Publikationen der SGO Stiftung* (pp. 115–126). Wiesbaden: Gabler Verlag Springer Fachmedien.

Lorenzo-Molo, C. F., & Siloran Udani, Z. A. (2013). Bringing back the essence of the "S" and "R" to CSR: Understanding the limitations of the merchant trade and the white man burden. *Journal of Business Ethics, 117*, 123–136.

Lounsbury, M. (2008). Institutional rationality and practice variation: New directions in the institutional analysis of practice. *Accounting, Organisations and Society, 33*, 349–361.

MacGregor, S. P., & Fontrodona, J. (2008). *Exploring the fit between CSR and innovation* (Vol. 759, pp. 1–20). Working Paper. Centre for Business in Society.

Mahmudur Rahim, M., & Wisuttisak, P. (2013). Corporate social responsibility oriented compliances and SMEs access to global market: Evidence from Bangladesh. *Journal of Asia-Pacific Business, 14*(1), 58–83.

Margolis, J. D., & Walsh, J. P. (2003). Misery loves companies: Rethinking social initiatives by business. *Administrative Science Quarterly, 48*(2), 268–305.

Martinet, A. C., & Payaud, M. A. (2008). Formes de RSE et entreprises sociales: une hybridation des strategies. *Revue Française de Gestion, 34*(180), 199–214.

Martinet, A. C., & Reynaud, E. (2001). Shareholders, stakeholders et stratégie. *Revue Française de Gestion, 136*, 12–25.

Matten, D., & Moon, J. (2008). 'Implicit' and 'explicit' CSR: A conceptual framework for a comparative understanding of corporate social responsibility. *Academy of Management Review, 33*(2), 404–424.

Mayr, S. (2015). Corporate social responsibility in SMEs: The case of Austrian construction company. Research Gate. Retrieved January 20, 2016, from https://www.researchgate.net/publication/279000462_CORPORATE_SOCIAL_RESPONSIBILITY_IN_SMES_THE_CASE_OF_AN_AUSTRIAN_CONSTRUCTION_COMPANY

McWilliams, A., & Siegel, D. (2000). Corporate social responsibility and financial performance: Correlation or misspecification? *Strategic Management Journal, 21*, 603–609.

McWilliams, A., & Siegel, D. (2001). Corporate social responsibility: A theory of the firm perspective. *Academy of Management Review, 26*, 117–127.

Mintzberg, H. (1983). The case for social responsibility. *Journal of Business Strategy, 4*(2), 3–15.

Morgan, D. L. (1988). *Focus group as qualitative research*. Thousand Oaks, CA: Sage.

Morgan, D. L. (1996). Focus groups. *Annual Review of Sociology, 22*, 129–152.

Morsing, M., & Perrini, F. (2009). CSR in SMEs: Do SMEs matter for the CSR agenda? *Business Ethics: A European Review, 18*(1), 1–6.

Nelling, E., & Webb, E. (2009). Corporate social responsibility and financial performance: The "Virtuous Circle" revisited. *Review of Quantitative Finance and Accounting, 32*(2), 197–209.

Neron, P., & Norman, W. (2008). Citizenship Inc.: Do we really want business to be good citizens. *Business Ethics Quarterly, 18*(1), 1–26.

Newman, A., & Sheikh, A. Z. (2012). Organisational commitment in Chinese small and medium-sized enterprises: The role of extrinsic, intrinsic and social rewards. *The International Journal of Human Resource Management, 23*(2), 349–367.

Nkiko, C. M. (2013). SME owner-manager as key drivers of corporate social responsibility in Uganda. *International Journal of Business Governance and Ethics, 8*(4), 376–400.

Nolan, J. (2013). The corporate responsibility to respect human rights: Soft law or not law? In S. Deva & D. Bilchitz (Eds.), *Human rights obligations of business: Beyond the corporate responsibility to respect?* (pp. 138–161). Cambridge: Cambridge University Press.

Orlitzki, M., Schmidt, F. L., & Rynes, S. L. (2003). Corporate social and financial performance: A meta-analysis. *Organisational Studies, 24*(3), 403–441.

Osterwalder, A., & Pigneur, Y. (2010). *Business model generation* (1st ed.). Hoboken, NJ: John Wiley & Sons.

Osuji, O. (2011). Fluidity of regulation-CSR Nexus: The multinational corporate corruption example. *Journal of Business Ethics, 103*, 31–57.

Pereira Pündrich, A., & Triki, D. (2014). *Les stratégies de RSE en temps des crises intraitables: le context. 8eme congrès de l'ADERSE*, March 2011, France, p. 16.

Philipsen, G. (1987). The prospects for cultural communication. In L. Kincaid (Ed.), *Communication theory: Eastern and Western perspectives* (pp. 245–254). San Diego: Academic Press.

Polonsky, M., & Wood, G. (2001). Can the over-commercialization of cause-related marketing harm society. *Journal of Macromarketing, 21*(8), 8–22.

Pondeville, S., Swae, V., & De Rongé, Y. (2013). Environmental management control systems: The role of contextual and strategic factors. *Management Accounting Research, 24*, 317–332.

Porter, M., & Kramer, M. (2006). Strategy and society: The link between competitive advantage and corporate social responsibility. *Harvard Business Review, 84*(12), 78–92.

Reidenbach, R. E., & Robin, D. P. (1993). Toward the development of a multidimensional scale for improving evaluations of business ethics. *Journal of Business Ethics, 9*(8), 639–653.

Richter, U. (2010). Liberal thought in reasoning on CSR. *Journal of Business Ethics, 97*, 625–649.

Rosenthal, U., & Koumin, A. (1993). Globalizing an agenda for contingencies and crisis management: An editorial statement. *Journal of Contingencies and Crisis Management, 1*(1), 1–12.

Roux-Dufort, C. (2003). *Gérer et décider en situation de crise*. Paris: Dunod.

RTS. (2015). *Mettez du bois dans votre chaudière*. 17 November 2009. Retrieved January 11, 2016, from http://www.rts.ch/emissions/ttc/934592-mettez-du-bois-dans-votre-chaudiere.html

Russo, A., & Perrini, F. (2010). Investigating stakeholder theory and social capital: CSR in large firms and SMEs. *Journal of Business Ethics, 91*(2), 207–221.

Samuel, O., Schilling, A., & Wehner, T. (2012). Corporate Volunteering aus der Perspektive schweizerischer NPO. In T. Wehner & G. C. Gentile (Eds.), *Corporate Volunteering. uniscope: Publikationen der SGO Stiftung* (pp. 101–114). Wiesbaden: Gabler Verlag Springer Fachmedien.

Sartre, V. (2003). *La communication de crise: anticiper et communiquer en situation de crise*. Paris: Editions Demos.

Scheuchzer. (2015). *Aujourd'hui*. http://www.scheuchzer.ch/web/guest/aujourd-hui

Schultz, F. (2013). Corporate social responsibility, reputation, and moral communication: A constructivist view. In C. E. Carroll (Ed.), *The handbook of communication and corporate reputation* (1st ed., pp. 362–375). Malden, MA: John Wiley & Sons.

Schur, L., Kruse, D., & Blanck, P. (2005). Corporate culture and the employment of people with disabilities. *Behavioral Science and the Law, 23*(1), 3–20.

Schwalbach, J., & Klink, D. (2012). Der Ehrbare Kaufmann als individuelle Verantwortungskategorie der CSR Forschung. In A. Schneider & R. Schmidpeter (Hrsg.), *Corporate social responsibility* (pp. 219–240). Heidelberg: Springer.

SECO State Secretariat for Economic Affairs. (2005). *The strategic focus of Swiss foreign economic policy*. Retrieved September 3, 2015, from http://www.seco.admin.ch/dokumentation/publikation/00008/00023/01742/index.html?lang=deanddownload=NHzLpZeg7t,lnp6I0NTU042l2Z6ln1acy4Zn4Z2qZpnO2Yuq2Z6gpJCDdYN2gmym162epYbg2c_JjKbNoKSn6A

SECO State Secretariat for Economic Affairs. (2010). *Annual Report 2009/2010 on Activities of the Swiss National Contact Point to the OECD Investment Committee*. Retrieved August 3, 2015, from http://www.seco.admin.ch/themen/00513/00527/02584/02586/index.html?lang=enanddownload=NHzLpZeg7t,lnp6I0NTU042l2Z6ln1ad1IZn4Z2qZpnO2Yuq2Z6gpJCFent5g2ym162epYbg2c_JjKbNoKSn6A--

SECO State Secretariat for Economic Affairs. (2011). *Ambitionen sind gut, Fokussierung ist besser*. 02.02.21011. Retrieved October 31, 2013, from http://www.kmu.admin.ch/aktuell/00524/02098/02122/index.html?lang=de

SECO State Secretariat for Economic Affairs. (2012a). *KMU-Portal: KMU-Themen*. Retrieved September 3, 2013, from http://www.kmu.admin.ch/themen/index.html?lang=de

SECO State Secretariat for Economic Affairs. (2012b). *KMU-Portal: Normierung*. Retrieved September 3, 2013, from, http://www.kmu.admin.ch/themen/01897/01898/index.html?lang=de

SECO State Secretariat for Economic Affairs. (2012c). *KMU-Portal: Zertifizierungen*. Retrieved September3,2013,fromhttp://www.kmu.admin.ch/themen/01897/01899/index.html?lang=de
SECO State Secretariat for Economic Affairs. (2012d). *KMU-Portal: Nachhaltige Entwicklung*. RetrievedSeptember3,2013,fromhttp://www.kmu.admin.ch/themen/02584/index.html?lang=de
SECO State Secretariat for Economic Affairs. (2012e). *KMU-Portal: Perspektiven der KMU-Politik*. Retrieved September 3, 2013, from http://www.kmu.admin.ch/politik/00152/00768/index.html?lang=de
SECO State Secretariat for Economic Affairs. (2012f). *CSR-Konzept des SECO*. Retrieved September 3, 2013, from http://www.seco.admin.ch/themen/00645/04008/index.html?lang=deanddownload=NHzLpZeg7t,lnp6I0NTU042l2Z6ln1acy4Zn4Z2qZpnO2Yuq2Z6gpJCFeXt3gGym162epYbg2c_JjKbNoKSn6A
SECO State Secretariat for Economic Affairs. (2012g). *Annual Report 2010/2011 on Activities of the Swiss National Contact Point for the OECD Guidelines for Multinational Enterprises*. Retrieved September 3, 2013, from http://www.seco.admin.ch/themen/00513/00527/02584/02586/index.html?lang=enanddownload=NHzLpZeg7t,lnp6I0NTU042l2Z6ln1ad1IZn4Z2qZpnO2Yuq2Z6gpJCFfH1,g2ym162epYbg2c_JjKbNoKSn6A--
SECO State Secretariat for Economic Affairs. (2012h). *Allgemeinverbindlich erklärte Gesamtarbeitsverträge*. Retrieved September 3, 2013, from http://www.seco.admin.ch/themen/00385/00420/00430/index.html
Spence, L. J., Schmidpeter, R., & Habisch, A. (2003). Assessing social capital: Small and medium sized enterprises in Germany and UK. *Journal of Business Ethics, 47*, 17–29.
SRF Schweizer Radio und Fernsehen. (2013). *Schweizer KMUs—Rückgrat der Wirtschaft*. 28 May 2013. Retrieved October 10, 2015, from http://www.srf.ch/news/infografik/schweizer-kmu-rueckgrat-der-wirtschaft
tel.search.ch. (2015). *search.ch*. URL: tel.search.ch [Accessed many times between 01.07.2015–23.01.2016].
Telle, K. (2006). "It pays to be green"—A premature conclusion? *Environmental and Resource Economics, 35*, 195–220.
Travail Suisse. (2015). *6 conséquences importantes de la hausse du franc suisse par rapport à l'euro*. Retrieved January 11, 2016, from http://www.travailler-en-suisse.ch/hausse-chf-eur.html
Van Marrewijk, M. (2003). Concepts and definitions of CSR and corporate sustainability: Between agency and communion. *Journal of Business Ethics, 44*(2), 95–105.
Vveinhardt, J., & Zygmantaite, R. (2015). Impact of social context on strategic philantropy: Theoretical insight. *Procedia: Social and Behavioral Sciences, 214*, 1165–1173.
Wagner, M. (2005). How to reconcile environmental and economic performance to improve corporate sustainability: Corporate environmental strategies in the European paper industry. *Journal of Environmental Management, 79*, 105–118.
Wagner, M., Schaltegger, S., & Wehrmeyer, W. (2001). The relationship between the environmental and economic performance of firms—What does theory propose and what does empirical evidence tell us? *Greener Management International, 34*, 95–108.
Wang, L., & Juslin, H. (2009). The impact of Chinese culture on corporate social responsibility: The harmony approach. *Journal of Business Ethics, 88*, 433–541.
Wartick, S. L., & Cochran, P. L. (1985). The evolution of the corporate social performance model. *Academy of Management Review, 10*(4), 758–769.
Westphalen, M. H. (1992). *Le Dicom*. Paris: Triangle Edition.
Zadek, S. (2004). The path to corporate social responsibility. *Harvard Business Review, 16*(2), 211–232.
Zefix (2014). *Central Business Names Index*. http://zefix.admin.ch/ [Accessed 05.02.2015, 14.02, 2015, 28.02.2015; 15.03.2015, 10.04.2015, 15.04.2015, 22.04.2015, 23.04.2015].
Zeyen, A., Beckmann, M., & Wolters, S. (2014). Actors and institutional dynamics in the development of multi-stakeholder initiatives. *Journal of Business Ethics, 135*(2), 341-360.
Zheng, Q., Luo, Y., & Maksimov, V. (2014). Achieving legitimacy through corporate social responsibility: The case of emerging economic firms. *Journal of World Business*, 389–403.

Authenticity: Is Corporate Social Responsibility the Key to Overcoming Crisis?

Mara Del Baldo

1 Introduction

The paper aims to develop within the theoretical and normative perspective the concept of the "golden rule" of authentic CSR-driven strategies as a necessary path to overcoming the financial crisis.

To this end, firstly a critical review is provided of the recent Encyclical Letter (Pope Francis, *Laudato Sì*, 2015) focusing on the care of our "common home" (our planet) that emphasizes the complexity of the crisis (financial and economic, preceded by a socially and anthropologically-based crisis) and suggests the path to overcoming the crisis through a renewed environmental, economic and social ecology. This last orientation involves the collective contribution to the common good, which presupposes the genuine responsibility of every person and organization belonging to the "near and far" socio-economic environment and the authentic orientation toward the integral development.

Secondly, the work provides the business context business context with a series of examples related to companies which in "hard times" voluntarily persist in investing in Corporate Social Responsibility (CSR) activities conceived as an effective and leading way to building the future and replying to the current and future crises.

The adopted research methodology was based on a content analysis of the encyclical document and on an empirical analysis which was developed through interviews addressed to the top management of ten Italian companies (both medium, large and small-sized) whose strategies are based on CSR and that have been studied for more than a decade.

M. Del Baldo (✉)
University of Urbino Carlo Bo, Urbino, Italy
e-mail: mara.delbaldo@uniurb.it

S.O. Idowu et al. (eds.), *Corporate Social Responsibility in Times of Crisis*,
CSR, Sustainability, Ethics & Governance, DOI 10.1007/978-3-319-52839-7_2

The implications of the work are twofold: on the one hand it intends to feed the scientific debate on the opportunity to focus the attention on the origins of entrepreneurial and managerial behaviors that have lead to the crisis and the need for a profound rethinking of the business conduct. On the other hand the work provides suggestions and examples to the corporate world by observing the validity of the strategic and operational behaviors founded on a durable CSR orientation and a genuine responsible decision-making process.

The paper is organized as follows. The following section introduces the theoretical framework, focusing on authenticity in CSR-driven strategies and on the business contexts where it is present. Third section presents a critical review of the "Laudato Sì" Encyclical Letter and discusses the main issues connected to the topic of sustainability and common good. Subsequently, after having introduced the methodological approach, fourth section describes the empirical findings while the final section synthesizes concluding remarks.

2 Authenticity and CSR

Authenticity has its roots in enlightenment philosophy. Rousseau relates authenticity to morality in the sense of pursuing the inbuilt voice of nature found within us (Taylor, 1992: 27). In contrast, Herder asserted that authenticity is more concerned with the idea that human beings have their own originality (Costas & Fleming, 2009). Authenticity as a driver of entrepreneurial and managerial behaviour relates to the expression of integrity (see Kaptein & Wempe, 2002 and "the diamond of integrity management model"). It is about being sincere, honest and genuine (Trilling, 1972). Entrepreneurship offers a mean to express one's authenticity and inwardly-derived values (Aragón Amonarriz & Iturrioz Landar, 2016; Cardon, Wincent, Singh, & Drnovsek, 2009). Entrepreneurial and managerial authenticity influences the nature and extent of CSR approaches and explains the orientation toward a genuine CSR and sustainability approaches claimed by the Encyclical Letter which forcefully invites to rethink the roots of the crisis and the values needed to overcome it by proposing a new social, economic, environmental and anthropological ecology.

"Organizational integrity means that the organization's values should guide its interactions with internal and external stakeholders " (Painter-Morland, 2006: 358). It reflects the organization's identity, goals and culture and is supportive of the ethical behavior. In other words, it requires an alignment and an ongoing interaction between an organization's values and its everyday practices, as well as between CSR programs and its ethics management processes (Painter-Morland, 2006). It can only exist where there is mutual trust between the organization and its stakeholders (Paine, 1994) which, in turn, depends on the manner in which the organization succeeds in living according to its stated values (Painter-Morland, 2006). Stakeholder engagement, organizational integrity and stakeholder activism (Alford & Signori, 2014) could be used to reconstruct a "social grammar" that would place business in a more sustainable relationship with society. A lack of integrity in the organization's functioning, as well as in organizational practices, constitutes "companies' vices" often diffused in large public companies which hinder an authentic CSR approach and innovative paths to facing the current and forthcoming crisis (Borgato, 2014; Malloch, 2008).

In the framework of leadership studies authenticity emerges as a key factor. Debating on the nature of CSR leadership, Visser (2011) points out that "We are convinced that without bold and effective leadership—at a political, institutional and individual level—we will fail to resolve our most serious social and environmental crisis" (Visser, 2011: 1). Moreover, as the latest Encyclical Letter clearly states, we will fail to resolve our deep anthropological crisis.

The various theories and approaches have set a frame for CSR leadership and relative key characteristics (i.e. systemic understanding, emotional intelligence and a caring attitude, values orientation, compelling vision, inclusive style, innovative approach and a long term perspective). A CSR leader is someone who inspires and supports actions toward a better word (Goffee & Jones, 2009). An authentic CSR leader orients the company toward the common good and respects the "golden rule"—typical of the Christian tradition—to do unto others what you would want for themselves.

Within the CSR debate, on the one hand CSR is either a commercial instrument and extrinsically motivated; on the other hand, it is based on intrinsic (or idealistic) reasons, which makes it a moral activity (see Looser & Wehrmeyer, 2015: 2).

The former—extrinsic CSR—is formalized, oriented to increasing legitimacy by improving image and market share and is aimed at external recognition (Matten & Moon, 2008). This opportunistic approach renders CSR a *façade* "or" emancipator rhetoric, giving rise to the so called "green-washing" phenomena (Mahoney, Thorne, Cecil, & LaGore, 2013; Visser, 2011).

The second one—intrinsic CSR—assumes that many companies are characterized by a business culture making CSR a moral duty (Looser & Wehrmeyer, 2015). In these contexts the underlying motivation which affects the entrepreneur/manager and the entire organization induces a stronger involvement in CSR, increases perceived sincerity, nurtures a positive relationship with employees and customers (Graafland & van de Ven, 2006). In these cases authentic leadership render social entrepreneurship a driver for social change and CSR has the power to generate "social goods".

Extrinsic CSR is mainly diffused among large companies while SMEs are more often intrinsically motivated by their tradition/family capitalism and rely on their long-standing informal networks rather than on formal, extrinsic policies (Looser & Wehrmeyer, 2015; Spence, Schmidpeter, & Habisch, 2003). Among the promising sustainability business and leadership models that have emerged, the "ideal-based company" (Capaldi, 2013; Del Baldo & Baldarelli, 2015; Malloch, 2008; Molteni, 2009) demonstrate a purest form (genuine/authentic) of CSR as it is ethical and supports CSR because that is "the noble way for corporations to behave" (Looser & Wehrmeyer, 2015: 3). Several studies in recent decades have in fact highlighted the existence of many companies, often little known, even silently, which are witnesses of authenticity in their CSR strategies and actions. Particularly, intrinsic motivations are diffused among community-based companies (Peredo & Chrisman,

2006), territorial companies (Del Baldo, 2010a), Economy of communion companies (Baldarelli, 2011), as well as family-based enterprises, whose social responsibility vision affects its responsible behavior (Aragón Amonarriz & Iturrioz Landar, 2016; Chua, Chrisman, & Sharma, 1999; Del Baldo, 2012).

All these companies form a widespread network that throughout the world, offers example of authentic CSR orientation and ethical-based connotation (Hoivik von Weltzien, 2014; Hoivik von Weltzien & Melé, 2009).

Accordingly, the "maturity model" proposed by Walker (2015) poses values and virtues at the foundation of practical wisdom in the European society. The set of cardinal virtues is relevant in the business context: justice has a strong relationship to the values of solidarity and integrity; temperance is related with respect, appreciation, freedom; fortitude has a strong connection to the value of subsidiarity, which represents a key value for innovations; prudence has a strong connection to the values of responsibility, accountability and liability. To overcome the crisis, the "typical" active entrepreneur orientation (aimed to increase equity capital and efficiency) is no longer enough. Hence, a proactive entrepreneurial behavior and new integrative management approaches are needed to manifest resilience through adaptability, diversity, efficiency and cohesion) which is a precondition for achieving sustainability (Walker & Salt, 2006). This approach includes fundamental pillars: stakeholder orientation, continuous improvement process, and the idea of a learning organization to foster knowledge and competences, ethical behavior, integrated reporting/accountability (see Walker, 2015: 5–6).

In short, what is needed is a genuine orientation to an integral (or holistic) development (Ketola, 2008; Sorci, 2007). The business world needs for "the application of successful intelligence and creativity toward the common good" and brave leadership capable of inventing actions which serve the common good through balancing values across diverse stakeholders and to contrast mainstream economics and business which neglect the common good, consider only monetary values, disregard non-market stakeholders, and discount the future (Zsolnai, 2015).

There has been an ongoing questioning concerning the lack of practical ethical training and the responsibilities and accountability of the business leader. Managers who hold a spiritual perspective on life are more ethical in their business conduct than those who do not hold a spiritual perspective (Bouckaert, 2011; Parry & Proctor-Thomson, 2002; Pruzan, 2011). Ethics literature has in fact emphasised management integrity, authenticity and virtues which are becoming widespread in the corporate context, giving rise to new models of governance and business aimed at constructing a more civil economy (Argandoña, 2003, 2011; Cortright & Naughton, 2002; Driscoll & Hoffman, 2000; Gui & Sugden, 2005; Luthans & Avolio, 2003; Pruzan, 2001; Zadek, 2006; Zamagni, 1995). These studies are a significant and emerging part of the theoretical framework of CSR and sustainability (Becker, 1998; Brown, 2005; Garriga & Melé, 2004; Hoivik von Weltzien, 2014; Malloch, 2008; Ruisi, 2010).

The "golden rule" translates concretely into principles which characterize purpose-led businesses: providing good and safe products and services; openly sharing knowledge and competences; building lasting relationships; being a good citizen; nurturing a responsible and responsive employer; being a guardian for future generations; and having a purpose which delivers long-term sustainable performances (Alford, 2015).

The Catholic social doctrine provides a relevant contribution in this field. Particularly in the last decades the encyclical letters have increasingly gone beyond the political and ecclesiastic/religious domain to impact significantly on the lives of every person (believer or non-believer) and the corporate world dimensions.

3 The Encyclical Letter "Laudato Sì": The Integral Ecology for People and Companies

Pope Francis' Encyclical on integral ecology provides an excellent opportunity for nurturing a conversation between the scientific, economic and spiritual "worlds" on the issues of responsibility, common good and sustainable development.

Integral ecology proposed by the Encyclical integrates concerns for people and the planet and suggests a path to sustainable development through a more frugal and moderate life-style, acknowledging the intrinsic value of nature and seeking holistic actionable knowledge. Frugality implies a low level of material consumption and a simple lifestyle (Bouckaert, Opdebeeck, & Zsolnai, 2007) and should precede efficiency in achieving sustainability. An integral and interdisciplinary understanding of the world sees ecology, economy, equity and justice, natural and social sciences, arts and humanities as systemically linked.

Reconsidering the role of business in society and nature is unavoidable as well as transforming it into a progressive social institution which respects nature, future generations and the common good of society (Zsolnai, 2015). This urgency has been previously pointed out by Patriarch Bartholomew (1997) who—drawing attention to the ethical and spiritual roots of environmental problems—claimed for solutions "not only in technology but in a change of humanity habits by replacing consumption with sacrifice, greed with generosity, wastefulness with a spirit of sharing and by learning to give, and not simply to give up" (See: Encycl. Letter, 2015: 8–9).

In the first section of the document ("what is happening to our common home?") Pope Francis underlines how the continued acceleration of changes affecting humanity and the planet (so-called "rapidification") has developed many contrasts with the naturally slow pace of biological evolution (pollution, waste and the throwaway culture): "Moreover, the goals of this rapid and constant change are not necessarily geared to the common good or to integral and sustainable human development"(Encycl. Letter, 20: 15). "The growth of the past two centuries has not always led to an integral development and an improvement in the quality of life. Some of these signs are also symptomatic of real social decline, the silent rupture of the bonds of integration and social cohesion" (Encycl. Letter, 46: 32).

Another issue addressed is the gap of leadership approaches and models capable of incorporating the orientation toward the common good and, consequently, to assume an ethical connotation (Olsen, 2010; Visser, 2011) "The problem is that we still lack the culture needed to confront this crisis. We lack leadership capable of striking out on new paths and meeting the needs of the present with concern for all and without prejudice toward coming generations" (Encycl. Letter, 53: 39). Only recently, in fact, has the issue of leadership been placed at the center of the international debate. For instance, both the accountability international guidelines for drafting sustainability/integrated report have included a specific section on leadership (GRI, 2013; IIRC, 2013).

Nevertheless, good examples and best practices of sustainability-oriented actions and strategies developed by companies, institutions, governments and individuals are increasing worldwide. "These achievements do not solve global problems, but they do show that men and women are still capable of intervening positively" (Encycl. Letter, 58: 42).

At the same time a "false or superficial ecology which bolsters complacency and a cheerful recklessness" is rising (Encycl. Letter, 59: 42–43). The Encyclical underlines in fact that every ecological approach needs to incorporate a social perspective, which—as mentioned in the previous sections—is typical of particular types of company, whose mission, governance and accountability are forged by the principle of "universal communion" and the common destination of goods (Del Baldo & Baldarelli, 2015). The principle of the subordination of private property to the universal destination of goods, is a golden rule of social conduct and "the first principle of the whole ethical and social order" (John Paul II, Encyclical Letter *Laborem Exercens*, John Paul, 1981, 626); see Encycl. Letter, 2015, 93: 68).

In Chapter three the Encyclical tackles the most profound aspect relative to "the human roots of the ecological crisis" (Ency. Letter, 101 and 102: 75) (Guardini, 1998).

The immense technological development has not been accompanied by a development in human responsibility, values and conscience (Encycl. Letter, 105: 77–78) and the object of holistic development has been missed since the idea of unlimited growth and the false notion of an infinite quantity of energy and resources has prevailed" (Encycl. Letter, 106: 78–79). Accordingly, the needs of "a distinctive way of looking at things" refers to the urgency of charismatic people and leaders capable of turning problems into opportunities and generating social innovation, starting from a different use of the immense potentialities of new technologies. In fact "the fragmentation of knowledge often leads to a loss of appreciation for the whole, for the relationships between things" (Encycl. Letter, 106: 82), and first and foremost, for relationships among people and the parts of the ecological systems which are a fundament of the holistic development (Gui & Sugden, 2005; Bruni, 2009). "We can put (technology) at the service of another type of progress, one which is healthier, more human, more social, more integral" (Encycl. Letter, 112: 84–85).

The Encyclical calls for a moral-based conduct oriented toward the daily implementations—included the business context—of fundamental values and virtues which have an authentic anthropological connotation (Argandoña, 2003, 2011; Melé, 2009, 2012). In fact the crisis of modern anthropocentrism is due to the underestimation of interpersonal relations. "There can be no renewal of our relationship with nature without a renewal of humanity itself. There can be no ecology without an adequate anthropology" (Benedict XVI, Benedict, 2009; Francis, 2015, 118: 96). We can read: "If the present ecological crisis is one small sign of the ethical, cultural and spiritual crisis of modernity, we cannot presume to heal our relationship with nature and the environment without healing all fundamental human relationships" (Encycl. Letter, 119: 89). The person is "the source, the focus and the aim of all economic and social life" (John Paul II, Encyclical Letter *Centesimus Annus*, John Paul, 1991). (See: Encycl. Letter, 127: 100).

An integral ecology (Chapter four) is the one that respects its human and social dimensions (Del Baldo, 2014b). Consequently, environmental and social crisis are not separate; rather they form one complex crisis which is both social and environmental and therefore "Strategies for a solution demand an integrated approach to combating poverty, restoring dignity to the excluded, and at the same time protecting nature"(Encycl. Letter, 138–139: 104–105).

The Pope claims the need for an "economic ecology" capable of appealing to a broader vision of reality, as we can read: "We urgently need a humanism capable of bringing together the different fields of knowledge, including economics, in the service of a more integral and integrating vision" (Encycl. Letter, 141: 105–106).

Moreover, he highlights the importance of a renewed cultural ecology, as well as an ecology of daily life. The first one is relative to the preservation of the historic, artistic and cultural heritage (Encycl. Letter, 143:107–108). The second one relates to efforts to bring about an integral improvement in the quality of human life which entails considering the setting in which people live their lives: "It is not enough to seek the beauty of design. More precious still is the service we offer to another kind of beauty: people's quality of life, their adaptation to the environment, encounter and mutual assistance" (Encycl. Letter, 151: 112).

The integral ecology is deeply connected to the common good which represents a central and unifying principle of social ethics[1] and should be concretely applied at a local level and extend to future generations (intergenerational solidarity) (Encycl. Letter, 159 and 196:118).

What emerges is the urgent need for dialogue and transparency in decision-making as well as an agreement on systems of governance for the whole range of the so-called "global commons" (Encycl. Letter, 174:128). As stated by

[1]The common good is "the sum of those conditions of social life which allow social groups and their individual members relatively thorough and ready access to their own fulfillment" Encycl. Letter, 156, p. 112) "Underlying the principle of the common good is respect for the human person. It has also to do with the overall welfare of society and the development of a variety of intermediate groups, applying the principle of subsidiarity" (Encycl. Letter, 157: 116–117).

Benedict XVI, "Love for society and commitment to the common good are outstanding expressions of a charity which affects not only relationships between individuals but also "macro-relationships, social, economic and political ones" (Encyclical Letter *Caritas in Veritate*, 2009; Encycl. Letter, 231: 156).

The reference to the responsibility at micro, meso and macro level and the proactive involvement of the community and the importance of a local shared orientation toward sustainability (territorial social responsibility and/or multilevel territorial governance) (Del Baldo, 2014a) "A community can break out of the indifference induced by consumerism" (Encycl. Letter, 167: 168).

Finally, the document recalls the importance of the dialogue between science and religion, and recommends a re-discovery of the ethical and spiritual treasures which could greatly contribute to providing replies to today's needs: "It needs educators capable of developing an ethics of ecology, and helping people, through effective pedagogy, to grow in solidarity, responsibility and compassionate care. Only by cultivating sound virtues will people be able to make a selfless ecological commitment" (Encycl. Letter, 210: 154). In this respective, the rich heritage of Christian spirituality can offer a precious contribution to the renewal of humanity because it proposes a growth marked by sobriety and moderation (Thomson & Eynikel, 2011). "When there is a general breakdown in the exercise of a certain virtue in personal and social life, it ends up causing a number of imbalances, including environmental ones. We have to dare to speak of the integrity of human life, of the need to promote and unify all the great values" (Encycl. Letter, 224: 163).

4 Authentic CSR-Orientation and Integral Ecology in the Daily Life of Companies

After having presented and briefly discussed the most important contents of the Encyclical Letter (Table 1), this section presents the first results relative to the empirical analysis aimed at verifying the presence of values, actions and strategies that consolidate these principles and are coherently characterized by an authentic CSR-orientation.

In order to verify whether these principles are applied in the business context, ten Italian companies have been selected from two different databases. The first is a national data base related to the companies monitored by the ISVI Italian Observatory (Istituto per i valori d'impresa—Institute for the Company's values). The second one is the international Observatory of the Economy of Communion companies (www.eoc-online.org). All the selected companies are CSR-oriented and over the years have implemented many actions, projects and accountability tools. Some of the companies were analyzed in previous stages of our research as best practices with reference to specific aspects of excellence (see: Del Baldo, 2010b, 2013a, b; Del Baldo & Baldarelli, 2015). The analysis was based on a qualitative

Table 1 Main and recurrent principles of the Encyclical Letter "Laudato Sì"

The Common Good
The urgency for "not weak responses"
Universal communion
The crisis and effects of modern anthropocentrism
Holistic/Integral development; Sustainable development
The question of leadership
The integral ecology: environmental ecology, economic ecology, social ecology, cultural ecology, and ecology of daily life
The centrality of the person and interpersonal relationships
Dialogue for innovative national and local policies
Transparency in decision-making
Subsidiarity
Collaboration, active involvement of the local community
Educating for the covenant between humanity and the environment
Ecological conversion
Toward a new lifestyle (sobriety, humility, frugality, spirituality)

research approach. Data have been collected through a variety of sources: interviews addressed to the entrepreneurs and the top and middle management team; direct observation during company visits and meetings (workshops, focus-groups, conventions); documental analysis (relative to company books, sustainability reports, company report) as well as technical, managerial and scientific publications refereed to the selected enterprises (Eisenhardt & Graebner, 2007; Naumes & Naumes, 2006; Yin, 1994).

The period of analysis is multi-year. Some of the companies have for years been involved in different forms of cooperation (interventions in entrepreneurship education initiatives, in meetings and congresses). Portions of the interviews are presented below. Moreover, some projects and strategies that concretize the Encyclical principles are briefly described. Table 2 summarizes the essential attributes of the ten companies.

Below, for each company some entrepreneurial and management statements grasped from informal conversations are cited.

4.1 Ariston Thermo Group Spa—Merloni

"In any entrepreneurial and business venture the value of the economic success does not exist if the commitment to social progress is not present" (A. Merloni, Founder of the Group, 1967).

Concrete and lasting values, with which the whole group identifies are: integrity, people, excellence, customer and Sustainability. Accordingly, among the shared key principles are: the centrality of the person; respect for employees' rights and

Table 2 Companies' profile

Company's name	Ownership and size	Economic sector and activity	Italian region	Year of foundation	Market
1. Ariston thermo group—Merloni group	Unlisted; Open-family companies Large-sized	Mechanics and electronics: production of appliances for heating water and environments	Marches	1930 More than 6,600 employees	Mainly international
2. Brunello Cucinelli Spa	Listed Open-family company Large-sized	Textile: production of high quality clothing, shoes, and accessories	Umbria	1978 1,300 employees	National and international
3. BoxMarche Spa	Unlisted; family based company (open to external members) small-sized company	Paper industry: design and production of packaging for the food and houseware sectors	Marches	1969 50 employees	Mainly national
4. Elica group	Listed: majority family-owned large company	Mechanics end Electrical equipment: motors for home appliances and central heating boilers	Marches	1970 Over 3,000 employees	Mainly international
5. FAAM Spa	Unlisted; medium-sized Not family owned	projects and solutions in the field of electrical accumulators and of environmentally friendly vehicles (i.e electric cars)	Marches	1974 300 employees	Mainly international
6. Loccioni group	Unlisted; family-owned medium-sized company	Electronic industry: electrical and automatic equipment; plants-design-robots; automotive; integrated technologies for environmental monitoring and quality control; biomedicine and medical	Marches	1968 More than 300 employees	Mainly international

(continued)

Table 2 (continued)

Company's name	Ownership and size	Economic sector and activity	Italian region	Year of foundation	Market
		equipment; equipment for the management of domestic energy (green energy); training courses and consultancy for technical and management education			
7. SGR group	Unlisted; family-based company medium-sized	Multi-utility Gas-energy sector: Distribution of natural gas and electric energy	Emilia Romagna	1956 330 employees	Regional, national and international
8. Insurance Agency "Ornella Seca"	Unlisted Individual entity EoC company small-sized	Insurance	Abruzzo	2002 6 employees and 7 collaborators	Local and regional
9. Bertola Srl	Unlisted; family-owned small-sized EoC company	Metalworking: galvanizing, grinding and polishing	Piemonte	1929 24 employees	National and international
10. Mercurio Ne:	Unlisted; Not family owned small-sized EoC company	ITC—Informatics solutions (hardware, software, services)	Marches	1994 7 employees and 10 external collaborators	Local and regional

needs, adoption of a participatory model of industrial relations; respect and care for the environment; innovation and quality; creation and distribution of the added value; continuous learning and training; transparency and fairness; integrity.

The Group, which was the first in Italy to introduce the social balance in the 1960s, is part of the international association "UCID"—Christian entrepreneurs for the common good. Starting from its foundation, the Group has greatly contributed to the development of the socio-economic conditions of the region, providing jobs to people in very poor and under-developed local areas and fighting depopulation caused by the flows of migrants to the Americas. Moreover, it has supported numerous projects aimed at enriching the cultural and social development of local territories, in Italy and abroad. It was part of the Copenhagen Communiqué on Climate Change and it was a pioneer in the development of high technologies aimed at preserving the environment and climate (i.e. The Energy Aware House", the ecological house for the future which obtained numerous award such as the Ecotech Award for being an eco-virtuous company).

4.2 Brunello Cucinelli Spa

Convinced that the peaceful atmosphere and the beauty of a local area enhance creativity and the respect of human values, Mr. Brunello Cucinelli set up the headquarters of the company in the fourteenth century village of Solomeo, in Umbria, in the same places where he had absorbed St. Francis's philosophy since his childhood. The entrepreneur is actively involved in working for the common good by producing a shared value, which is the fruit of his involvement, together with his employees and the local community.

"I dream of a form of modern capitalism with strong ancient roots. I believe in a humanistic enterprise, which always places human values in first place. I think that there can be no quality without humanity. In my organization the reference point is the common good" (B. Cucinelli, 2015).

"All revenues are divided in four equal parts: the first is attributed to the company, in order to consolidate future investment; the second one to the entrepreneur; the third to the employees, whose salaries are about 20% higher than the national contract, while the fourth part is to beautify humanity, that is, the creation of services, i.e. a theater or a hospital, that can benefit the entire community" (B. Cucinelli, 2015)[2]".

"I would like to convey our land and the place from which we come through our artifacts, trying to work with dignity, tolerance and respect. Our daily work becomes a partaker of spirituality" (B. Cucinelli, 2015).

4.3 BoxMarche Spa[3]

"The true roots of BoxMarche and of many Marchegian entrepreneurs, can be traced back 400 years to the culture surrounding the harvest, which prepared these populations for the world of entrepreneurship. The agrarian culture embraced profound values that were based on Christian doctrine, and which became the foundation for sustainable, authentic and humane socio-economic development. The management of BoxMarche is the classic demonstration of this category of entrepreneur who has embraced the precious education inherited from one's

[2]www.brunellocucinelli.com/it/filosofia#/page1

[3]For a more detailed analysis of this case, see: Del Baldo (2010b).

ancestors, from whom they inherited the fundamental values of family, faith, work, commitment, and courage. The family unit was and is today at the forefront of BoxMarche's hierarchy of values; it is the reason why the company stands out, a badge of honor for our land" (Don Lamberto Pigini, President of Pigini Group, partner, 2013).

"We have an emotional tie to our territory. We want to use our abilities to sustain the local economy. Our activities are not only business choices but are networks made, first and foremost, by Human Beings (T. Dominici, Managing director of BoxMarche, May 23rd, 2012)."

"Perhaps it's a little presumptuous, but we love to define ourselves as the agents of civilization. The small entrepreneur is a "builder" (of activities, of men, of wealth). Our firm embodies a narrative identity and tells a story. For this reason, it has a soul linked to the spirit and dignity of the persons and is called to a great responsibility, inasmuch we have inherited the land from our fathers, but we also have it on loan from our children" (T. Dominici, Managing director of BoxMarche April 5th, 2011).

4.4 *Elica Group*

Elica has obtained a number of awards and recognitions that formally acknowledge the genuine commitment to human resources. Since 2007 it has been among the leading attractors of talent in Italy. For three consecutive years it was included in the Top Employers Italy and in 2011 acknowledged as "a company excelling in the management of human resources" by the CRF Institute (CRF, 2012). Moreover, it was recognized as "Best place to work in 2011 in Italy and Europe" by the Great Place to Work Institute, which evaluates over 1000 businesses throughout Europe to identify the best working environments. In Mexico, it has been awarded by the Instituto Nacional para la Educaciòn de los adultos for its work in improving the scholastic education of those who work for ElicaMex. The Ermanno Casoli Foundation, in the memory of the founder of Elica, was established in 2007 in order to strengthen the link between the worlds of the arts and industry, by promoting initiatives in the field of contemporary art.

"Elica is for us an engine through which passion, experience, innovation, well-being and listening to internal and external needs can create an impression on our surrounding environment" (F. Casoli, President of Elica Group, May 27th 2011).

"We are very proud of these awards, which recognize the bounty and continuity of the work that we're doing. The first objective of human resources management is to give support to the company in order to create the best conditions for the personal and professional development of its people and provide the necessary tools for them to become the main actors in their own growth with the same willingness dedicated to the group's growth" (E. Zampetti, HR Manager, June 12th, 2011).

"Excellence cannot be created if we do not live in a workplace which practices excellence in the care of the individual, both within and outside the workplace environment. From this starting point, Elica Life was created, whose objective is to provide employees with a series of services to improve their lifestyle and work life balance and establishing more than a mere professional relationship" (CEO, September 10th, 2011).

4.5 FAAM Spa

FAAM aims to be a proper, innovative and transparent company, mainly founded on human resources and, more generally, on social and relational capital, which are the basis of growth projects. It has created a junior academy for internal training, adopted a code of ethics, the balanced scorecard, and several environmental certifications. The company has obtained numerous awards for its products quality, innovations, and the promotion of women in all hierarchical positions.

"Each stakeholder who has with us any kind of relationship, looks for trust in our relationships (. . .) If you want to get out of the crisis, the only way is to re-orientate toward CSR authentic practices. FAAM will continue the path already taken because we believe in it and in the future we will be more and more geared toward CSR" (F. Vitali, Past President, February 21st, 2012).

4.6 Loccioni Group[4]

Enrico Loccioni's entrepreneurial venture represents an excellent example of an evolutionary path (started in the 1960th) taken toward a knowledge-based business, centred on the principles of tradition and innovation. At the same time it exemplifies the industrial processes of many Italian small towns launched first by entrepreneurs who knew how to recuperate and give value to the heritage of customs, traditions, civil conditions that their sharecropper ancestors had left. An inheritance of co-responsibility that is deeply entrenched in the Marchegian spirit and culture: solidarity, good sense, wisdom, prudence, work ethics and neighbourliness. Nowadays the Loccioni Group is included among the Italian best performing companies (Marchegian excellent companies—Istao, 2014) and has been recognized as "the best place to work" and "top employers company" 2014 by the CFR and the Great Place to Work Institutes. Moreover, a multitude of recognitions have been attributed both to the founder (Enrico Loccioni) and to his Group: for the excellent level of innovation, the authentic implementation of CSR and sustainability-oriented projects and the genuine ethical orientation (i.e.: the Sodalitas Social Award in 2005, 2008 and 2009, for "Internal Processes of CSR and network enterprise model", the "Metalmezzadro project in the knowledge-based business" and "Sustainability Projects"); the "Business and Culture" Award in 2003; the Legambiente award (for the "Leaf Community

[4]For a more detailed analysis see: Del Baldo (2013a).

Project: Leaf Energy and Future" and for being a partner of the European Commission in the Sustainable Energy Europe Campaign); and the "Olivettiano Business of the year 2008" by Ernst & Joung.

"I had the opportunity to meet him (Enrico Loccioni) the first time in the early Sixties, as a supplier and installer of electrical equipment at our plant. I followed his ' take-off" and his high flying" with admiration (F. Merloni, Ariston Thermo Group, Preface; see: Bartocci, 2011: 11).

The culture of the whole Group is based on the following shared values: Imagination (being capable of creating); Energy (the capability to dream and to accomplish one's dreams); Responsibility (for the air that we breathe, the land that we walk on, the resources that we utilize, and the trust that we earn); and Tradition & Innovation.

"I had and I have the dream to create a model of a company aware of its social role, of its future in the territory and the world; we desire to be the creators of the future rather than just mere spectators. To nurture this dream, people must understand the future they want for themselves and for the group" (E. Loccioni, President of the Loccioni Group, March 23rd, 2014).

"Take care of this piece of land where the company is located, to stay here where our roots are, trying to add value to people and the environment, is my (and our) great enterprise. We want to spread a new work culture based on passion, enjoyment and beauty. To network with the territory, with its institutions, its cultural and economic community is essential for me and for us. Our young people (employees and external collaborators) are passionate about doing their job well. Seeing them with a smile in their eyes, seeing their respect and sense of fair play in the work place is something which fills my heart with joy and affection. There must be a passion for continuous improvement because as a friend of mine, a village priest, said: the best thing we can do is leave things in a better way than how we found them as we have received more than we have given". (E. Loccioni, President of the Loccioni Group, July 20th, 2013).

From these speeches the charismatic personality of the leader emerges, his ability to interpret the reality through a different outlook and to transform problems into opportunities for the community and for business (Bruni & Sena, 2013). The afore stated values and principles have driven the Loccioni Group to take care of the Esino river, which twice in the past damaged the company when it flooded. A specific investment project called "Flumen" (which involves five towns and the local public institution) has been implemented to reinstate the river course to its original position, to drain and clean up the dunes, preserve the fauna and flora biodiversity. At the present time the 2 km of "river auction" are a fluvial laboratory for the measurement and prevention of ecological disasters. It is an example of good practice in "social imagination" which refers to the love of the territory and the fusion of innovation and tradition (Varvelli & Varvelli, 2014).

"Why spend money on something that is not mine? I saw an opportunity in the river, not only a threat. With the biomass (hydropower) energy we will repay our investments and the benefits will be shared by the entire community. And, most importantly, I returned the territory to its history and its beauty of 50 years ago" (E. Loccioni, January 26th, 2014).

A further examples of the authentic orientation to the common good is provided by many other projects, such as the Leaf Community and the Chemo Apothecary.

"The leaf community is something like a philosophical current and religious faith. The leaf house is inhabited on the top floor by transient guests, and the other two floors are permanently inhabited by young people working in the Group. However, this house does not produce an ounce of the dreaded CO2. It does not consume a single watt of electricity, nor does it waste even half a litre of water" (Bartocci, 2011: 89 and 91).

"As a first point, the centrality of the Humancare project lies in the word human, that is the centrality of the man, a fundamental asset of this company. And, so the health line of the Loccioni Group has already taken consistency in the Chemo Apothecary, a system for automatic dispensing of chemiotherapy drugs, first in the world and produced by Loccioni researches". (Bartocci, 2011: 105).

4.7 SGR Group

The importance that SGR has attributed to authentic relationships comes from the past; going back 20–30 years to the history of the group's business activities, the supply of methane gas to the area and the country, represents a strong relationship with the territory. The group is in fact a "territorial company" (Baldarelli, Del Baldo, & Nesheva-Kiosseva, 2014) which spreads the culture of sustainability through a wide variety of initiatives. It actively contributes to building a model of sustainable local governance, promoted by a network of public and private operators (universities, institutions and non profit organizations) which activate mechanisms of participation in the socio-economic fabric aimed at the common good.

The President of the Group is a woman—Dionigi M.—who acts as a charismatic leader and reference point for the company, whose values have been inherited from the founders and interpreted in coherence with the changed internal and external environmental context. Throughout difficulties and challenges, she has combined humility with tenacity, determination, the spirit of sacrifice and energy. Her relational approach can be translated into the principle of the "door being open" to each collaborator. Democratic participation, trust and relationships characterize SGR governance.

"We are known as an innovative and dynamic multi-utilities company, respectful of the environment which is greatly tied to the territory and the community. Our sustainability report is a process of dialogue with all the protagonists of context in which the SGR Group operates and which contains challenging objectives on which we will concentrate our efforts. It is the story of a live experience with the territory, the community and our stakeholders (M. Dionigi, President, May 2012).

4.8 Insurance Agency Ornella Seca

Ornella (the founder) decided to enter the project Economy of Communion from the start of the company, because her life has always been based on the sound values of brotherhood, love of neighbor and justice, which she also wants to integrate in her work. Each year she allocates part of the profits to the Association United World, a NGO aimed at spreading the culture of dialogue and unity among peoples and to realize sustainable activities for the development of poor countries (i.e. in Bolivia).

Ornella is always the first to take responsibility for the good of others—whether it is an employee or a customer—and this is perceived as a concrete example of being able to implement the EoC values (the willingness to dialogue, clarification and training) even in difficult times.

4.9 Bertola Srl

The founder (Mr. Livio Bertola) introduced the basic concepts of the Economy of Communion, proposing to live the "golden rule" in his daily life: "whatever you wish that men would do to you, do you to them". He progressively became interested in his employees, their daily difficulties and personal interests, as well as their family environment. This knowledge allowed him to take care of everyone's needs, making them feel part of a big family, near (the company and the local community) and far (the entire world). Among the main project promoted by Bertola Srl, the company created schools in Sudan and Senegal that provide the training and education of young people and women.

"One day an unemployed person came looking for work. He looked like a man in pain; he was tired and had no family. I knew that at that moment I could not hire anyone, because there was not enough work and that person seemed the less suitable. But I could not bear the thought of leaving him on the road. I decided to take the young man reserving a warm welcome. Subsequently I found out he was a drug addict. I helped him. Today this young man is still among our employees. He has given up his drug addiction, is happy and works with the same community to help others overcome drug" (L. Bertola, October 2nd, 2014).

4.10 Mercurio Net Srl

The company is part of the EoC project, based on the principles of communion and universal fraternity applied to the business environment, following the charisma of Lubich (2001) who founded the Catholic Movement called "Focolare". It is a concrete example of how the spiritual and moral values can actually influence daily economic life. All corporate profits are divided into three parts: one for future investments, one for the poor living near and far, and the third one to fund projects for the formation of "new men" (through professional and spiritual education and training).

"The EoC entrepreneur must have a strong charisma, so as to become a guide for himself and for others and act as educator, aware that the relationships that are built during life are much more important than the company itself " (R. Ruffini, November, 11th, 2014).

"Every Monday morning we have a meeting with all our employees, to share objectives, politics, problems, results and outcomes, i.e. the periodical revenues and the budget. Everyone is called to express ideas, in other words we put the activity and the business planning in communion" (R. Ruffini October 9th, 2014).

"Since 2008, with the outbreak of the economic crisis, there has been a worsening of the company finances and I decided to waiver my compensation, until the end of the crisis. At this point all the employees spontaneously decided to reduce their salary" (R. Ruffini October 9th, 2014).

"We should never compromise on the ethics issue, making unfair competition (i.e. selling unoriginal software). Even at the cost of risking the closure of the company, we have always found different solutions. For example a few years ago a bank accidentally credited 35,000 euro onto the current account of the company; we immediately decided to inform the bank and return the entire amount, although we had serious cash flow problems. A few days later our company was selected by a major customer to handle an important project. Our silent partner (the Providence) could find space in our company" (R. Ruffini, November, 11th, 2014).

The above mentioned narrative statements and principles are summed up in the following table (Table 3) where for each company the main values are linked to the Encyclical principles.

5 Concluding Reflections

Francis (Encycl. Letter, 2015) urges a major attempt to integrate all dimensions of ecology and suggests that the main task is to incorporate nature as primordial stakeholder in the functioning of human organizations, including businesses, public administration and civil society organizations. Hence, the crucial question concerns how to develop the ecological sensitivity and responsiveness of people at different levels of organizations, and how then to translate the emerging ecological consciousness into effective and caring organizational practices that organizations can develop into a culture of Earth Citizenship.

Starting from this assumption, the work highlights that both scholars and companies demonstrate that resilience and a creative response to the deep economic, social and anthropological crisis can be experienced and testified. The case studies are in fact examples of how entrepreneurs/managers can promote cultural reorientation, helping others to unlearn the bad habits inspired by 'turbo-capitalism' (Matacena, 2010) and valorize humanity, relationships and the territory in which they do business. They show not so much original business experiences, but rather exceptional stories of humanity in which it is difficult, if not impossible, to separate the reasons of the heart from those of the mind (F. Merloni; see: Bartocci, 2011, Prologue). These companies can be considered virtuous and "happy companies" where the gap between the declared and shared values is reduced, and people are

Table 3 Summary of the case study analysis

Encyclical principle	Company
The common good	Mercurio Net Srl Bertola Srl, insurance agency O. Seca, Brunello Cucinelli, BoxMarche, Loccioni
The urgency for "not weak responses"	Mercurio Net Srl, Bertola Srl, insurance agency O. Seca, Loccioni group
Universal communion	Mercurio Net Srl, Bertola Srl, insurance agency O. Seca
The crisis and effects of modern anthropocentrism	Mercurio Net Srl Bertola Srl, insurance agency O. Seca, Ariston thermo group, Brunello Cucinelli, BoxMarche, Loccioni group
Holistic/Integral development; Sustainable development	Mercurio Net Srl, Bertola Srl, insurance agency O. Seca, Brunello Cucinelli, Ariston thermo group, BoxMarche, Elica, Faam, Loccioni, SGR group
The question of leadership	Mercurio Net Srl, Bertola Srl, insurance agency O. Seca, Brunello Cucinelli, Loccioni group, SGR group
The integral ecology: environmental ecology, economic ecology, social ecology, cultural ecology, and ecology of daily life	Mercurio Net Srl, Bertola Srl, insurance agency O. Seca Ariston thermo group, Brunello Cucinelli, BoxMarche, Loccioni group
The centrality of the person and interpersonal relationships	Mercurio Net Srl, Bertola Srl, insurance agency O. Seca, Ariston thermo group, Brunello Cucinelli, Box Marche, Elica, Faam, Loccioni group, SGR group
Dialogue for innovative national, local and international policies	Mercurio Net Srl, Bertola Srl, insurance agency O. Seca., Ariston thermo group, Box Marche, Elica, Loccioni group
Transparency in decision-making	Mercurio Net Srl, Bertola Srl, insurance agency O. Seca. Box Marche, Elica, Faam, Loccioni group, SGR group
Subsidiarity	Mercurio Net Srl, Bertola Srl, insurance agency O. Seca., BoxMarche, Elica, Loccioni group, SGR group
Collaboration, active involvement of the local community	Mercurio Net Srl, Bertola Srl, insurance agency O. Seca., Ariston thermo group, Brunello Cucinelli, BoxMarche, Elica, Faam, Loccioni, SGR group
Educating for the covenant between humanity and the environment	Mercurio Net Srl, Brunello Cucinelli, BoxMarche, Elica, Faam, Loccioni group, SGR group
Ecological conversion	Brunello Cucinelli, BoxMarche, Loccioni group, SGR group
Toward a new lifestyle (sobriety, humility, frugality, spirituality)	Mercurio Net Srl, Bertola Srl, insurance agency O. Seca, Brunello Cucinelli, BoxMarche, Loccioni group

really engaged in the pursuit of the common good (Borgato, 2014). Accordingly, we can summarize the following lines of reflection.

Fist, a courageous and ethics-driven leadership render themselves capable of influencing the socio-economic terrain they belong to and calling others to imitate those virtues. A common trait of all entrepreneurs/managers considered in this study is that they possess moral depth (often related to the family nature of the business, and a high level of social cohesion and social capital), which generates charisma, enthusiasm, the capacity to communicate the vision of a company where it is a pleasure to work. Entrepreneurs also present themselves as innovators, who—starting from 'an act of faith'—are able to inspire in others the positive will to emulate their choices.

Second, the conviction that the bases for integral development are in the diffusion of authentic anthropological values, characterize the companies which "adhere" to the Encyclical Letter's principles and render them multicultural companies engaged in the local and global community.

Finally, the capacity to stimulate and contribute to the progress of individuals and the socio-economic contexts derives to a rootedness which is not an anchor, but rather a jumping off point for confronting the world armed with a collection of authentic values (Del Baldo & Demartini, 2012).

We can thus conclude drawing from A prayer for our earth (Pope Francesco): (we should) Bring healing to our lives, that we may protect the world and not prey on it, that we may sow beauty, not pollution and destruction (Encycl. Letter, 2015: 178).

References

Alford, H. (2015). *Against the "Hollowing-out" of meaning. Virtue ethics in the "Blueprint for Better Business"* (pp. 1–15). Lecture presented at the SPES Congress "Virtues and vices in economics and business", Centre for Economics and Ethics of the Catholic University of Leuven in cooperation with the European SPES Institute, June 19–20, 2015, Leuven, Belgium. Retrieved December 11, 2015, from http://eurospes.org/content/virtues-and-vices-economics-and-business#sthash.0kVF1jSz.dpuf

Alford, H., & Signori, S. (2014). Brief considerations on the effectiveness of shareholder activism. A virtue ethics approach. *Impresa Progetto-Electronic Journal of Management, 3*, 1–10.

Aragón Amonarriz, C., & Iturrioz Landar, C. (2016). Responsible family ownership in small- and medium-sized family enterprises: An exploratory study. *Business Ethics: A European Review, 25*(1), 75–93.

Argandoña, A. (2003). Fostering values in organizations. *Journal of Business Ethics, 45*, 15–28.

Argandoña, A. (2011). Beyond contracts: Love in firms. *Journal of Business Ethics, 99*(1), 77–85.

Baldarelli, M. G. (2011). *Le aziende dell'economia di comunione. Mission, Governance e Accountability*. Roma: Città Nuova.

Baldarelli, M. G., Del Baldo, M., & Nesheva-Kiosseva, N. (2014). Implementing sustainability reporting: (Neo) institutional theory insights in the analysis of SGR group Italy and CityGas Bulgaria. *Journal of Modern Accounting and Auditing, 10*(11), 1067–1104.

Bartholomew P. (1997, November 8). *Address in Santa Barbara, California*. Cfr. Chryssavgis, J., *On earth as in heaven: Ecological vision and initiatives of Ecumenical Patriarch Bartholomew*. Bronx, NY, 2012.

Bartocci, M. (2011). *Animal spirits in Vallesina Enrico Loccioni e "l'impresa come gioco"*. Milano: Luiss University.

Becker, T. (1998). Integrity in organizations: Beyond honesty and conscientiousness. *Academy of Management Review, 23*(1), 154–161.

Benedict, X. V. I. (2009). *Enciclical Letter Caritas in Veritate, 29 giugno 2009*. Libreria Editrice Vaticana: Città del Vaticano.

Borgato, R. (2014). *L'impresa felice. La responsabilità sociale come impulso alla crescita*. Milano: Franco Angeli.

Bouckaert, L. (2011). Spirituality and economic democracy. In L. Zsolani (Ed.), *Spirituality and ethics in management* (2nd ed., pp. 41–52). Dordrecht: Springer.

Bouckaert, L., Opdebeeck, H., & Zsolnai, L. (Eds.). (2007). *Frugality. Rebalancing material and spiritual values in economic life*. Oxford: Peter Lang Academic Publishers.

Brown, M. T. (2005). *Corporate integrity. Rethinking organizational ethics and leadership*. Cambridge: Cambridge University Press.

Bruni, L. (2009). *The economy of communion. Toward a multi-dimensional economic culture*. USA: New City Press.

Bruni, L., & Sena, B. (Eds.). (2013). *The charismatic principle in social life*. New York: Routledge.

Capaldi, N. (2013). How American spiritual capital inform business and affects the common good. In S. Groschl (Ed.), *Uncertainty, diversity and the common good. Changing norms and new leadership paradigms* (pp. 25–40). Surrey: Gower.

Cardon, M. S., Wincent, J., Singh, J., & Drnovsek, M. (2009). The nature and experience of entrepreneurial passion. *Academy of Management Review, 34*(3), 511–532.

Chua, J. H., Chrisman, J. J., & Sharma, P. (1999). Defining the family business by behavior. *Entrepreneurship Theory and Practice, 23*(4), 19–39.

Cortright, S. A., & Naughton, M. J. (2002). *Rethinking the purpose of business. Interdisciplinary essays from the catholic social tradition*. Notre Dame, IN: University of Notre Dame Press.

Costas, J., & Fleming, P. (2009). Beyond dis-identification in contemporary organizations. *Human Relations, 62*(3), 353–378.

CRF. (2012). *Top Employers Italia 2012*. Milano: Franco Angeli.

Del Baldo, M. (2010a). Corporate social responsibility and corporate governance in Italian SMEs: Toward a 'territorial' model based on small 'champions' of CSR. *International Journal of Sustainable Society, 2*(3), 215–247.

Del Baldo, M. (2010b). CSR and sustainability: Mission, governance and accountability in Italian Smes. The experience of BoxMarche, a "Convivial" and "Territorial" enterprise: Toward a "territorial model of sustainability". In M. G. Baldarelli (Ed.), *Civil economy, democracy, transparency and social and environmental accounting research role* (pp. 289–345). Milano: MacGraw-Hill Italia.

Del Baldo, M. (2012). Familial and territorial values for sustainable entrepreneurship in Italy: Some reflections in theory and practice. *Journal of Modern Accounting and Auditing, 8*(8), 1185–1203.

Del Baldo, M. (2013a). CSR-oriented SMEs: A question of entrepreneurial virtues in action? Reflections in theory and practice. In J. O. Okpara & S. O. Idowu (Eds.), *Corporate social responsibility. Challenges, opportunities and strategies for 21st century leaders* (pp. 145–170). Heidelberg: Springer.

Del Baldo, M. (2013b). Corporate social responsibility, human resource management and corporate family responsibility. When a company is "the Best Place to Work": Elica group, the Hi-Life company. Ekonomska Istraživanja. *A Journal of Economic and Social Research, 1*(Special Issue), 201–224.

Del Baldo, M. (2014a). Sustainable entrepreneurship: Next stage of responsible business. In C. Weidinger, F. Fischler, & R. Schmidpeter (Eds.), *Sustainable entrepreneurship. Business success through sustainability* (pp. 135–153). Heidelberg: Springer.

Del Baldo, M. (2014b). Developing businesses and fighting poverty: Critical reflections on the theories and practices of CSR, CSV, and inclusive business. In L. Pate & C. Wankel (Eds.), *Emerging research directions in social entrepreneurship, Advances in business ethics research* (pp. 191–223). Dordrecht: Springer Science, Business Media.

Del Baldo, M., & Baldarelli M.G. (2015). *From weak to strong CSR: The experience of the EoC (Economy of Communion) industrial parks in Germany and Italy* (pp. 1–14). Berlin, Heidelberg: UWF, UmweltWirstschaftssForum, Springer. doi:10.1007/s00550-015-0364-5. Published on line 26th October 2015. www.springer.com/home/SGWID=0-0-1003-0-

Del Baldo, M., & Demartini, P. (2012). Small business social responsibility and the missing link: The local context. In W. D. Nelson (Ed.), *Advances in business management* (Vol. 4, pp. 69–94). New York: Nova Science Publishers, Inc.

Driscoll, D. M., & Hoffman, W. M. (2000). *How to implement values-driven management.* Walham, MA: Bentley College.

Eisenhardt, K. M., & Graebner, M. E. (2007). Theory building from cases: Opportunities and challenges. *Academy of Management Journal, 50*(1), 25–32.

Francis, P. (2015). *Laudato Sì, Encyclical Letter on care for our common home, May 24th, 2015.* Rome: Vatican Press.

Garriga, E., & Melé, D. (2004). Corporate social responsibility theories: Mapping the territory. *Journal of Business Ethics, 53*(1), 51–71.

Goffee, R., & Jones, G. (2009). Authentic leadership: Excite others to exceptional performance. *Leadership Excellence, 26*(7), 17.

Graafland, J. J., & van de Ven, B. (2006). Strategic and moral motivation for corporate social responsibility. *Journal of Corporate Citizenship, 22*, 111–123.

GRI—Global Reporting Initiative. (2013). *Sustainability reporting guideline-G4*, May 2013. Retrieved September 2, 2015 http://www.globalreporting.org

Guardini, R. (1998). *The end of the modern world* (9th ed.). Wilmington, DE: ISI Books.

Gui, B., & Sugden, R. (2005). *Economics and social interactions. Accounting for interpersonal relations.* Cambridge: Cambridge University Press.

Hoivik von Weltzien, H. (2014). The heart of leadership is ethics. *Impresa Progetto-Electronic Journal of Management, 11*, 1–9.

Hoivik von Weltzien, H., & Melé, D. (2009). Can a SME become a global corporate citizen? Evidence from a case study. *Journal of Business Ethics, 88*, 551–562.

IIRC—International Integrated Reporting Council. (2013). *The International <IR> Framework.* December 2013.

Istao. (2014). *Marchegian excellent companies.- classifica delle principali imprese marchigiane anno 2013.* Ancona: Istao, Fondazione Aristide Merloni. Retrieved November 11, 2015 http://www.fondazione-merloni.it

John Paul, I. I. (1981). *Encyclical letter Laborem Exercens, 14 September 1981.* Libreria Editrice Vaticana: Città del Vaticano.

John Paul, I. I. (1991). *Encyclical Letter Centesimus Annus, 1 May 1991.* Libreria Editrice Vaticana: Città del Vaticano.

Kaptein, M., & Wempe, G. (2002). *The balanced company: A theory of corporate integrity.* Oxford: Oxford University Press.

Ketola, T. (2008). A holistic corporate responsibility model: Integrating values, discourses and actions. *Journal of Business Ethics, 80*, 419–435.

Looser, S., & Wehrmeyer, W. (2015). *Doing well or doing good? Extrinsic and intrinsic CSR in Switzerland* (pp. 1–14). UWF. Heidelberg: Springer. doi:10.1007/s00550-015-0360-9. Published on line August 14, 2015.

Lubich, C. (2001). *L'Economia di Comunione-Storia e profezia.* Roma: Città Nuova.

Luthans, F., & Avolio, B. J. (2003). Authentic leadership: A positive developmental approach. In K. S. Cameron, J. E. Dutton, & R. E. Quinn (Eds.), *Positive organizational scholarship: Foundations of a new discipline* (pp. 241–261). San Francisco, CA: Berrett-Koehler.

Mahoney, L. S., Thorne, L., Cecil, L., & LaGore, W. (2013). A research note on standalone corporate social responsibility reports: Signaling or greenwashing? *Critical Perspectives on Accounting, 24*(4–5), 350–359.

Malloch, T. R. (2008). *Spiritual enterprises. Doing virtuous business*. New York: Encounter Books.

Matacena, A. (2010). Corporate social responsibility and accountability: Some glosses. In M. G. Baldarelli (Ed.), *Civil economy, democracy, transparency and social and environmental accounting research role* (pp. 7–59). Milano: McGraw-Hill.

Matten, D., & Moon, J. (2008). 'Implicit' and 'explicit' CSR: A conceptual framework for a comparative understanding of corporate social responsibility. *Academy of Management Review, 33*(2), 404–424.

Melé, D. (2009). *Business ethics in action. Seeking human excellence in organizations*. Hampshire, GB: Palgrave MacMillan.

Melé, D. (2012). *Management ethics. Placing ethics in the core of good management*. New York: Palgrave McMillan.

Merloni, F. (2011). Preface. In M. Bartocci (Ed.), *Animal Spirits in Vallesina. Enrico Loccioni e "l'impresa come gioco"* (Vol. 11). Milano: Luiss University.

Molteni, M. (2009). Aziende a movente ideale. In L. Bruni & S. Zamagni (Eds.), *Dizionario di Economia Civile* (pp. 65–75). Roma: Città Nuova.

Naumes, W., & Naumes, M. J. (2006). *The art and craft of case writing* (2nd ed.). London: M.E. Sharpe Inc.

Olsen, O.K. (2010). *Are good leaders moral leaders? The relationship between effective military operational leadership and morals* Dissertation for the philosophiae doctor degree (PhD) at the University of Bergen.

Paine, L. S. (1994). Managing for organizational integrity. *Harvard Business Review, 72*(2), 106–117.

Painter-Morland, M. (2006). Triple bottom-line reporting as social grammar: Integrating corporate social responsibility and corporate codes of conduct. *Business Ethics: A European Review, 15*(4), 352–364.

Parry, K. W., & Proctor-Thomson, S. B. (2002). Perceived integrity of transformational leaders in organizational settings. *Journal of Business Ethics, 35*, 75–96.

Peredo, A. M., & Chrisman, J. (2006). Towards a theory of community-based enterprise. *Academy of Management Review, 31*(2), 309–328.

Pruzan, P. (2001). The question of organizational consciousness: Can organizations have values, virtues and visions? *Journal of Business Ethics, 29*, 271–284.

Pruzan, P. (2011). Spirituality as the context for leadership. In L. Zsolani (Ed.), *Spirituality and ethics in management* (2nd ed., pp. 3–22). Dordrecht: Springer.

Russi, M. (2010, 9–11 September). *Measure entrepreneurial virtues. Towards a new perspective for the indicators of corporate success*. Paper presented at the 23rd Eben Annual Conference which values for which organisations? Trento.

Sorci, C. (Ed.). (2007). *Lo sviluppo integrale delle aziende*. Milano: Giuffré.

Spence, L. J., Schmidpeter, R., & Habisch, A. (2003). Assessing social capital: Small and medium sized enterprises in Germany and the U.K. *Journal of Business Ethics, 47*(1), 17–29.

Taylor, C. (1992). *The ethics of authenticity*. Cambridge, MA: Harvard University Press.

Thomson, M. G., & Eynikel, J. (2011). *Leading with wisdom. European SPES Cahier, 7*. Antwerpen: Garant.

Trilling, L. (1972). *Sincerity and authenticity*. London: Oxford University Press.

Varvelli, M. R., & Varvelli, R. (2014). *2Km di futuro. L'impresa di seminare bellezza*. Milano: Il Sole 24 Ore.

Visser, W. (2011). The nature of CSR leadership. Definitions, characteristics and paradoxes. *CSR International Paper Series, 4*, 1–10.

Walker, T. (2015, July 29–31). *Entrepreneurial wisdom*. Paper presented at the 2nd International Conference on CSR, Sustainability, Ethics and Governance, Nanjing, China, 2015_01_Entrepreneurial_Wisdom_Walker_final.docm, pp 1–12. Walk-on Institute for sustainable solutions. www.walk-on.co.at

Walker, B., & Salt, D. (2006). *Resilience thinking: Sustaining ecosystems and people in a changing world*. Washington, DC: Island Press.
Yin, R. K. (1994). *Case study research: Design and methods* (2nd ed.). Sage: Thousand Oaks.
Zadek, S. (2006). Responsible competitiveness: Reshaping global markets through responsible business practices. *Corporate Governance, 6*(4), 334–348.
Zamagni, S. (Ed.). (1995). *The economics of altruism*. Hants: E. Elgar.
Zsolnai, L. (2015, June 19–20). *Prudence in management and economic wisdom*. Lecture presented at presented at the SPES Congress "Virtues and vices in economics and business", Centre for Economics and Ethics of the Catholic University of Leuven in cooperation with the European SPES Institute, Leuven, Belgium.

Developing Fourth Generation Sustainability-Oriented Business Models: Towards Naïve, Native, and Narrative Intelligence

Frans Melissen and Lars Moratis

1 Introduction: The Failure of CSR

Ranging from poverty to biodiversity and from climate change to the need to develop cross-sector partnerships to tackle some of the world's most pressing sustainability problems, the Sustainable Development Goals have presented business and society with a clear and up to date global sustainability agenda. At the same time, the severity and the interconnectedness of the world's most pressing sustainability problems also indicate that corporate social responsibility (CSR) seems to have failed: despite the initiatives firms have taken over the years to contribute to a better world in an ecological and social sense through directing their resources and competencies towards this goal, the world has been degrading on many important sustainability-related indicators (cf. Visser, 2011). By implication, firms need to resort to other ways of integrating societal goals into their strategies, organizational architecture, and decision-making processes.

Sustainability-oriented business models (SOBMs) may present a way to turn the tides and provide an alternative to relying on firm's CSR activities for creating a better world. The need for developing and descriptions of SOBMs has been increasingly recognized by both academics and practitioners over the past few years (e.g., Stubbs & Cocklin, 2008; SustainAbility, 2014). This conceptual and discursive chapter aims to contribute to this discourse by identifying three generations of SOBMs and their limitations in realizing sustainable development as well as by presenting an interpretation of fourth generation SBOMs. It differentiates

F. Melissen
NHTV Breda University of Applied Sciences, Breda, The Netherlands

L. Moratis (✉)
NHTV Breda University of Applied Sciences, Breda, The Netherlands

Antwerp Management School, Antwerp, Belgium
e-mail: lars.moratis@ams.ac.be

S.O. Idowu et al. (eds.), *Corporate Social Responsibility in Times of Crisis*, CSR, Sustainability, Ethics & Governance, DOI 10.1007/978-3-319-52839-7_3

SBOMs by analyzing them in terms of the extent to which they approach sustainability in holistic ways and the extent to which they are built on instrumental approaches towards sustainability. In presenting fourth generation SOBMs, it integrates insights from evolutionary psychology and identifies three types of 'sustainability intelligence' firms need to develop in order to be successful in developing SOBMs. The chapter start with a systemic view on sustainability transformation and addressing the role of business in sustainable development. It then turns to describing three generations of SOBMs and presents a fourth generation SOBM. After discussing some leadership implications of SOBMs it identifies three types of sustainability intelligence that SOBMs require: naïve, native, and narrative intelligence. The chapter ends with a brief discussion of implications and future research suggestions.

2 The Old and the New System

In recent years, some encouraging signs of the early stages of what some refer to as a transformation or transition towards a more sustainable society and economic system can be witnessed (Loorbach, 2014). Others have described this as a co-evolutionary process (Laukkanen & Patala, 2014) in which new, more sustainable, technologies, social practices and institutions are slowly but surely being created. Loorbach writes that a gradual and creeping revolution is taking place "in which citizens, consumers, [...] entrepreneurs, civil servants, researchers and activists are changing the way in which we live, consume, [and] produce in small but certain steps" (Loorbach, 2014: 35). Obvious examples include AirBnB and Uber, the most well-known representatives of the sharing economy. Other examples relate to the opportunities to buy your renewably energy directly from farmers and other individuals that have installed solar panels or windmills that generate electricity.

Interestingly, most of these initiatives and new practices would not have been possible without the very same system that put society on an unsustainable course in the first place—the system that still dominates the everyday lives of most people. This system based on free markets and promoting private property, autonomy and economic growth has provided many of us with the material means, technologies such as the internet and social media, knowledge, desires and imagination needed to set up these initiatives and participate in these practices (Varul, 2013). However, the invisible hand of the free market and society's pre-occupation with economic growth and material wealth not only damages the natural environment and promotes inequality, it also slows down or even destroys a number of initiatives and practices, especially promising business practices, that are crucial to resolving, repairing, and preventing these problems. This resistance of the old system could seriously delay the build-up of an alternative system—a system that supports sustainable development. Furthermore, obviously, the longer the delay, the more severe the consequences will be, especially if a continued focus on promoting private property, autonomy and economic growth creates a situation in which the associated negative outcomes lead to an uncontrolled, chaotic collapse of the old system before we have managed to build up the new one (Loorbach, 2014).

3 Business and Sustainable Development

In order to realize the transition to a new system, it should be acknowledged that our current socio-economic systems have been mutually shaped by various actors, including governments, nongovernmental organizations, civil movements, and business. Consequently, all of these actors, and firms in particular, have a role to play in striving for sustainable development and overcoming the resistance of the current system (see e.g. Doane, 2005; Laukkanen & Patala, 2014; Stubbs & Cocklin, 2008). And ensure a relatively smooth transition to a sustainable socio-economic system. However, it is important to realise that the importance of businesses is not just the result of other actors consciously choosing not to step up to the plate. It is also linked to the fact that other actors are very much influenced in their behaviour by the way business fulfils its role. Given that the functioning of our current socio-economic system is highly dependent on the various functions business has in society, firms represent a potential powerful force in societal transformation.

This means that firms not only bear a huge responsibility in adjusting the course of our society to a more sustainable one, their potential to do so is equally substantial—and for this latter reason alone there is role for business in making the transition towards a more sustainable society (Davis, 1973). If businesses manage to create and apply the capabilities, technologies and practices needed to support sustainable development, there seems to be every reason to be optimistic about our future.

The actual functioning and contribution of individual businesses can best be described by reviewing what products and services they are offering or providing to which customers, the way in which they deliver these products and services to their customers and, finally, the way in which they deal with costs and revenues. Together these three elements shape the value creation processes that constitute the so-called business model that is applied by a specific firm (see e.g., Bocken, Short, Rana, & Evans, 2014; Schaltegger, Lüdeke-Freund, & Hansen, 2011; Teece, 2010; Zott, Amit, & Massa, 2011).

The next sections focus on the development of business models from the perspective of sustainable value creation, which we call sustainability-oriented (SOBMs), and categorize these into different development stages.

4 Generations of Sustainability-Oriented Business Models

4.1 First Generation Sustainability-Oriented Business Models

Efforts by firms to reduce their negative impact from a sustainability perspective have usually started by trying to adjust the way in which products and services are created and delivered (cf. Visser, 2011), looking for ways to create and deliver value to their customers that use fewer resources and that generate less waste, emissions and pollution. In fact, through redesigning products and services and applying circular economy principles such as recycling, Cradle-to-Cradle, and leasing instead of buying, many firms have already taken significant steps in controlling and avoiding pollution, as well as maximizing material and energy efficiency of their value creation processes. We can also see an increased effort of many firms to stimulate and engage in the production and use of renewable energy. In other words, the business world has already made progress in developing technologies, products, services, and practices that minimize or even eliminate negative environmental consequences associated with its business models.

4.2 Second Generation Sustainability-Oriented Business Models

In contrast to first generation SOBMs, second generation SOBMs are applied by firms that not only address their environmental impact but also their social impact. In other words, second generation SOBMs explicitly account for and try to optimize all consequences of their functioning from a people, planet, and profit perspective (Elkington, 1997). Whereas a few decades ago it was still more or less common practice to stick to the mantra of Milton Friedman (1962) stating that the only responsibility of business is to increase its profits, nowadays many firms consider their responsibilities to include more than merely making money. Numerous firms focus on minimizing their environmental impact, yet also on making a significant positive contribution to the social component of sustainable development. While some firms choose to invest in the health and overall wellbeing of their employees, others focus on improving living conditions in local communities or set up charity projects across our globe. Some firms probably only do this to persuade potential customers of their good intentions and ethics, hoping that this will increase sales and thus profits (Marques & Mintzberg, 2015); others undoubtedly do this based on genuine good intentions of the people involved. Unfortunately though, regardless of what motivates these practices, the invisible hand of the free market, weighted by passive and oftentimes sceptical consumers, usually finds a way to temper ambitions and obstruct long-term meaningful progress.

The explanation for this mechanism is as simple as it is sobering: ultimately, the business model applied by almost all of these firms still revolves around capturing as much (monetary) value from the market as possible. Given that the dominant coordination mechanism within our current socio-economic system is that of the market, this is hardly a surprise. In order to be successful and to survive in the free market, firms have to try and increase its market share and profits (Boons & Lüdeke-Freund, 2013). Ultimately, this means that firms that want to engage in CSR have to take on the challenge of merging two competing paradigms: creatively trying to synthesize the social and economic purpose of the business (Wilson & Post, 2013). Consequently, more and more of these firms are looking for ways to engage in CSR that simultaneously improve their economic performance, thus increasingly treating environmental and social practices as a business case (e.g., Carroll & Shabana, 2010; Dohrmann, Raith, & Siebold, 2015).

Ultimately, treating these practices as a business case—as a means to an end, instead of an end in itself—implies judging sustainability initiatives on their contribution to meeting the requirements of the market, such as cost reduction, profit margin, risk reduction, reputation, and brand value (Moratis, 2014; Schaltegger et al., 2011). Only through operating based on meeting these requirements can firms ensure the returns on investment demanded by its shareholders, meet the expectations of stock market analysts, generate the financial means needed to acquire their material and financial resources, and continue to satisfy their current customers.

Combined with the reluctance of governments and consumers to support and stimulate sustainable business practices, this explains how our current socio-economic system has created self-reinforcing mechanisms (Bohnsack, Pinkse, & Kolk, 2014) with respect to the way in which business operates and the business models they use. Even though more and more firms claim to account for people, planet, and profit, these second generation SOBMs still treat sustainability initiatives as an add-on or plug-in (Marques & Mintzberg, 2015).

4.3 *Third Generation Sustainability-Oriented Business Models*

The fact that sustainability initiatives have been treated at add-ons or plug-ins is why scholars have been calling for the development of different types of business models and new management paradigms (Stubbs & Cocklin, 2008). A promising development in relation to this call is the recent rise of so-called social or societal business models, which we refer to as third generation SOBMs. In contrast to the first and second generation, these business models have been purposely designed to also generate environmental and social value instead of just financial value. In fact, they aim at creating economic value through creating societal value (Porter & Kramer, 2011) and some of these businesses even consider profitability as a means to an end rather than an end in itself (Wilson & Post, 2013). Those values and ends are oftentimes defined in terms of solving specific environmental problems, community development, or helping out the world's poor.

Many of the entrepreneurs responsible for social businesses have made a conscious choice to create social value through operating on the market as a business. Applying a self-sustaining or even profit-focused business model is referred to by some as an act of positive protest as an attempt to reform capitalism (Wilson & Post, 2013), by others as an attempt to empower capitalism to deal with some of the challenges our society is facing (Yunus, Moingeon, & Lehmann-Ortega, 2010), or even as the way to create a "higher form of capitalism" (Porter & Kramer, 2011: 75). Even though many of these firms acknowledge that operating on the market place can lead to conflicts, the reasoning is that these simply represent "facts of life" (Yunus et al., 2010: 318) and involve "conscious trade-offs rather than deep conflicts of interest" (Wilson & Post, 2013: 727).

For these firms, making a contribution to sustainable development is the purpose. However, these SOBMs need to be backed up by owners and shareholders that will accept (s)lower financial returns on their investment to support the creation of higher environmental or social value. In today's marketplace, this might make it rather difficult for them to acquire the financial and material resources needed to scale up and, consequently, to capture a sizable market share. This, in turn, makes these businesses "fragile organizations that must walk a fine line between achieving a social mission" (Santos, Pache, & Birkholz, 2015) and participating in the free market game, despite the protests of some of these players, is still played within the context of regulatory and tax systems that favour the less ethical ones; a game in which firms applying traditional profit-focused business models can still count on the strong support of consumers; and a game in which the winner takes all (Doane, 2005).

Ultimately, third generation SOBMs do not fundamentally change the role of firms in our socio-economic system or the reference points of the system as a whole (Crane, Palazzo, Spence, & Matten, 2014). Therefore, it is not surprising to see that many societal entrepreneurship initiatives have not survived or have never managed to move beyond the status of a niche player. Those that have survived often have made significant concessions to their original mission and objectives with respect to creating societal value. While there are exceptions to this rule, one cannot but conclude that these third generation SOBMs models are not game changers either.

4.4 Fourth Generation Sustainability-Oriented Business Models

Firms that want to divert from the path of the free market (Bohnsack et al., 2014) and escape the lock-in of our current socio-economic system in their efforts to contribute to a more sustainable course of our society, thus need to apply business models that move beyond the ones discussed so far. Ideally, these fourth generation SOBMs incorporate mechanisms that can truly change the rules of the game (Bocken et al., 2014). Perhaps the best way to identify these mechanisms is to review the current behaviour of one of the main players and, especially, possible and promising avenues for influencing that behaviour.

If firms are to play the role of catalysts for a relatively smooth transition towards a more sustainable system, they need the support of consumers. Sustainable products and services—produced and delivered through sustainable processes—can only have a positive influence on our society if consumers are willing to adopt them as appropriate means to satisfy their needs. Firms applying first, second and third generation SOBMs have so far found it rather difficult to convince consumers to do so. Interestingly, research has shown that most people do consider climate change and sustainability to represent urgent and important problems (Gifford, 2011). However, people are inclined to prefer short-term personal and usually economic pay-offs over long-term, collective, and non-economic pay-offs, such as a liveable planet for current and future generations. Such a situation is often referred to as a social dilemma or social trap (Dawson, 1980).

In order to develop the full potential of fourth generation SOBMs, insights from evolutionary psychology into the workings of human behaviour, partnerships and leadership should be integrated into the conceptualization of these models. The next sections will deal with these issues respectively.

4.4.1 The Role of Human Nature

The most logical starting point to explore the contours of such mechanisms is looking at the fact that one of the main causes for the pickle the human species finds itself in is that it has become too good at extracting and exploiting the Earth's resources to satisfy our needs (Sennett, 2008). In their paper on the evolutionary bases for sustainable behaviour, psychologists Griskevicius, Cantú, and van Vugt (2012) explain that this tendency to self-destruct is clearly linked to our ancestors' struggle to survive and the resulting dominance of selfish genes in humanity's current gene pool. However, they also explain that this is not necessarily a bad thing because "selfish genes can and do build organisms that are capable of behaving in ways that are kind, charitable and sustainable" (p. 116, referring to Dawkins, 1976). This suggests that mechanisms designed to convince consumers to accept sustainable products, services and practices need to tap into the same evolutionary processes that created these genes.

One example would be to make use of people's desire for status, especially relative status. Costly signalling theory tells us "natural selection favours people who engage in activities that are increasingly costly—involving significant resources, energy, risk, or time" (Griskevicius et al., 2012: 120–121). Within the context of promoting sustainable behaviour, one could purposely make use of this tendency because it means that consumers are likely to consider people and organizations that risk spending a lot of resources on environmental and social causes to represent interesting potential friends, leaders, and co-creators of their lives. Based on the same logic, fourth generation SOBMs should also offer additional value to consumers in presenting them with clear signs or tags that allow them to display their sustainable behaviour to others, thus increasing their own status as well.

Another example would be to make use of people's tendency to copy the behaviour of others, especially those considered to possess leadership features. In fact, by setting the right example and inviting others to join, for instance by investing in community development, firms and their leadership can not only use people's desire for status and propensity to copying behaviour as a means to build a market, they could also tap into various other human tendencies, such as reciprocal altruism and social obligation. The effect of these tendencies has been proven to be stronger in small communities and dense, interdependent social networks. They originally relate to the inclination of people trying to survive in hunter-gatherer groups to help out others, if those others return the favour—especially if those others take first turn. This explains why businesses that have managed to create a community feeling linked to their business model and that donate to environmental and social causes before asking consumers to accept and buy their products and services are far more successful than those that promise to make a donation once the consumer has paid the bill (Goldstein, Griskevicius, & Cialdini, 2012).

However, trying to make selfish genes serve a good cause through creating group identity and trust is not the only reason why fourth generation SOBMs should focus on and make use of communities and social networks, including those shaped by modern technologies such as social media. A transition towards a sustainable alternative for our current socio-economic system entails much more than sustainability practices at the level of an individual business (Stubbs & Cocklin, 2008). Entrepreneurial actions at this level can probably provide us with some of the physical technologies needed to solve specific environmental problems, but fully tackling the challenge of sustainable development requires more than that. It requires a holistic approach that combines advanced physical technologies with new social technologies (Laukkanen & Patala, 2014). These new social technologies refer to new ways of organising ourselves and our society, including the firms that co-create it—possibly even to redefining what we consider to represent society's needs in relation to what the Earth can provide for and the search for an alternative to capitalism as the system assisting in meeting those needs. Obviously, rising to this challenge will require risk taking and experimenting, breaking old rules, and developing new ones as we go along. Obviously, it would not be wise for individual firms to take on this challenge in isolation, without the support and assistance of some of the other players in the game. The required innovations and the context in which they need to be implemented are simply too complex and reach beyond the boundaries of an individual firm (Bocken et al., 2014; Rohrbeck, Konnertz, & Knab, 2013).

4.4.2 The Role of Partnerships

The second type of mechanisms that need to be incorporated in fourth generation SOBMs are hence linked to a focus on and the means to set up collaborations with other firms and stakeholders involved in producing and consuming the products and services that these models deliver, and defining and redefining the needs that they satisfy (Doane, 2005). Also by applying the lessons learned from the evolutionary processes discussed earlier, these mechanisms need to create partnerships and infrastructures that provide access to the resources and expertise needed to solve a specific problem or satisfy a specific need (Dahan, Doh, Oetzel, & Yaziji, 2010). Once again, the most promising and logical level to create these partnerships and infrastructures is at the community or social network level. This not only represents the level where needs are formulated and social norms are shaped, this is also the level that has already proven to be the most resistant to the invisible hand of the free market. It thus offers the most promising environment to tap into and join the already on-going but gradual and still mostly hidden revolution referred to earlier. Joining forces with this movement of engaged citizens, entrepreneurs, civil servants, researchers, and activists not only opens the door to an invaluable collection of resources, imagination, and knowledge, but also provides firms applying sustainable business models with a strengthened license to operate. Together with these stakeholders, firms could explore ways to co-create viable alternatives to our current socio-economic system—alternatives that are based on equality, inclusiveness, and a responsible and sustainable way of interacting with our natural environment (Loorbach, 2014).

4.4.3 The Role of Leadership

Installing these mechanisms and applying fourth generation SOBMs places high demands on firms and their leadership. It requires courage and vision to do things differently and to base corporate decisions on more than reducing costs, increasing profits, and satisfying the needs of current customers. It requires viewing firms as a community of civilized craftsmen who aspire to use their resources for a collective good (Sennett, 2008). It also requires ethical leadership allowing to not only ask the question whether things are done in the right way, but also whether the right things are done for the right reasons (Ciulla, 2005).

Joining existing or collaboratively shaping new communities and networks that focus on redefining and satisfying specific needs and tackling specific environmental and social problems through experimenting with alternatives to our current socio-economic system also requires accepting the concept of distributed control (Loorbach, 2014). Given that such communities and networks place those specific needs and problems at the centre implies that firms, at least to some extent, need to give up autonomy and to share power. Ultimately, achieving the goal of the community or network is a collective effort and thus requires involving others in decision-making processes, even those beyond the boundaries of the firm.

Finally, it requires conviction, enthusiasm and particular skills to convey one's intentions and what firms can bring to the table. This may very well involve lobbying and teaming up with influential stakeholders to persuade policy makers and public authorities to adjust regulatory systems or to at least allow firms to not follow current regulations to the letter. It definitely requires firms to create a clear and appealing vision, to be able to share that vision, and to follow it up with actions that show they mean it. Somehow, leaders will have to be able to convey to others that they have got what it takes to be part of a community or network that is based on long-term commitment and true reciprocity rather than short-term gains and opportunism.

5 The Composition of Sustainability Intelligence

The perspective on (the need for) the development of fourth generation SOBMs that are truly sustainability-oriented as portrayed in this chapter puts an important question on the table: what is the type of 'intelligence' that firms should have in order to be successful in developing these business models?

Our conception of sustainability intelligence is based on a typology that differentiates between naive intelligence, native intelligence, and narrative intelligence. Before we elaborate on these three dimensions of sustainability intelligence, it is important to understand what we mean and do not mean by intelligence.

We prefer to use the concept of intelligence and avoid using related concepts such as knowledge, competencies, and skills. The reason for this is that we consider the concept of intelligence to represent a broader orientation encompassing a range of characteristics, including awareness, attitude, aptitude, and ability. In doing so, we draw on a somewhat shared definition of the contested concept of intelligence as "[a] very general mental capability that, among other things, involves the ability to reason, plan, solve problems, think abstractly, comprehend complex ideas, learn quickly and learn from experience. It is not merely book learning, a narrow academic skill, or test-taking smarts. Rather, it reflects a broader and deeper capability for comprehending our surroundings—'catching on', 'making sense' of things, or 'figuring out' what to do" (Gottfredson, 1997: 18). Intelligence relates to the understanding of complex ideas, achieving a successful fit with the surrounding environment, experiential learning, and the engagement in various forms of reasoning in order to overcome barriers by taking thought (Neisser et al., 1996). Although the concept of intelligence may seem rooted in cognition and may appear dominantly cerebral, our interpretation of intelligence is versatile and includes affective and conative dimensions (cf. Basu & Palazzo, 2008; Gardner, 1993). This approach towards intelligence corresponds with our conception of sustainability as set out in this chapter, which stresses an involvement and engagement with all kinds of stakeholders, including future generations. In addition, it aligns with the need to understand and influence human behaviour towards sustainability, which involves applying insights from social and evolutionary psychology.

Our conception of sustainability intelligence as we elaborate in the next sections also differs from Goleman's popular concept of ecological or green intelligence (Goleman, 2009). This notion predominantly focuses on becoming aware and obtaining knowledge about the impact of the products we use in order to reduce the ecological footprint of our consumption patterns. Goleman's idea of intelligence is rooted in the need for gauging the environmental externalities for any given action, be it a human behaviour, a way of value creation, or a business decision, and relies on the need for full transparency to optimize ecological footprint reduction and practice patterns of sustainable living. While this may be seen as an integral part of sustainability intelligence, the concept that we propagate is more encompassing and surpasses the idea of merely acquiring data to improve a firm's information position in the field of sustainability.

It should be noted that various scholars, for instance in the field of systems thinking and artificial intelligence, tend to use naive intelligence and native intelligence as synonyms referring to common sense and general cognitive capacity as a human capability for problem-solving as opposed to calculus-based approaches (e.g., Simon, 1990). However, within the context of the train of thought presented here, naive, native, and narrative intelligence represent related though categorically distinctive types of intelligence.[1]

5.1 *Naive Intelligence*

We consider naive intelligence as crucial for the development towards fourth generation SOBMs.

Firms need naive intelligence to escape the pathway formed by current and generally applied, mostly unsustainable, business models and envision new, truly SOBMs that do not exist or have been applied yet. In our view, business should embrace rather than reject a certain level of naivety in their strategic and organizational decision-making to avoid the sort of systemic and cognitive lock-in created by operating in (market) environments characterized by taken-for-granted elements, such as the continuous pursuit of maximizing profits, ever larger market shares, and providing the highest possible return on investment for shareholders. Naivety then includes a freshness, an openness, and an intuitiveness towards ideas that may shape a future that can be, rather than a future that is simply a replication, projection, or extrapolation of the past. Through not letting firms be constrained by

[1] Although our focus is particularly on businesses, it should be noted that the types of sustainability intelligence we identify here apply to other organizations as well. This includes organizations that want to explore SOBMs themselves, but also those organizations that are stakeholders of firms that want to co-create these business models.

experience or historical facts, naive intelligence may build on entrepreneurial and intrapreneurial bricolaging (Halme, Lindeman, & Linna, 2012) as an antidote for the processes of institutional isomorphism that dictate patterns of reproduction of business strategies and organizational forms.

Naive intelligence is needed to break the sustainability barriers that we are confronted with due to ingrained conceptions of, for instance, the purpose of business, organizational architecture, and leadership styles (Visser, 2014). Against the background of an environment that has gained complexity due to the increasing intertwinement of business and society and the expanding relationships and consequent feedback loops between them it can be argued that naive intelligence and intuition may be the best tools for coping with a very complex and non-deterministic future (Ayres, 1999).

While one may relate the concept of naivety to blessed ignorance, naive intelligence should not be mistaken for ignorance, as it actually allows for learning, exploring the unknown, and the generation of new visions, whereas ignorance usually amounts to shielding oneself against learning, defying new visions, and raving the known. Naive intelligence is not so much an uninformed mode of intelligence, but reflects a certain level of healthy and deliberate unprejudiced appreciation of old and new ideas, and the assumptions behind existing and new business models, through a truly reflexive process. As such, naive intelligence cultivates the adaptive capacity firms need to foster resilience.

5.2 *Native Intelligence*

The second type of sustainability intelligence, native intelligence, draws on several approaches towards and aspects of the value creation processes that are characteristic of fourth generation SOBMs. A first aspect of our conception of native intelligence relates to the role of evolutionary psychological processes. In order to be successful in developing fourth generation SOBMs, firms need to understand the evolutionary roots of human behaviour and their role in day-to-day decision-making. Referring to the work of Griskevicius et al. (2012), this chapter observed that there are built-in barriers towards sustainability in the human psychology (cf. Marshall, 2015; Stoknes, 2015). However, while this 'native dimension' of human behaviour seems to limit us in our quest to develop sustainable practices, a recognition of this nativity may enable firms to work with evolutionary processes rather than be constrained by them. In other words, when human behaviour is guided by, among other things, self-interest, short-termism, obsessive status orientation, as argued by evolutionary psychologists, firms should realize that they can promote sustainable behaviour by using these dimensions of human behaviour to their benefit as well. Native intelligence then means that firms not only are aware of the psychological limitations of humans in developing sustainable behaviour, but also have the ability to turn these limitations into opportunities by learning to work with them. It should be noted that possessing native intelligence would also enable and urge firms to reflect on their own behaviour and assumptions behind it from this viewpoint.

Second, it draws on a topological approach towards value creation—emphasizing the need for firms to get a grip on local aspects of doing business and coordinating activities with stakeholders. The word 'native' is generally used in contexts of indigenous people, for instance referring to the age-old knowledge and capacity of Indians and Aboriginals to solve problems based on ancient wisdom and without relying on modern technologies in doing so. Our conception of native intelligence is somewhat different, but draws on the underlying principles of rootedness, culture, and craftsmanship.

Fourth generation SOBMs rely on forging (new) coalitions between firms and their stakeholders. The composition of the stakeholder network of firms in a sense represents the roots of the company. Stakeholder thinking is characterized by the identification and recognition of dependencies between firms and their stakeholders. As such, one can only understand firms through understanding their stakeholder network. Although normative approaches towards stakeholder thinking are seen as a separate strand of stakeholder thinking (Donaldson & Preston, 1995) in fact all strands of stakeholder thinking have ethical implications since firms are confronted with demands, interests, and expectations from their stakeholders and need to decide how to respond to these.[2] Fourth generation SOBMs are not only rooted in a stakeholder culture, but also propagate an ethical culture in which they do not only recognize the interdependencies between themselves and their stakeholders but also commit themselves towards cultivating the success of the entire stakeholder network, which includes society, nature, and future generations. This, in turn, means that stakeholder relationships go beyond thinking in terms of agenda's (cf. Polonsky, Maignan, & Ferrell, 2005), but also build on the wisdom and intelligence that is available within the stakeholder network and use this for the overarching goals of the company, which includes those of all stakeholders, not only those whose goals align best with that of the firm.

Related to the above, native intelligence also refers to an orientation on increasingly local patterns of value creation processes that have recently been identified as a trend in business practice (Moratis, 2016). Local production and consumption clearly seems to appeal to people and it may be argued from a psychological viewpoint that we have a deeply ingrained tendency towards and preference for local (cf. Scruton, 2012). Local value creation processes enable the development of closed loop solutions that knit together responsible use of natural resources and waste recovery strategies that provide the foundation of the circular economy. Such local approaches towards creating value are also spurred as citizens and entrepreneurs alike are trying to respond to the questions raised by anonymity of globalized capitalism and try to rediscover and revive fundamental values people share by taking matters into their own hands again. This trend towards 'sustainability going

[2]This response may vary from neglecting demands, interests, and expectations to honouring demands, furthering interests, and cultivate expectations.

native' manifests itself partly through the creation of bottom-up, grassroots initiatives and social entrepreneurship which contain elements of modern craftsmanship (ibid.).

5.3 Narrative Intelligence

Narrative intelligence is grounded in the deeply ingrained human tendency and ability to organize experience into a narrative form (Blair & Meyer, 1997). As such, it reflects the fact that people are 'narrative animals' and thus acknowledges an important aspect of human psychology in the progress of thought (Mateas & Sengers, 1999). However, narrative intelligence is more than simply being open and able to translate and transfer experience into and through stories respectively. As author Stephen Denning writes in his book 'The secret language of leadership', narrative intelligence "reflects a recognition that the narrative aspects of the world matter because human goals matter, and narratives encapsulate human goals" (Denning 2007: 45).

Narrative intelligence, in our conception, is fundamentally about sensemaking, defined as an interpretive process in which people assign meanings to ongoing occurrences (Wagner & Gooding, 1997; cf. Weick, 1995). It aligns with Schön's idea of the reflective practitioner as it entails "the cyclical process of taking action, extracting information from stimuli resulting from that action, and incorporating information and stimuli from that action into the mental frameworks that guide further action" (Seligman, 2006; cf. Schön, 1983). Fourth generation SOBMs need narrative intelligence as well in the sense that they build on recursive and iterative process of exploring and experimenting with possible futures (cf. naive intelligence) and that they require firms not only to recognize stakeholders and their characteristics, but also truly engage with them in this process (cf. native intelligence).

Bearing this in mind, narrative intelligence goes well beyond communicating and even storytelling. Through a co-creation process with their stakeholders it enables firms that want to develop fourth generation SOBMs to create joint interpretation mechanisms and attach shared meaning to the interface of business and society. Whereas the norm in storytelling may be called story-yelling, narrative intelligence cultivates storycrafting. Drawing on the work of Morsing (Christensen, Morsing, & Thyssen, 2013; Morsing & Schultz, 2006), it can be argued that fourth generation business models not only rely on sensemaking rather than sensegiving processes, but also accelerate their development through aspirational talk. By means of aspirational talk, firms challenge both their status quo as well as that of society and paint their idea of a desired future that breaks with more obvious future directions for business and society to develop into. It is this desired future that then becomes the compass for corporate decision-making.

6 Conclusion

This chapter has argued that sustainability-oriented business models (SOBMs) provide a way of developing business into an institutional force for a better world beyond what the CSR concept has sofar been capable of. Focusing on value creation processes that encompass social and environmental dimensions it has distinguished between different generations of SOBMs—each with their own characteristics and limitations. By presenting fourth generation SOBMs, which rely on insights from evolutionary psychology and partnership thinking, and which transcend Porter and Kramer's (2011) idea of creating shared value we hope to have contributed to the developing discourse on business models from a sustainability perspective. In particular, by identifying three types of what we have called sustainability intelligence (naïve, native, and narrative intelligence) we hope to have shed light on the knowledge, skills, attitudes, and experiences that are necessary to develop fourth generation SOBMs.

From an academic viewpoint, it may not only be interesting to further conceptualize fourth generation SOBMs, the value creation processes they involve, and the trade-offs they may engender between the different performance dimensions that are inherent in the holistic interpretation of sustainability residing in these SOBMs. As such, SOBMs may be viewed as hybrid organizations, operating under different value systems at the same time, and fall victim to the risks in institutional pluralism (Pache & Santos, 2010). Conceptual and empirical research may hence focus on the way in which firms that are implementing SOBMs deal with competing institutional demands and to what extent the experience the further hybridization of the value creation processes as implementation barriers. From a practical viewpoint, one of the main questions that result from chapter is in what ways naïve, native, and narrative intelligence may be developed by organizations, and what role corporate leadership plays in the development of sustainability intelligence. Research into corporate change initiatives in the sustainability realm for instance suggests that well-equipped and well-positioned so-called social intrapreneurs (e.g., Grayson, McLaren, & Spitzeck, 2014) may play an essential role in this as they may pressingly bring new and innovative perspectives on the company-society interface to the table, are able to find and develop new stories around these perspectives, and have a mandate to develop new narratives into organizational action.

References

Ayres, R. (1999). *Turning point: The end of the growth paradigm*. London: Earthscan.

Basu, K., & Palazzo, G. (2008). Corporate social responsibility: A process model of sensemaking. *Academy of Management Review, 33*, 122–136.

Blair, D., & Meyer, T. (1997). Tools for an interactive virtual cinema. In R. Trappl & P. Petta (Eds.), *Creating personalities for synthetic actors: Towards autonomous personality agents* (pp. 83–91). Berlin: Springer.

Bocken, N., Short, S., Rana, P., & Evans, S. (2014). A literature and practice review to develop sustainable business model archetypes. *Journal of Cleaner Production, 65*, 42–56.

Bohnsack, R., Pinkse, J., & Kolk, A. (2014). Business models for sustainable technologies: Exploring business model evolution in the case of electric vehicles. *Research Policy, 43*, 284–300.
Boons, F., & Lüdeke-Freund, F. (2013). Business models for sustainable innovation: state-of-the-art and steps towards a research agenda. *Journal of Cleaner Production, 45*, 9–19.
Carroll, A., & Shabana, K. (2010). The business case for corporate social responsibility: A review of concepts, research and practice. *International Journal of Management Review, 12*, 85–105.
Christensen, L., Morsing, M., & Thyssen, O. (2013). CSR as aspirational talk. *Organization, 20*, 372–393.
Ciulla, J. (2005). The state of leadership ethics and the work that lies before us. *Business Ethics: A European Review, 14*, 323–335.
Crane, A., Palazzo, G., Spence, L., & Matten, D. (2014). Contesting the value of 'Creating shared value'. *California Management Review, 56*, 130–153.
Dahan, N., Doh, J., Oetzel, J., & Yaziji, M. (2010). Corporate-NGO collaboration: Co-creating new business models for developing markets. *Long Range Planning, 43*, 326–342.
Davis, K. (1973). The case for and against business assumption of social responsibilities. *Academy of Management Journal, 16*, 312–322.
Dawkins, R. (1976). *The selfish gene*. New York: Oxford University Press.
Dawson, R. (1980). Social dilemmas. *Annual Review of Psychology, 31*, 169–193.
Denning, S. (2007). *The secret language of leadership: How leaders inspire action through narrative*. San Francisco: Jossey-Bass.
Doane, D. (2005). Beyond corporate social responsibility: Minnows, mammoths and markets. *Futures, 37*, 215–229.
Dohrmann, S., Raith, M., & Siebold, N. (2015). Monetizing social value creation: A business model approach. *Entrepreneurship Research Journal, 5*, 127–154.
Donaldson, T., & Preston, L. (1995). The stakeholder theory of the corporation: Concepts, evidence, and implications. *Academy of Management Review, 20*, 65–91.
Elkington, J. (1997). *Cannibals with forks: The triple bottom line of 21st century business*. Oxford: Capstone.
Friedman, M. (1962). *Capitalism and freedom*. Chicago: University of Chicago Press.
Gardner, H. (1993). *Frames of mind: The theory of multiple intelligences*. New York: Basic Books.
Gifford, R. (2011). The dragons of inaction: Psychological barriers that limit climate change mitigation and adaptation. *American Psychologist, 66*, 290–302.
Goldstein, N., Griskevicius, V., & Cialdini, R. (2012). Reciprocity by proxy: Harnessing the power of obligation to foster cooperation. *Administrative Science Quarterly, 44*, 441–473.
Goleman, D. (2009). *Ecological intelligence: How knowing the hidden impacts of what we buy can change everything*. New York: Broadway Books.
Gottfredson, L. (1997). Mainstream science on intelligence (editorial). *Intelligence, 24*, 13–23.
Grayson, D., McLaren, M., & Spitzeck, H. (2014). *Social intrapreneurism and all that jazz—How business innovators are helping to build a more sustainable world*. Sheffield: Greenleaf Publishing.
Griskevicius, V., Cantú, S., & van Vugt, M. (2012). The evolutionary bases for sustainable behavior: Implications for marketing, policy, and social entrepreneurship. *Journal of Public Policy & Marketing, 31*, 115–128.
Halme, M., Lindeman, S., & Linna, P. (2012). Innovation for inclusive business: Intrapreneurial bricolage in multinational corporations. *Journal of Management Studies, 49*, 743–784.
Laukkanen, M., & Patala, S. (2014). Analysing barriers to sustainable business model innovations: Innovation systems approach. *International Journal of Innovation Management, 18*, 1440010.
Loorbach, D. (2014). *To transition! Governance panarchy in the new transformation*. Rotterdam: Erasmus University.
Marques, J., & Mintzberg, H. (2015). Why corporate social responsibility isn't a piece of cake. *Sloan Management Review, 56*, 7–11.
Marshall, G. (2015). *Don't even think about: Why our brains are wired to ignore climate change*. New York: Bloomsbury.

Mateas, M., & Sengers, P. (1999). *Narrative intelligence. Proceedings of the AAAI fall symposium on narrative intelligence* (pp. 1–10). North Falmouth: AAAI Press.
Moratis, L. (2014). The perversity of business case approaches towards CSR: Nuancing and extending the critique of Nijhof and Jeurissen. *International Journal of Sociology and Social Policy, 34*, 654–669.
Moratis, L. (2016). *Connecting the dots: Towards integrated value. CSR Trend Report 2016*. Antwerp: Antwerp Management School.
Morsing, M., & Schultz, M. (2006). Corporate social responsibility communication: Stakeholder information, response and involvement strategies. *Business Ethics: A European Review, 15*, 323–338.
Neisser, U., et al. (1996). Intelligence: Knowns and unknowns. *American Psychologist, 51*, 77–101.
Pache, A., & Santos, F. (2010). When worlds collide: The internal dynamics of organizational responses to conflicting institutional demands. *Academy of Management Review, 35*, 455–476.
Polonsky, M., Maignan, I., & Ferrell, O. (2005). A stakeholder model for implementing social responsibility in marketing. *European Journal of Marketing, 39*, 956–977.
Porter, M., & Kramer, M. (2011). Creating shared value. *Harvard Business Review, 89*, 62–77.
Rohrbeck, R., Konnertz, L., & Knab, S. (2013). Collaborative business modelling for systemic and sustainability innovations. *International Journal of Technology Management, 63*, 4–23.
Santos, F., Pache, A.-C., & Birkholz, C. (2015). Making hybrids work: Aligning business models and organizational design for social enterprises. *California Management Review, 57*, 36–58.
Schaltegger, S., Lüdeke-Freund, F., & Hansen, E. (2011). *Business cases for sustainability and the role of business model innovation: Developing a conceptual framework*. Lüneberg: Centre for Sustainability Management.
Schön, D. (1983). *The reflective practitioner*. New York: Basic Books.
Scruton, R. (2012). *How to think seriously about the planet: The case for an environmental conservatism*. New York: Oxford University Press.
Seligman, L. (2006). Sensemaking throughout adoption and the innovation-decision process. *European Journal of Innovation, 9*, 108–120.
Sennett, R. (2008). *The craftsman*. New Haven: Yale University Press.
Simon, H. (1990). Invariants of human behavior. *Annual Review of Psychology, 41*, 1–19.
Stoknes, P. (2015). *What we think about when we try not to think about global warming: Toward a new psychology of climate action*. White River Junction: Chelsea Green Publishing.
Stubbs, W., & Cocklin, C. (2008). Conceptualizing a 'sustainability business model'. *Organization & Environment, 21*, 103–127.
SustainAbility. (2014). *Model behaviour: 20 business models innovations for sustainability*. London: SustainAbility.
Teece, D. (2010). Business models, business strategy and innovation. *Long Range Planning, 43*, 172–194.
Varul, M. (2013). Towards a consumerist critique of capitalism: A socialist defence of consumer culture. *Ephemera—Theory & Politics in Organization, 13*, 293–315.
Visser, W. (2011). *The age of responsibility: CSR 2.0 and the new DNA of business*. London: Wiley.
Visser, W. (2014). *Sustainable frontiers: Unlocking change through business, leadership and innovation*. Sheffield: Greenleaf Publishing.
Wagner, J., & Gooding, R. (1997). Equivocal information and attribution: An investigation of patterns of managerial sensemaking. *Strategic Management Journal, 18*, 275–286.
Weick, K. (1995). *Sensemaking in organizations*. Thousand Oaks: Sage.
Wilson, F., & Post, J. (2013). Business model for people, planet (& profits): Exploring the phenomena of social business, a market-based approach to social value creation. *Small Business Economics, 40*, 715–737.
Yunus, M., Moingeon, B., & Lehmann-Ortega, L. (2010). Building social business models: Lessons from the Grameen experience. *Long Range Planning, 43*, 308–325.
Zott, C., Amit, R., & Massa, L. (2011). The business model: Recent developments and future research. *Journal of Management, 37*, 1019–1042.

Corporate Social Responsibility in Times of Crisis: Evidence from Romania

Catalina Sitnikov and Claudiu Bocean

1 Introduction

The economic and financial crisis that hit the world during 2008–2013 was characterized by a unique intensity, a high complexity and difficulties. They had to be faced and overcome by national governments and international bodies. Social responsibility programs, which consume large financial resources, were questioned. However, some authors as Escobar and Miras (2013) believe that the difficult financial situation provides an opportunity to test the organizations commitment to social responsibility principles and values. According to Fernández-Feijóo Souto (2009), the crisis can be viewed as both a threat and an opportunity for any organization's social responsibility activities. The crisis is seen as a threat since, in order to survive, organizations are forced to reduce their costs, with social responsibility programs being affected first due to the high implementing costs. The crisis can also be seen as an opportunity for social responsibility programs, as CSR is a mean to manage the difficult financial conditions and to support organizations in overcoming the consequences of the crisis. The result will drive to positive effects both in terms of organization's reputation and the advantages brought for the local communities (Miras-Rodríguez et al, 2013).

Arevalo and Aravind (2010) stated that the degree of social responsibility integration into the core strategy of the organization influences the power of economic and financial crisis impact on an organization. Organizations that have a proactive CSR strategy will succeed in better coping with difficult and challenging financial and economic conditions (Lee, 2008), while organizations with responsive social responsibility programs will experience problems (Mateo-Perez, Martinez-Roman, & Domenech-Lopez, 2015).

C. Sitnikov (✉) • C. Bocean
University of Craiova, Craiova, Romania
e-mail: catalinasitnikov@yahoo.com

S.O. Idowu et al. (eds.), *Corporate Social Responsibility in Times of Crisis*,
CSR, Sustainability, Ethics & Governance, DOI 10.1007/978-3-319-52839-7_4

Hsu (2006) believes that social responsibility has a positive effect on crisis management within an organization. Social responsibility programs have also an important role in the development of responsible behaviour within organization, which will be an advantage during potential future crises (Banerjee, 2008). When an organization is facing a crisis, in addition to poor economic outcomes, reputation suffers the most. Social responsibility programs and activities can prevent this loss of reputation (Coombs, Frandsen, Holladay, & Johansen, 2010). It can therefore be mentioned that CSR has a preventive feature, being not only a reactive tool used to reduce the negative effects of a crisis. According to Haigh and Dardis (2012), responsible behaviour can support an organization to safely navigate and substantially improve its reputation during crisis.

Fernández-Feijóo Souto (2009) states that if any organization would implement CSR programs and would include social responsibility in organizational strategy, financial and economic crisis had not manifested or, at least, had been more manageable. A crisis has a lot of implications for organizations, consequences ranging from profits decrease to bankruptcy (Miras-Rodríguez et al., 2013). The crisis has led many organizations to change their business patterns, implementing austerity policies in order to decrease losses (Mateo-Perez et al., 2015). Austerity policies may have a bad influence on the financing of social responsibility programs, particularly within those organizations where social responsibility is not a strategic issue. Nevertheless, in times of economic and financial crisis social needs are higher than ever (Karaibrahimoglu, 2010). Those being the case, social responsibility measures are more than necessary to ensure the welfare of society as a whole and to avoid degradation of living standards and quality of life. It is apparent that, during economic and financial crises there are increased requirements from society and local community, to organizations supporting social and environmental causes (Gregory, 2011).

However, there are opinions arguing that implementing social responsibility programs in times of crisis could threaten the organization. Orlitzky, Schmidt, & Rynes, (2003) considers it normal for organizations to behave in more conservative and defensive ways in times of crisis. Giannarakis and Theotokos (2011) believe that, in order to be able to meet shareholders' expectations in periods of crisis (reducing losses and maximizing profits) organizations must go beyond the limits of a socially responsible behaviour. In times of crisis, organizations' activities run under financial restrictions, shareholders requiring cost reductions in order to maintain financial performance at an acceptable level (Yelkikanat & Kose, 2012). The immediate consequences may be the delay or cancellation of many CSR initiatives (Njoroge, 2009).

Giannarakis and Theotokos (2011) consider that, in times of economic and financial crisis, the effects manifested after implementing CSR strategy and initiatives are more important than ever for the survival of any organization. Social responsibility supports organizations differentiate their products or services and increases stakeholders' confidence in the organization (Thome, 2009). Organizations tend to invest in social responsibility programs to enhance organization's reputation, the brand, consumer's confidence and to redefine the relationship between organizations and society. That way, an economic and financial crisis provides organizations with the opportunity to turn CSR into an opportunity.

Jacob (2012) shows that 2008 economic and financial crisis influenced CSR initiatives in many companies, the most common results being the cost reduction for targeting community involvement and programs restructuring which resulted in employees layoffs. However, there is a range of benefits focused on improving organizational governance, setting priority on environmental policy and restructuring wage policies to increase reward equity. As regards the stakeholders, key stakeholders affected by the crisis were employees, followed by investors and customers (Jacob, 2012), while labour practices were significantly affected. Jacob (2012) concludes that the most important issues for an organization are those targeted by most influential stakeholders or on those that do not incur a huge reputational risk. During economic and financial crisis, when the whole economy is restructured, employees' dismissal as a result of organizational programs restructuring or work conditions and labour relations impairment will not be perceived as a threat to the organization's reputation (Sharma & Narwal, 2006). Organizations prefer to focus on environmental and brand issues, because they could significantly affect their reputation (Gotsi & Wilson, 2001). Emeseh, Ako, Okonmah, and Obokoh (2010) confirm that economic and financial crisis has an impact on social responsibility programs due to the pressure and restrictions, which organizations must meet in order to survive. However, the impact is not all negative, as social responsibility finds new opportunities of development and, with the crisis ending it will revive stronger than before.

Simionescu and Dumitrescu (2014) analyse the effect of CSR practices on financial performance of 19 companies listed in Romania during the financial crisis. Although the financial crisis has caused many organizations to reduce their social responsibility programs because of high costs, benefits arising from the implementation of such programs are undeniable. As a result of research, Simionescu and Dumitrescu (2014) found that there is a positive connection between CSR and financial performance. The positive effects of social responsibility programs influence not only financial performance, but also organizational reputation (Thome, 2009). However, there are empirical studies underlining that the connection between social responsibility programs and financial performance has altered as a consequence of turbulent economic environment through the financial crisis (Miras et al., 2013; Ducassy, 2013).

Regarding CSR in times of economic crisis, the situation in Romania is similar to the global one. In Ernst & Young study "CSR Trends and Realities in Romania" there were interviewed 113 top managers from companies residing in Romania (Ernest & Young, 2014). Most respondents (74%) felt that CSR practices are developed especially to acknowledge, raise the visibility and notoriety. CSR is not an end in itself or an independent strategy but is a part of the public relations strategy. The study concludes that, in Romania, CSR practices aim at commercial and not ethic objectives, rooted in the company responsible behaviour. Romanian companies focus their efforts particularly towards meeting the needs of external stakeholders to ensure better recognition on the market. These CSR market characteristics make companies develop rather formal CSR activities that are not sustainable and do not produce significant effects. Although, lately, due to financial and economic crisis there is a need for companies to turn towards CSR, developing CSR programs, in Romania CSR is developed more to achieve commercial objectives and not as a result of socially responsible behaviour.

Crisan-Mitra and Borza (2015) believes that the economic crisis has had a major impact on corporations' willingness to use resources in areas not directly related to their core business, avoiding investments in complex CSR programs with long-term effect. They showed that in Romania social responsibility actions do not significantly influence customers' purchase decision, as they focus more on goods or services price and characteristics. Such buying behaviour is characteristic to countries with low living standards as, due to low income, consumers are reluctant to pay higher prices to reward companies' socially responsible behaviour (Crisan-Mitra & Borza, 2015).

Obrad, Petcu, Gherheş, and Suciu (2011) shows that CSR practices have been implemented in Romanian corporate environment as the result of importing values and principles from European business environment. Local companies, particularly subsidiaries of multinational companies, have begun to adopt similar practices and adapted to the specific local organizational culture. The economic crisis has affected to a greater extent Romanian CSR market, due to the mimetic stage CSR is in Romania.

In 2016, a consulting company, The Azores, which publishes Romania CSR Index, found that top performing companies on Romanian social responsibility market are multinational companies that usually publish local CSR reports. Most of organizations in Romania involve too little stakeholders in decision making on CSR investment and do not perform detailed analyses to determine their needs. The effect of the crisis on CSR in Romania is influenced by multinational companies' availability to invest in CSR programs.

In a comparative study, Zaharia and Grundey (2011) show that large companies in Romania seem to develop national CSR projects on a larger scale, unlike large companies in Lithuania, more focused on the local community. The two authors recommended for the two countries to provide greater transparency, implement good practice guides, raise awareness regarding CSR issues, and enhance cooperation with stakeholders, especially in times of crisis.

Based on the findings from the few papers that explore the relationship between economic crisis and social responsibility programs the authors addressed the effect of economic and financial crisis on CSR in Romania at both microeconomic level, examining the issues concerning social responsibility in OMV Petrom during 2009–2014, and at macroeconomic level analysing the National Strategy for Promoting Social Responsibility developed by Romanian Government and the effects it has on society. The authors have chosen OMV Petrom to study the effect of the crisis on CSR at the microeconomic level due to the fact that OMV Petrom is the largest integrated oil group from South-Eastern Europe, it is a leading company in terms of performance and corporate governance in Romania, its annual turnover is about 5% of Romania's GDP, in 2013 it spent over 150 million euro for environmental projects and for community (OMV Petrom, 2014), representing a substantial part of the whole CSR market in Romania.

2 Evolution of CSR Within OMV Petrom in Time of Economic and Financial Crisis

OMV Petrom is the largest oil and gas integrated group in South-Eastern Europe, with activities in the exploration, production, transport and sale of fuels branches. Since 2006, OMV Petrom has undergone through one of the most complex and successful transformation processes of a Romanian state-owned company, currently being a competitive private company, a major player on the European energy market. Following a responsible approach regarding restructuring, massive investment, efforts to boost competitiveness and sustainable development, OMV Petrom has developed into a leading company in terms of performance and corporate governance, which generates value for all stakeholders (OMV Petrom, 2012). Regarding social responsibility OMV Petrom is a pioneer in the Romanian business environment. Since 2007, the company has integrated the principles of corporate responsibility within its business strategy and has allocated more than 40 mil euros for the Romanian communities' development, focusing on environmental protection, education, health and local development. Since 2009, "Andrei's Country" has become a platform for community involvement of OMV Petrom and the best known CSR program in Romania (OMV Petrom, 2013). In 2012, OMV Petrom has passed from CSR concept to sustainability concept and introduced the "resourcefulness" in its business strategy, a concept that encourages the organization to profitable grow through responsible use of natural, human and financial resources. 2014 marked the eighth year of social and community involvement, during which OMV Petrom developed community projects that targeted one million people. OMV Petrom received recognition from both the general public and key opinion leaders as the most involved company in Romania in social and environmental issues (OMV Petrom, 2015).

OMV Petrom approach is designed to establish mutually beneficial relationships with stakeholders. It brings together all responsible commitments related to the management of environmental issues, new energy sources, education and development, health and safety, diversity, business ethics, human rights and relationships with community stakeholders, all melted in a unique and comprehensive strategy. OMV Petrom commitments are organized around three key areas (the three E—Fig. 1):

- eco-efficiency—limiting environmental impact;
- eco-innovation—the adoption of alternative energy sources and opportunities to generate sustainable innovations;
- education for development—encouraging skills that people need in order to succeed.

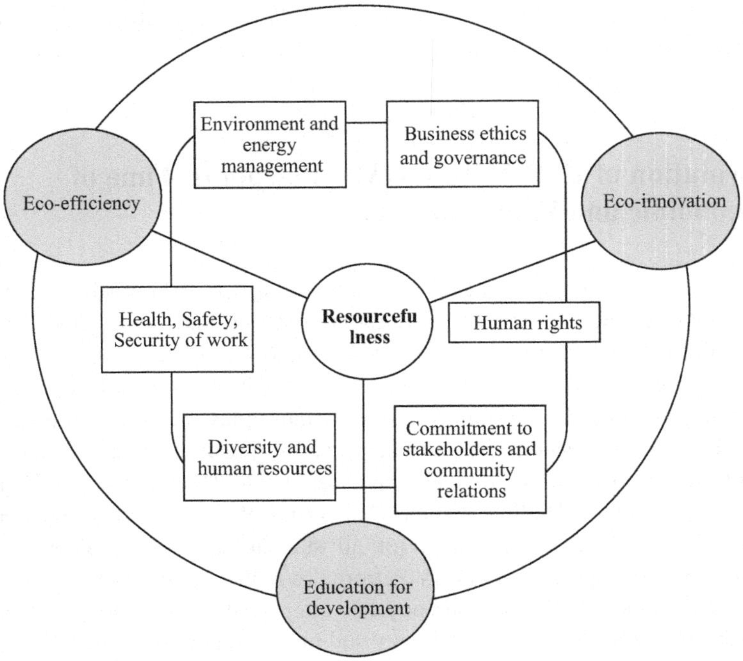

Fig. 1 Key areas and key topics of OMV Petrom social responsibility strategy

Regarding commitment to Eco-efficiency, OMV Petrom aims at paying attention to the management company's impact on the environment and the safety and efficiency of its operations. Regarding commitment to Eco-Innovation, OMV Petrom focuses on identifying new ways to meet growing global energy needs, given the challenges of climate change and reducing fossil fuel reserves. Regarding the commitment to education for development, OMV Petrom aims at developing skills that people need to succeed. OMV Petrom invests in economic development of the communities through "Andrei's country" programme, providing know-how and financial supporting people and sustainable projects that help developing Romanian communities in the long term.

In 2014, OMV Petrom set up the Steering Committee for the concept of "resourcefulness", an internal body with the objective to conduct, inform and align all functions and business units towards sustainable development. This committee is held every quarter and is led by OMV Petrom CEO (OMV Petrom, 2015).

OMV Petrom has identified the following key topics of social responsibility that influence performance and are of great interest to stakeholders (OMV Petrom, 2014):

- Environment and energy management,
- Health, safety, work safety,
- Human rights,
- Diversity and human resources,

- Business ethics and governance,
- Commitment to stakeholders and community relations.

Environment and Energy Management. OMV Petrom monitors the environmental impact along entire value chain, efficiently using natural resources and aiming at minimizing waste and greenhouse gases and water consumption used for productive operations.

Health, Safety, Security of Work. Health, safety and security are basic requirements for the success of OMV Petrom. The main concern is employees' physical safety and health.

Human Rights. Human rights are a central topic for OMV Petrom. In December 2013 OMV Petrom joined the platform of the United Nations Global Compact (UNGC) and included the specific principles in human rights policy and Code of Conduct. In 2013, there were established new key performance indicators related to human rights.

The Diversity and Human Resources. OMV Petrom workforce comprises around 51 different nationalities, with diversity, inclusion and equal opportunities being high on the agenda. OMV Petrom aims at attracting the best people from diverse groups and providing the tools and skills they need to perform at their maximum potential (OMV Petrom, 2015).

Business Ethics and Governance. OMV Petrom aims at building a culture of ethics and accountability based on trust and integrity. Governance procedures are based on a code of business ethics that includes clear rules on conflicts of interest, gifts offering, as well as third-party assessment.

Commitment to Stakeholders and Community Relations Management. OMV Petrom interacts with a large number of stakeholders. The objective is to have a regular dialogue to build sustainable and fruitful relations with them—to support in understanding the visions, interests and needs to improve activities (OMV Petrom, 2015).

Following the analysis of OMV Petrom annual reports and sustainability reports (OMV Petrom, 2012, 2013, 2014, 2015) the authors have focused on the evolution of the main social responsibility indicators that characterizes the key topics on which OMV Petrom built CSR strategy (Table 1). 2014 was not included in the analysis because for this year OMV presented consolidated reports for the entire group only. In the analysis there were introduced two performance indicators (sales revenue and net profit distributed to shareholders). Annual net profit distributed to shareholders, personnel costs, expenses for environmental protection and funds allocated to "involvement in community" projects enable an analysis of funds distribution by category of stakeholders.

It can be seen that throughout the crisis period net profit distributed to shareholders and environment costs registered a steady upward trend reflecting the fact that OMV Petrom targeted these categories of stakeholders. The net profits distributed to shareholders registered sustained growth also as a result of the higher sales

Table 1 OMV Petrom social responsibility indicators in period 2009–2013

	2009	2010	2011	2012	2013
Sales revenues (mil RON)	16090	18616	22614	26258	24185
Net profit attributable to equity holders (mil RON)	860	2201	3757	3953	4821
Staff (employees)	28984	24662	22912	21650	19619
Staff costs (mil RON)	2344.25	1919.30	1784.3	2037.83	2030.44
Lost-time injury rate (LTIR) (employees)	0.67	0.64	0.47	0.41	0.37
Total recordable injury rate (TRIR) (employees)	0.87	0.90	0.74	0.78	0.67
Lost-time injury severity (LTIS) (employees)	26.82	24.79	28.09	20.08	15.74
Power consumption (PJ)	53.2	50.6	45.4	42.9	49.5
Total water consumption (mil. Cubic meters)	43.8	39.94	33	26	24
GHG (mil. Tones CO2 equivalent)	4.98	4.73	4.3	4.5	5.2
Expenses for environmental protection (mil RON)	458.49	472.47	488.11	507.23	661.19
Funds allocated to "involvement in community" projects (mil RON)	19.15	18.14	17.80	21.78	17.38

Source: data from annual reports and sustainability reports of OMV Petrom (2012, 2013, 2014, 2015)

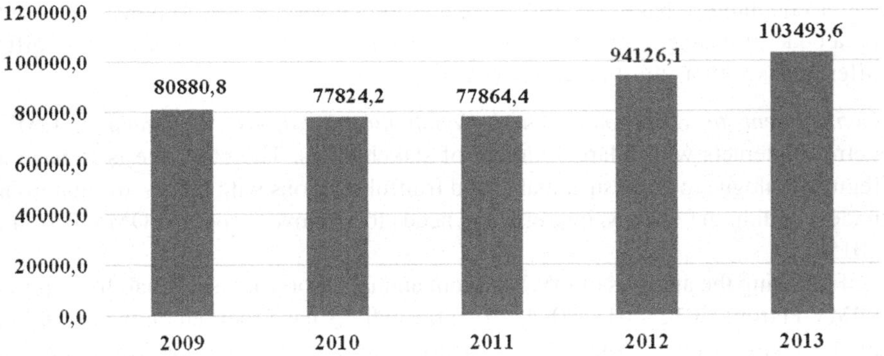

Fig. 2 Average staff expenses per year including social spending. Source: data from annual reports (OMV Petrom, 2015)

revenues alongside the beneficial effects of restructuring and fossil fuel price increases in the studied period. Although development in personnel expenses is not spectacular, if one considers the number of employees one can see that OMV Petrom pays particular attention to this type of stakeholders Fig. 2.

2010 and 2011 were the years when the crisis peaked in Romania and Europe, fact reflected in the decline of personnel expenses. The funds allocated to projects related to community involvement also experienced a similar trend.

Besides financial indicators illustrating the distribution of resources by stakeholders, the authors have introduced in the analysis indicators showing the effects social responsibility policies produce: health, safety, work security indicators (Lost-time injury rate, Total recordable injury rate, Lost-time injury severity) and environment protection indicators (Power consumption, Total water consumption, greenhouse gases emissions). Regarding health, safety, work security indicators, a continuous improvement it can be seen, showing a constant interest in terms of working conditions at OMV Petrom. Regarding environment protection indicators it can also be seen a sustained improvement, exclusive 2013. Decrease of crisis effects in 2013 caused a large increase in fossil fuel production, which made the GHG emissions and water consumption used in the production process to increase compared to previous years.

To deepen the analysis the authors calculated the correlations (Pearson coefficient) among social responsibility indicators at OMV Petrom from 2009 to 2013 (Table 2).

Analysing the correlations, one can see the interest paid to shareholders (an almost perfect correlation between the trend of sales revenues and trend of net profits distributed to shareholders) and environmental protection (Pearson coefficient showing the correlation between the trend of sales revenues and the trend of environmental expenditures recorded a value of 0.535).

Increased sales revenue is the result of the rise in fossil fuel prices and of staff reductions (Pearson coefficient showing the correlation between the trend of sales revenues and changes in the number of employees recorded a value of −0.938).

Staff professional development programs and dismissal of those who do not meet requirements causes a decrease in indicators that illustrate workplace accidents. One can mention a strong trend of these indicators correlated with the number of employee's trend (LTIR 0.926, TRIR 0.808, and LTIS 0.739).

Another correlation that can be mentioned is the one between the trend of environmental protection expenditures and that of greenhouse gas emissions (Pearson coefficient recorded a value of 0.570).

Summarizing, at OMV Petrom level, economic and financial crisis had a limited and mixed impact on the social responsibility issues. OMV Petrom has given a particular care to shareholders and environment protection, while employees have not been affected considerably by the economic and financial crisis effects. Staff reduction was performed under restructuring programs implemented due to the transition from state ownership to private ownership and in order to create a strong and modern company able to cope with competitive energy environment.

3 National Strategy on Social Responsibility

In Romania, the concept of social responsibility is still in the early stages of development. While multinationals and large Romanian companies have started to develop gradually their own social responsibility culture, social responsibility

Table 2 Correlations among OMV Petrom social responsibility indicators in period 2009–2013

	SR	NP	S	SC	LTIR	TRIR	LTIS	PC	TWC	GHG	EEP	FAICP
Sales revenues	1	0.912*	−0.938**	−0.816*	−0.798	−0.642	−0.503	−0.710	−0.812*	−0.314	0.534	−0.087
Net profit	0.912*	1	−0.983**	−0.581	−0.967**	−0.886*	−0.676	−0.671	−0.960**	−0.102	0.755	−0.068
Staff	−0.938**	−0.983**	1	0.587	0.926*	0.808*	0.739	0.627	0.948**	0.045	−0.767	0.064
Staff costs	−0.816*	−0.581	0.587	1	0.400	0.323	−0.064	0.572	0.363	0.613	−0.093	0.311
LTIR	−0.798	−0.967**	0.926*	0.400	1	0.917*	0.701	0.700	0.985**	0.073	−0.743	−0.093
TRIR	−0.642	−0.886*	0.808*	0.323	0.917*	1	0.595	0.450	0.859*	−0.094	−0.817*	0.219
LTIS	−0.503	−0.676	0.739	−0.064	0.701	0.595	1	0.179	0.793	−0.539	−0.871*	−0.093
Power consumption	−0.710	−0.671	0.627	0.572	0.700	0.450	0.179	1	0.676	0.717	−0.058	−0.504
Total water consumption	−0.812*	−0.960**	0.948**	0.363	0.985**	0.859*	0.793	0.676	1	0.005	−0.769	−0.148
GHG	−0.314	−0.102	0.045	0.613	0.073	−0.094	−0.539	0.717	0.005	1	0.570	−0.329
Expenses for environmental protection	0.534	0.755	−0.767	−0.093	−0.743	−0.817*	−0.871*	−0.058	−0.769	0.570	1	−0.338
Funds allocated to "involvement in community" projects	−0.087	−0.068	0.064	0.311	−0.093	0.219	−0.093	−0.504	−0.148	−0.329	−0.338	1

Significance of *: correlation is significant at the 0.05 level
Significance of **: correlation is significant at the level 0.01

concept is little known in the area of Small and Medium Size Enterprises, while in many cases, public authorities at central, regional and local level are not sufficiently involved in activities promoting the concept of social responsibility for the benefit of communities. Thus, involvement in social responsibility campaign is still very weak at this level, but in the context of globalization, Romanian companies and organizations must adopt a socially responsible attitude to withstand on a competitive market.

Based on the Integrated Guidelines for Growth and Jobs (2005–2008), which recommends that the EU Member States encourage organizations to develop own social responsibility vision and ISO 26000 International Standard on Social Responsibility, which creates prerequisites to implement social responsibility by both companies, multinationals, state-owned companies and small and medium enterprises, and NGOs, trade unions, employers and government institutions, Romania has developed a National Strategy for the Promotion of Social Responsibility. In applying this strategy, Romanian government envisages active involvement of public sector, private and non-governmental sectors creating synergies that will facilitate favourable conditions for developing the concept of social responsibility.

Under the National Strategy for the Promotion of Social Responsibility are presented principles, objectives and measures for an active and coherent policy to promote application of social responsibility concept in companies, organizations and public authorities in Romania (Romanian Government, 2011). In developing the strategy there were involved governmental authorities with direct responsibility in the implementation of measures to promote social responsibility, civil society representatives, and foreign experts with experience in social responsibility.

The main aim of the strategy is to increase the awareness of social responsibility importance and benefits and increase public sector involvement in the application of social responsibility in Romania. The goal of involving public authorities is to support the development of conditions for the implementation of corporate social responsibility in companies and other organizations through a series of active measures. All these in order to increase motivation and the ability to exercise social responsibility, including advice, openness to cooperation and partnerships, dialogue and exchange of experience among authorities, the private sector and civil society. Regarding private companies and other organizations, the National Strategy for the Promotion of Social Responsibility aims to promote the concept of social responsibility in their activities, public authorities having the role to create the necessary conditions and to generate social dialogue on this subject. The strategy aims also at promoting a responsible approach to the Romanian consumer multinationals.

The objectives of the National Strategy for the Promotion of Social Responsibility are economic, social, environmental, educational, and research objectives. Economic and financial objectives aimed at strengthening organizations competitiveness and position, maintaining a responsible relationship with employees, consumers, shareholders and other stakeholders. Social objectives are considering informing the population about the concept of social responsibility and increasing the capacity of public authorities and companies to address social needs.

Environmental objectives are focused on increasing accountability of organizations on preservation and environmental protection as to ensure minimal negative impact on the environment. The objectives of education and research targeted at enhancing knowledge and training regarding social responsibility so that organizations can express interest in the research and development (Romanian Government, 2011).

National Strategy on Social Responsibility provides an overview of Romanian authorities' initiatives and expectations regarding social responsibility of organizations. First, authorities and public sector as a whole are responsible for maintaining high standards in their activities. Authorities also play an important role as owner, investor and purchaser, role that provide the opportunity to promote direct application of the social responsibility concept. The strategy also focuses on the active involvement of the public sector in promoting social responsibility in companies and other private organizations by providing advice, information and training in social responsibility area.

The project of National Strategy for the Promotion of Social Responsibility aimed at increasing awareness and training on the concept of social responsibility. State-owned companies should develop their own strategies to support them overcome the crisis more simply. Though it has begun to be implemented since 2011, a year before the end of the implementation period, the National Strategy for the Promotion of Social Responsibility has not had much impact on the central and local authorities or on private companies. Because of misunderstandings, lack of tools and the failure in defining clear responsibilities in terms of implementing, monitoring and evaluating, the effects and impact on Romanian society and economy has been limited. Designed as a tool to promote social responsibility in the Romanian economy and society, to support the social responsibility programs of private organizations and public institutions in crisis and post crisis in the period 2011–2016, the National Strategy for the Promotion of Social Responsibility failed to meet these goals. Multinational companies continued their social responsibility programs, with fewer resources than before, small and medium enterprises have ignored the problem except punctual sponsorships and charitable projects (Anca, Aston, Stanciu, & Rusu, 2011), while public institutions have not been involved in social responsibility programs, except some.

Despite the fact that the progress of integrating responsible practices into core business organizations achieved more through organizations individual efforts, a number of challenges continue to be present in the Romanian economy, challenges that must be addressed by all stakeholders, such as: raising awareness of social responsibility and its integrated approach importance; increasing awareness of the needs and benefits associated with the use of responsible practices in core activities to ensure organization sustainability; increasing transparency, monitoring and evaluation of the effects of social responsibility activities on all stakeholders (Anca et al., 2011). In Romania, the challenge is to overcome the current belief that social responsibility is not only a tool to maximize the value of the organization by improving the image and reputation, but an essential element of long-term success, which is directly linked to the performance of social and environmental community (Barbuta, Hentea, Mesteru, & Cociorvei, 2014).

4 Conclusions

Given the dynamic and turbulent environment in which organizations must carry out their activities, many professionals and scientists are concerned about the evolution of CSR in times of economic and financial crisis because of reduced organizational performance and austerity policies that are promoted in such times. Economic and financial crises enable a better and clearer identification of the real motivations or interests organizations promote through programs of social responsibility. CSR activities and budgets will be significantly affected by the crisis if organizations run social responsibility actions only for increasing external reputation or getting direct benefits in the short term. However, if organizations are truly attached to the values of social responsibility and have integrated into their core strategy the social responsibility dimension, they can turn the crisis into an opportunity. Therefore, the economic and financial crises do not entail the disappearance of CSR actions, but affects budgets dedicated to social actions and their magnitude.

In recent years, Romanian organizations have made progress in the implementation of social responsibility programs. Despite an extension of the strategic approach of social responsibility, most organizations still focus on sponsorship and charity projects. Although the role of sponsorship and donations is beneficial in resolving the issues concerning environmental protection and development of local communities, it is also necessary an integrated proactive approach of social responsibility programs. This will be done by integrating the principles and generally accepted practices in the core activity and paying special attention to stakeholders. OMV Petrom is a success story in terms of social responsibility strategic approach by including CSR strategy in its basic strategy. As regards the economic and financial crisis effects on social responsibility, the authors noticed that OMV Petrom was provided with opportunities to integrate further the social responsibility dimension in the company's core strategy. In 2012, OMV Petrom has passed since the concept of social responsibility to sustainability concept, integrating "resourcefulness" in its business strategy, a concept that encourages the organization to profitable grow through responsible use of natural, human and financial resources. The quantitative and qualitative analyses underlined the fact that OMV Petrom paid a special attention to shareholders and environmental protection, but neither employee has being impacted too much by the economic and financial crisis effects. Staff reduction was performed under restructuring programs implemented due to the transition from state ownership to private ownership and in order to create a strong and modern company to cope with competitive energy environment.

Regarding macroeconomic analysis, the authors studied the components of National Strategy for the Promotion of Social Responsibility and the effects it had and has on society. National Strategy on Social Responsibility was a tool that could provide support and opportunities for all stakeholders and any organization that wishes to adhere to the principles of social responsibility for adopting such an integrative approach. Romanian Government focused on committing business and institutional communities towards correlating the values and principles of social responsibility with stakeholder expectations. National Strategy on Social Responsibility had a limited impact on Romanian society and economy despite the opportunities provided by the economic and financial crisis. It was rather a formal document, an attempt to align with EU policies on social responsibility.

Following the analysis undertaken at microeconomic and macroeconomic level the authors concluded that strategic integration of CSR into the core business led to the adoption of responsible behaviour that is less vulnerable to external factors such as economic and financial crises.

References

Anca, C., Aston, J., Stanciu, E., & Rusu, D. (2011). *(Corporate) Social responsibility in Romania—Situation analysis and current practices review*. Report prepared as part of the project "Strengthening the capacity of the Romanian companies to develop social partnerships-CSR", November 2011. Retrieved December 17, 2015, from http://undp.ro/libraries/projects/CSR/deliverables/Analiza%20Situatiei%20RSC%20in%20Romania%20ENG.pdf

Arevalo, J. A., & Aravind, D. (2010). The impact of the crisis on corporate responsibility: The case of UN global compact participants in the USA. *Corporate Governance: The International Journal of Business in Society, 10*(4), 406–420. doi:10.1108/14720701011069641.

Barbuta, R., Hentea, M., Mesteru, S.D., & Cociorvei D. (2014). Rolul Consiliilor Economice şi Sociale şi al instituţiilor similare din UE în promovarea conceptului de Responsabilitate Socială a Companiilor, CES Romania. Retrieved December 19, 2015, from http://www.ces.ro/newlib/studii-ces/STUDIUL-Responsabilitate-Sociala-Companii.pdf

Coombs, W. T., Frandsen, F., Holladay, S. J., & Johansen, W. (2010). Why a concern for apologia and crisis communication? *Corporate Communications: An International Journal, 15*(4), 337–349. doi:10.1108/13563281011085466.

Crisan-Mitra, C., & Borza, A. (2015). Approaching CSR in Romania: An empirical analysis. *Procedia—Social and Behavioral Sciences, 207*, 546–552.

Ducassy, I. (2013). Does corporate social responsibility pay off in times of crisis? An alternate perspective on the relationship between financial and corporate social performance. *Corporate Social Responsibility and Environmental Management, 20*(3), 157–167. doi:10.1002/csr.1282.

Emeseh, E., Ako, R. T., Okonmah, P., & Obokoh, L. O. (2010). Corporations, CSR and self regulation: What lessons from the global financial crisis? *German Law Journal, 11*(2), 230–259.

Ernest and Young. 2014. *Trends and realities CSR in Romania*. Retrieved May 17, 2016, from http://www.ey.com/Publication/vwLUAssets/CSR_Survey_15_May_15/$FILE/CSR%20Survey%202015_RO.pdf. Accessed Febuary 20, 2015.

Fernández-Feijóo Souto, B. (2009). Crisis and corporate social responsibility: Threat or opportunity? *International Journal of Economic Sciences and Applied Research, 2*(1), 36–50.

Giannarakis, G., & Theotokas, I. (2011). The effect of financial crisis in corporate social responsibility performance. *International Journal of Marketing Studies, 3*(1), 2–10. doi:10.5539/ijms.v3n1p2.

Gotsi, M., & Wilson, A. M. (2001). Corporate reputation: Seeking a definition. *Corporate Communications: An International Journal, 6*(1), 24–30. doi:10.1108/13563280110381189.

Grigore, F. G. (2011). Corporate social responsibility and marketing. *Developments in Corporate Governance and Responsibility, 2*, 41–58. doi:10.1108/S2043-0523(2011)0000002006.

Haigh, M. M., & Dardis, F. (2012). The impact of apology on organization—Public relations and perceptions of corporate social responsibility. *Public Relations Journal, 6*(1), 1–16.

Hsu, Y.S. (2006). *Corporate social responsibility and crisis communication: Nike Taiwan Jordan Crisis vs. Paolyta Bullwild Crisis*. Gainesville, FL.

Jacob, C. K. (2012). The impact of financial crisis on corporate social responsibility and its implications for reputation risk management. *Journal of Management and Sustainability, 2* (2), 259–275. doi:10.5539/jms.v2n2p259.

Karaibrahimoglu, Y. Z. (2010). Corporate social responsibility in times of financial crisis. *African Journal of Business Management, 4*(4), 382–389.

Lee, M. D. P. (2008). A review of the theories of corporate social responsibility: Its evolutionary path and the road ahead. *International Journal of Management Reviews, 10*(1), 53–73. doi:10.1111/j.1468-2370.2007.00226.x.

Mateo-Perez, M.-A., Martinez-Roman, M.-A., & Domenech-Lopez, Y. (2015). Social services in times of economic and social crisis: The case of Spain. *Revista de cercetare și intervenție socială, 50*, 96–110.

Miras-Rodríguez, M.M., Escobar-Pérez, B., & Carrasco-Gallego, A. (2013a, September 25–27). *How is the crisis affecting the socially responsible behavior of Spanish firms?* Presented in XVII Congreso AECA, Pamplona, Spain.

Miras-Rodríguez, M.M., Escobar-Pérez, B., & Carrasco-Gallego, A. (2013b, September 28–30). *The impact of the economic crisis on the environmental responsibility of the companies*. Proceedings of the 2013 International Conference on Environment, Energy, Ecosystems and Development, Venice, Italy.

Munilla, L. S., & Miles, M. P. (2005). The corporate social responsibility continuum as a component of stakeholder theory. *Business and Society Review, 110*(4), 371–387. doi:10.1111/j.0045-3609.2005.00021.x.

Njoroge, J. (2009). Effects of the global financial crisis on corporate social responsibility in multinational companies in Kenya, Covalence Intern Analyst Papers. Retrieved December 16, 2015, from http://www.covalence.ch/docs/Kenya-Crisis.pdf

Obrad, C., Petcu, D., Gherheş, V., & Suciu, S. (2011). Corporate social responsibility in Romanian companies—Between perceptions and reality. *Amfiteatru Economic, XIII*(29), 43–55.

OMV Petrom. (2012). *2011- Dezvoltare durabilă Petrom: fapte și cifre*. Retrieved December 21, 2015, from https://www.omvpetrom.com/SecurityServlet/secure?cid=1255739406493&lang=ro&swa_id=1411688294303.22&swa_site=

OMV Petrom. (2013). *Raport privind dezvoltarea durabilă pe anul 2012*. Retrieved December 21, 2015, from https://www.omvpetrom.com/SecurityServlet/secure?cid=1255753974868&lang=ro&swa_site=&swa_nav=&swa_pid=&swa_lang=

OMV Petrom. (2014). *Resourcefulness 2013 Fapte și cifre*. Retrieved December 21, 2015, from m https://www.omvpetrom.com/SecurityServlet/secure?cid=1255762043258&lang= ro&swa_site=&swa_nav=&swa_pid=&swa_lang=

OMV Petrom. (2015). *Annual reports, 2008–2014*. Retrieved December 21, 2015, from https://www.omvpetrom.com/portal/01/petromcom/petromcom/OMV_Petrom/Relatia_cu_investitorii/Rapoarte_si_prezentari_pentru_investitori/Rapoarte_anuale

Orlitzky, M., Schmidt, F. L., & Rynes, S. L. (2003). Corporate social and financial performance: A meta-analysis. *Organization Studies, 24*(3), 403–441. doi:10.1177/0170840603024003910.

Romanian Government. (2011). *National strategy on social responsibility*, 2011–2016. Retrieved December 23, 2015, from http://www.sgg.ro/nlegislativ/docs/2011/05/2x4bdfjnwskv89n17pzq.pdf

Sharma, T., & Narwal, M. (2006). Managing business crisis: The CSR perspective. *Social Responsibility Journal, 2*(2), 124–130. doi:10.1108/eb059242.

Simionescu, L., & Dumitrescu, D. (2014). Corporate social responsibility and financial crisis. *Journal of Public Administration, Finance and Law, 1*(Special Issue).

The Azores. (2016). *Romania CSR index* 2015, Retrieved May 18, 2016, from http://www.altreileasector.ro/wp-content/uploads/Romania_CSR_Index_2015.pdf

Thome, F. (2009). *Corporate responsibility in the age of irresponsibility: A symbiotic relationship between CSR and the financial crisis?* International Institute for Sustainable Development. Retrieved December 18, 2015, from http://www.iisd.org/pdf/2009/csr_financial_crisis.pdf

Yelkikalan, N., & Köse, C. (2012). The effects of the financial crisis on corporate social responsibility. *International Journal of Business and Social Science, 3*(3), 292–300.

Zaharia, R. M., & Grundey, D. (2011). Corporate social responsibility in the context of financial crisis: A comparison between Romania and Lithuania. *Amfiteatru Economic, XIII*(29), 195–206.

The Practice of Internal Corporate Social Responsibility in SMEs in Cyprus

Ioanna Papasolomou

1 Introduction: What Is CSR?

Corporate social responsibility (CSR) has gained significant ground as a strategic approach (Carroll & Buchholtz, 2006; Blowfield & Murray, 2008). Corporations are faced with heightened pressure to demonstrate their responsibility towards the society and transform into more ethical organizations that are able to balance their financial interests with the long term interests of the society (Christensen, Morsing & Thyssen, 2011). More and more businesses are realizing that by adopting CSR they can gain a positive reputation and strengthen their relationships with different stakeholder groups such as the government, customers, investors, suppliers and their employees (McWilliams & Siegel, 2001; McWilliams, Siegel & Wright, 2006; Weber, 2008; Ali, Rehman, Ali, Yousaf & Zia, 2004). Successful CSR practices can also reduce conflicts with stakeholders and contribute to increased loyalty and sustainable growth (Weber, 2008).

The existent literature suggests that there is an overemphasis on the externally oriented and often PR-driven approach to CSR in contrast to the internally integrated and employee-driven CSR practice which has been rather neglected (Vogel, 2005; Kuhn & Deetz, 2008; Mory, Wirtz & Gottel, 2015). However, employees are considered to be one of the most important stakeholder groups and play a significant role in the success and growth of an organization. Internal CSR programs can enhance employee morale and productivity, attract potential employees and enhance corporate reputation (Lee, Park & Lee, 2013; Turban & Greening, 1996). Internal CSR involves employee oriented social responsibility practices in the area of safety, health and well-being, training and participation in the business, equality of opportunities and

I. Papasolomou (✉)
Marketing Department, University of Nicosia, 46 Makedonitissas Avenue, P.O. Box 24005, 1700 Nicosia, Cyprus
e-mail: Papasolomou.i@unic.ac.cy

S.O. Idowu et al. (eds.), *Corporate Social Responsibility in Times of Crisis*,
CSR, Sustainability, Ethics & Governance, DOI 10.1007/978-3-319-52839-7_5

work-family relationship (Vives, 2006). Internal CSR is directly related with the physical and psychological working environment of employees (Turker, 2009). Vives (2006) suggests that internal CSR practices demonstrate a concern for the health and well-being of employees, their training and participation in the business, equality of opportunities, and the work-family relationship.

There is a growing need to shed more light onto the impact of internal CSR practice on employees and their behaviour in the workplace (Aguilera, Ruth, Rupp, Williams, & Ganapathi, 2007; Lee & Bruvold, 2003; Turker, 2009). The chapter draws empirical evidence from a research study that was carried out with the aim of exploring the perceptions of 180 employees working within 12 SMEs in Cyprus across four different sectors. The study was aimed at exploring the employees' perceptions in relation to the implementation of four dimensions of internal CSR practice. These dimensions are: (1) health and safety in the workplace, (2) organisational justice and fairness, (3) employee training and development, and (4) work-life balance.

CSR has been studied extensively by academics and practitioners alike. It's a term associating businesses with corporate citizenship and self-regulation. It enables businesses to comply with legal frameworks, public policy and ethical standards. At the heart of CSR is the notion of benefiting the society (McWilliams & Siegel, 2001; McWilliams, Siegel, & Wright, 2006). The EU definition of CSR is: "A concept whereby companies integrate social and environmental concerns in their business operations and in their interaction with their stakeholders on a voluntary basis" (Commission Green Paper, 2001). Corporations engage in activities which impact several internal and external stakeholder groups and CSR highlights the importance of ensuring that these activities do not harm and instead create mutual benefit and value for both the business and all the stakeholders involved. Roberts and Markley (2011) posit that the aim of CSR is to "maximizing the creation of shared value for...owners/ shareholders and for their other stakeholders and society at large."

Indeed, CSR has been studied by researchers since the 1950s (Carroll, 1999), but there is a lot of confusion in the way CSR is defined, why and whether it should be implemented and the way in which it should be practised within corporations (Welford, 2004). It is rather difficult to provide a universally accepted definition since the topic has been described as elusive, broad and complex (Mohr, Webb, & Harris, 2001; Smith & Langford, 2009). The difficulty in developing a single, universally accepted definition for CSR is due to the plethora of views within the academic literature, which create confusion (Godfrey & Hatch, 2007).

Ali, Rehman, Ali, Yousaf, and Zia (2010) suggest that CSR has internal and external practices. Internal CSR addresses the need of employees by highlighting diversity in the workplace, by promoting the health and safety of the workforce, as well as the provision of monetary and non-monetary rewards. These areas have been traditionally incorporated into the organizational HRM practice but what is new about them today is that they are integrated under a general effort to be responsible and ethical. They are practiced by businesses not only merely to comply with legal and governmental frameworks but also in order to responsibly recognize and satisfy the interest and well-being of internal stakeholder groups. External CSR practices focus on the local community, business partners and suppliers, customers, public authorities and NGOs representing local communities, and the environment such as philanthropy, volunteerism and environmental protection (European Commission, 2001).

Several authors have expressed their views in terms of what constitutes internal CSR. Internal CSR practices are primarily directed towards employees and highlight the importance of investing in human capital (Ali et al., 2010). Martinuzzi, Gisch-Boie, and Wiman (2010), stipulate that internal CSR integration is founded on fair and efficient HRM practices such as workforce health and safety, employee respect and freedom, lack of discrimination and diversity in the workplace. Turker (2009) suggests that internal CSR practices are directly related to the physical and psychological working environment of employees whilst Vives (2006) posits that internal CSR reflects a concern for the health and well-being of employees, their training and participation in the business, equality in job opportunities, and addressing work-family relationships.

It can be argued that the main difference between internal CSR and HRM goes at the heart of the two concepts. On the one hand, internal CSR utilises HRM activities in order to responsibly and ethically address the needs, interests and well-being of the employees. On the other hand, HRM can be viewed as a management led effort to meet the legal requirements in terms of the workforce. Papasolomou, Koundouros and Kitchen (2012) suggest that CSR is a process directed towards stakeholder engagement within a context which can nourish the necessary conditions for viewing stakeholders such as employees not as obstacles to business progress but as valuable partners in achieving business success. CSR concepts such as collaboration, dialogue, understanding/negotiation, ethical behaviour, codes of conduct, accountability and ethical values can be developed within procedures of employee involvement with significant benefits to the corporation (Papasolomou et al., 2012). The existent literature suggests that CSR activities have a direct positive impact on human resource management practices such as improving trust and understanding, increasing employee commitment, motivation, efficiency, and supporting organizational culture and communication effectiveness (Jenkins, 2001). A study carried out by Kramar (2004) across 25 countries showed that employees felt greater loyalty, satisfaction and motivation when their organizations were practicing internal CSR. Internal CSR (e.g., human resources practices such as training and labor participation) can improve firm's labor productivity (Calveras, 2013). Aguilera, Rupp, Ganapathi, and Williams (2006) found that CSR can have a positive impact on job performance and Valentine and Fleischman's (2008) showed that there is a positive perception of the influence of CSR on job satisfaction.

2 Employees as Stakeholders

From a stakeholder perspective, corporations are expected to meet their duties and responsibilities towards their stakeholder groups (Simmons, 2004). Stakeholder theory holds that the needs of shareholders cannot be met without satisfying to some degree the needs of other stakeholders. Even when a firm primarily aims at meeting the needs of its shareholders, its success in doing so is likely to be affected by other stakeholders (Foster & Jonker, 2005; Hawkins, 2006). The stakeholder approach was adopted by Papasolomou-Doukakis, Krambia-Kapardis, and Katsioloudes (2005) in a research study carried out in Cyprus. They identified six key stakeholder groups namely employees, customers, investors, suppliers, the community and the environment and relevant CSR actions vis-a-vis each cluster. Their findings showed that Cypriot firms give priority to meeting the needs of their employees and consumers in their pursuit of CSR whereas they exhibit moderate attention to their community, suppliers, investors, and the environment.

Barnett (2007) claimed that CSR is a form of corporate investment characterized by a dual orientation towards the improvement of *social welfare* and of *stakeholder relations*. This focus on *stakeholder relations* explains why employees, as a stakeholder group, impact CSR policy. The employees are one of the primary stakeholders of a corporation to whom the firm has a duty and a responsibility to behave in an ethical and responsible manner (Lee, Park, & Lee, 2013). Firstly, employees can act as agents for social change when they push corporations to adopt socially responsible behavior (Aguilera, Ruth, Rupp, Williams, & Ganapathi, 2007). Secondly, environmental policy demonstrates that the employees' support is necessary to secure effective CSR programs and policies (Ramus & Steger, 2000). Thirdly, employees as a stakeholder group perceive, evaluate, judge and react to CSR programs and actions (Rowley & Berman, 2000; Rupp, Gananpathy, Aguilera, & Williams, 2006; Wood & Jones, 1995).

The literature also suggests that when the employees feel that their employer practices CSR activities there will be heightened employee loyalty and commitment and decreased employee turnover rate within the firm (Lee et al., 2013). Riordan, Gatewood, and Bill (1997) used external corporate image as a proxy for social performance, and found that there is a positive correlation between job satisfaction, and a negative impact on the employee turnover rate and employee turnover intentions. There is also a strong empirical evidence to support the positive impact of CSR on employee (or organizational) commitment (Aguilera et al., 2006; Brammer, Millington, & Rayton, 2007; Maignan & Ferrell, 2001; Maignan, Ferrell, & Hult, 1999; Peterson, 2004).

The concept of "organizational commitment" has significant impact on work performance, job attraction and employee retention (Armstrong & Green, 2013). The existent literature reviewed suggests that if CSR programs are successfully implemented they have the potential to generate positive organizational reputation and contribute towards producing a positive perception for both existing and potential employees (Lee et al., 2013; Turban & Greening, 1996). Similarly, an ethical work environment within an organization can become a company's vehicle for strengthening the employees' organizational commitment, decreasing the rate of employee turnover, increasing employee productivity and performance and generating positive attitudes among employees towards their work (Lee et al., 2013; Sims & Keon, 1997).

CSR has been found to enhance the perceived trustworthiness of a corporation for the job seeker (Turker, 2009). The literature suggests that there is a positive correlation between CSR and employee commitment. Specifically, Maignan et al. (1999) claim that a philanthropic culture and a strong reputation for corporate citizenship have the potential to improve the levels of employee commitment, customer loyalty and corporate performance. Similarly, Peterson (2004) supports that there is a positive relationship between corporate citizenship and organizational commitment whilst Brammer, Millington, and Rayton (2005) showed a positive relationship between socially responsible behaviour and organizational commitment. Turker (2008) and Brammer et al. (2005) argue that a sense of belonging is more likely to develop when employees perceive the organization to engage in socially responsible behaviour.

3 Some Empirical Data from the SME Sector in Cyprus

CSR practice varies significantly across EU countries. The UK has a leading position in terms of n CSR in Europe largely due to being the 'home' of the world's leading academics, practitioners and campaigners in the field. A number of reasons have contributed to the growth of CSR in the UK primarily, a dedicated business community to CSR principles, a strong legal framework designed to support voluntary CSR initiatives and a dedicated NGO community (Mullerat, 2013). French corporations are increasingly embracing and adopting CSR practices (Berthoin & Sobczak, 2007) although the state regulations and agreements governing labour relations are counter-productive of CSR growth (Mullerat, 2013). In Germany, CSR principles and initiatives are rather slow in terms of gaining ground. CSR in Italy is driven by the financial and banking sectors and in general the approach towards CSR is influenced by the guidelines of the EC Green Paper 2001. The Baltic countries seem to be more experienced when it comes to CSR practice. In Sweden, CSR is effectively implemented by the business sector which is prepared to invest financial resources in socially CSR initiatives. A study carried out by Energy Dalen (2005) revealed that despite the heightened competition and difficult market conditions the Swedish business sector has directly linked CSR to long-term profitability. Recently there has been a shift towards the less developed EU countries to further the development of CSR especially within SMEs. Specifically, the EC funded the Leonardo da Vinci project "Sustainability and Social Responsibility through learning in SMEs" with the aim to transfer and further develop an innovative methodology based on a strategic performance management tool to develop a CSR strategy within SMEs in Europe between 2008 and 2010 in Bulgaria, Cyprus, Latvia, Lithuania, Portugal, Spain and Romania. In the EU 2020 Strategy the EC stresses the importance of CSR among SMEs for growth, job creation and sustainable development. Due to the fact that CSR was created primarily by and for larger companies, several EU public policy initiatives aim to familiarize SMEs with the CSR concept.

The chapter presents some empirical evidence from the business sector in Cyprus. Data has been collected from 180 employees working within 12 SMEs that operate in four different sectors namely retailing, professional services, healthcare services and education. The study investigated the employees' perceptions, attitudes and experiences regarding the implementation of internal CSR and specifically the practice of four tools which emerged in the literature reviewed as being among the key elements of the internal CSR practice. These activities were:

(1) employee training and development, (2) organisational justice and fairness, (3) work-life balance, and (4) health and safety standards in the workplace. The employees were asked to reveal their personal perceptions and experiences in terms of these four dimensions in an effort to shed some understanding onto the degree in which the SMEs studied practice internal CSR.

The methodological approach that guided the specific study was the *deductive* since the aim was to investigate the relationship between employee organizational commitment and internal CSR practices namely, employee training and development, justice and fairness in the workplace, health and safety at work, work-life balance. The deductive approach is related to developing theories and hypotheses that will be used in a research strategy for testing the relationships between these variables. This requires a collection of quantitative data and sampling of an adequate population size (Saunders, Lewis, & Thornhill, 2015). Descriptive research aims to capture and provide an accurate and detailed picture of the environment. Therefore, it is important to have a clear overview of the phenomena that is going to be examined, prior to collecting the data (Collins & Hussey, 2003). Descriptive research aims at addressing "what?" questions and utilizes quantitative and statistical techniques in the process of data collection (Collins & Hussey, 2003; Saunders et al., 2015). The use of a questionnaire enables the researchers to perform a statistical analysis of data which is structured and formalized and allows a high degree of control. Since the primary was to gain an overall understanding of whether the local business sector has adopted internal CSR activities and whether these appear to have an impact on the employees' level of commitment the use of a quantitative research study was justified.

The study's sample was determined by using the convenience sampling method. A total of 180 usable questionnaires out of 200 were collected from employees working within 12 local SMEs across a number of sectors in order to gain a more representative picture of the internal CSR practice in Cyprus. The questionnaire consisted of measurement statements relating to the four internal CSR dimensions which were based on a five-point Likert-scale ranging from "strongly disagree" to "strongly agree". The questionnaire was pilot tested before it was used. The data was analyzed by using frequency distribution statistics. Descriptive statistics were computed for each item and the data was processed and analyzed by using SPSS.

4 Research Findings

The study revealed some descriptive, preliminary findings in terms of the implementation of internal CSR in the SMEs that participated in the study. From a total of 180 employees, who completed and returned the questionnaires, 104 were females (57.85%) and 76, were males (42.2%). The majority of participants (45%) were 26–35 years old. 91.7% were full-time employees with a Bachelor's degree. Approximately, 40% of the participants were working in organizations with more than 100 employees but less than 250, followed by 28% working in organizations with 51–100 employees. Approximately, 8% of the participants were employed in firms with more than 150 employees and less than 250. The remaining 34% were employed in firms with less than 50 employees. 30.56% of the employees worked in the retailing industry, 60.56% in professional services, 5.56% in healthcare services and 3.33% in education. Less than 37.2% of the sample, have been employed for less than a year, 28.9% of the sample have been employed between 1 and 3 years, 16.7% have been employed for 4–6 years and 17% for 7–10 years.

The organisations studied use primarily the four human resource management tools that follow bellow to demonstrate a socially responsible behaviour towards their employees and practice internal CSR. The business practices that are primarily targeted at employees are in essence human resource management activities. These four tools are supported by the literature as being some of the key internal CSR dimensions (Calveras, 2013; Vives, 2006). Calveras (2013, p. 7) claims that: "As internal CSR we may consider high commitment human resources practices such as training, labor stability, employee participation, high wages, etc.). Polak-Weldon, Balogh, Bogdany, and Csizmadia (2013) after reviewing the relevant literature and social responsibility standards and guidelines they identified five groups of performance indicators of internal CSR activities: employment, management-employee relationship, employee training and development, health and wellbeing, and workplace inclusion.

4.1 *Employee Training and Development*

In the Cyprus study, seven items were used in the questionnaire based on the literature reviewed to measure the extent to which the employees are offered training and development opportunities as part of the company's internal CSR practice. Table 1 presents the mean scores in relation to the development and training statements.

Table 1 Employee training and development mean scores

Descriptive statistics		
	N	Mean
My organization trains employees on skills that prepare them for future jobs and career development	180	3.8222
My organization allows employees to have the time to learn new skills that prepare them for future jobs	180	3.1667
My organization provides support when employees decide to obtain ongoing training	180	3.8389
My organization supports employee learning and development, and uses a blend of classroom, e-learning, coaching, job shadowing and mini training sessions	180	4.0167
My organization organizes activities, events and/or after work happenings to enhance the relationship/bonding between employees	180	3.5944
The management keeps me up to date with the company's strategic goals and ambitions	180	3.5389
My organization appreciates and praise hard work	180	3.4556
Valid N (listwise)	180	

When the respondents were asked to express their perceptions in terms of whether they felt that their organization encourages and supports them in participating in training and personal development programs aimed at improving their skills for future development, 60% agreed and 15% strongly agreed. In relation to whether there are opportunities for ongoing training and development, 84% had a positive perception. When though they were asked whether their organization allows the employees time to learn new skills that will prepare them for future jobs 35% had a positive view, 19% had a negative view and 46% were neutral. This finding in essence is counter-productive to the ability of the employees to take advantage of any opportunities for training and personal development.

The study also attempted to explore the employees' perceptions in relation to whether the organization invests in activities aimed at enhancing the quality of employee relationships. 52.8% of the employees that participated in the study had a positive perception. In relation to the effectiveness of the top-down internal communication, 52.8% agreed. When the employees studied were asked whether their managers praise them and show appreciation for their efforts, 52.2% were neutral and 47.8% felt that they were appreciated for their efforts.

In the present study, it seems that the "training and development" dimension of CSR is perceived to be successfully implemented by the participants who in their majority feel satisfied with the opportunities given to them for personal development and training. There is also a general positive perception in relation to the quality of employee relationships, internal communication and employee recognition. These findings show that the SMEs studied appear to invest in this specific area of internal CSR in an effort to demonstrate a caring and socially responsible behavior towards their employees. This finding is in agreement with the current literature in the area. Aguinis and Kraiger (2009) believe that training is an area that makes a clear contribution to improving human well-being and performance both in the organization and the society in general. In the United Kingdom, the government wishes to improve the skills of the workforce and encourages the development of lifelong learning practices through a variety of organizations and initiatives (Lee, 2004). In a study carried out by Polak-Weldon et al. (2013) one of the indicators used to measure the activity level of organisational internal CSR practices was *employee training and development.* In the same study, the researchers used a number of performance indicators to measure the degree to which employee training and development was implemented in the business that participated in the study, one of which was the "The Investors in People" (IIP) program which gives awards to organizations that implement excellent practices in the training of individuals to achieve business goals. The program provides a framework to help organizations find the most suitable means for achieving success through people (Investors in People Standard, 2005).

4.2 Work-Life Balance

Five items were used in the Cyprus study based on the themes that emerged from the literature reviewed to measure the extent to which the participants perceive and experience a work-life balance (Table 2).

When the employees studied were asked whether they have achieved a balance between work and their personal life the findings showed that there is a significant negative response. Thirty percent of the employees studied had strong negative perceptions and another 30.6% had negative perceptions. Overall, this statement received a mean score 2.25 which implies a strong negative employee perception in terms of balancing work and personal/social life. Only 15.5% of the sample had positive perceptions and the remaining 23.9% of the sample were neutral.

The majority of the respondents (55%) were neutral when they were asked whether their supervisor shows an understanding if they experience any personal/family issues that might influence their work. In addition, 28.3% of the sample had negative perceptions and only a relatively small percentage (16.7%) had positive perceptions. Forty percent of the employees studied "*look forward to being with their colleagues every day*", 32% were neutral and 28% disagreed. Only 10% of the sample stated that their organization takes initiatives to help them manage and balance work-life. This statement received a high negative response with almost 50% of the respondents having a strong negative perception. The study's findings revealed that the remaining 40% of the respondents showed a neutral view regarding the statement. It seems that local businesses have failed to demonstrate an understanding towards the importance of helping their employees to balance work and family responsibilities. The study also revealed that the majority of participants (86%) are happy with the insurance coverage offered to employees, this statement actually received the highest mean score of the index (4.05), and also 73% of the respondents agreed in that their organization provides a comfortable working environment to its employees. The study revealed that the SMEs studied have not managed to create an organizational culture and structure that creates the right 'infrastructure' that will allow work-life activities to be adopted and nourished.

Table 2 Mean score of descriptive statistics for work-life balance

Descriptive statistics		
	N	Mean
I feel that I have the right balance between work and life outside of work	180	2.2500
My supervisor is understanding when I talk about personal or family issues that affect my work	180	2.9833
I look forward to being with the people I work with each day	180	3.1389
I work for long hours/overtime or even on holidays	180	3.3389
My organization takes initiatives to manage work life of its employees (i.e. flexible working hours, mother scheme, etc.)	180	2.5667
Valid N (listwise)	180	

The present study shows that this specific dimension of internal CSR lacks behind in terms of effectiveness and success. The employees studied share strong negative perceptions in terms of their organisation enabling them to achieve a balance between their life and work commitments and responsibilities. In general there is a shared negative perception among the employees in relation to the inability of their supervisors to show a genuine understanding towards the employees' need to find a balance between their work and family responsibilities. In terms of the quality of the working environment the employees were satisfied as they were in terms of the provision of employee insurance cover schemes.

Work-life balance has received significant attention from employers, employees, government, researchers, and the media (McPherson & Reed, 2007, p. 13). Employees often have to balance an array of non-paid activities or commitments such as: study, travel, sport, voluntary work, personal development, leisure or eldercare. They have to deal with increased work and home responsibilities such as a changing family structure (a rise in the number of single parents), a growing reluctance on behalf of employees to accept longer hours at work, and the rise of the 24/7 society. These changes have resulted in conflicts between the different roles that individuals have. Increasingly corporations have to respond by designing various kinds of practices, intended to enable employees to balance work-life responsibilities. The "work-life balance" factor has been recognized as being critical for the internal dimension of CSR by Welford (2008): "Work-life balance is an essential part of Corporate Social Responsibility. Corporations are increasingly recognizing that an inadequate work-life balance can have detrimental impacts on staff performance, satisfaction and retention. When employers target good work-life balance, they can see that reducing stress and frustration resulting from poor work-life balance can be beneficial to both parties." (p. 3)

Morison and Thurnell (2012) carried out a research study in a large New Zealand construction company and the results showed that company provision of a wide variety of work-life benefits from which employees can choose during different stages in their life and career is ideal. Being perceived as having innovative work-life balance practices allows organizations to attract and retain employees. For example, Arup Laboratories, a Medical and Testing Reference Laboratory with 1789 employees in Salt Lake City, USA, has offered flexible work scheduling to its employees which has in turn enabled them to more than double their employee base from 700 in 1992 to 1700 employees in 2004 and reduce employee turnover from 22 to 11% (Hartel, Fujimoto, Strybosch, & Fitzpatrick, 2007). According to Worklifebalance.com (2015) the *Work-Life Balance Benefits to Organizations* are multiple: increases in individual productivity, accountability and commitment, better teamwork and communication, improved employee morale, increased engagement and commitment levels and less negative organizational stress. According to the same source, the work-life balance benefits for individual employees are: better understanding of what their best individual work-life balance is, increased productivity, improved relationships both on and off the job and reduced stress.

4.3 Organisational Justice and Fairness

The researcher used a four-item index Table 3 in order to study the perceived organisational justice and fairness in the workplace.

Sixty percent of the respondents agreed with the statement that the "process used to conduct my performance appraisal is fair". When the participants were asked whether the "*process used to make a decision about promotion is fair*", 65% of the respondents had a neutral view. When the participants were asked whether the "*process used to determine their salary is fair*" again there was also a relatively high neutral view (56.7%). When the participants were asked whether the "process used to address any complaints about the firm is fair" 46.1% of the sample agreed with the statement.

Although in the present study 60% of the respondents agreed that the performance appraisal processes are fair in the businesses they worked for, 65% shared a negative view when it came to the company's promotion procedures and remuneration programs. This finding sends rather conflicting messages in relation to the degree of effectiveness of this specific CSR dimension. There is a need for businesses to identify any problem areas in relation to the 'organisational justice and fairness' dimension of internal CSR by enabling and encouraging the employees to raise their concerns and dissatisfaction. If they fail to do that there is the risk of employee dissatisfaction and discontent. On the contrary one could argue that if employees feel that they are being treated fairly then they will be happy and satisfied and will be likely to show a stronger commitment towards the organisation.

The literature reviewed suggests that in order to ensure ethical behaviour and fairness in the workplace there is a need to ensure that the organizational policies and processes treat employees fairly and equally (Collier & Esteban, 2007). Carroll (1979) suggests that procedural justice can be placed within the domain of ethical citizenship (in Meyer, Allen, & Gellatly, 1990) and hence, be an integral part of internal CSR. In fact, Brammer et al. (2007) has identified procedural justice and employee training as two aspects of internal CSR. Procedural justice refers to the employees' perceived fairness of organisational processes (Colquitt, 2001; Saks, 2006). It is important that corporations today adopt transparency, justice, integrity and fairness both externally and internally. Kim, Lin, and Leung (2013) collected survey data from 151 employees from businesses in Hong Kong operating in

Table 3 Mean scores for Organisational Justice and Fairness in the workplace

Descriptive statistics		
	N	Mean
The process used to conduct my performance appraisal is fair	180	3.3500
The process used to make decision about promotion is fair	180	2.9833
The process used to determine my salary is fair	180	2.8778
The process used to address any complaints about the firm is fair	180	3.2167
Valid N (listwise)	180	

various sectors such as the finance, service, information technology, manufacturing, and education sector. In particular, they measured fairness perceptions, job satisfaction and affective organizational commitment. They found that changes in distributive, procedural, and interactional justice significantly explained variances in job satisfaction. Higher perceptions of procedural justice by employees are more likely to reciprocate with greater organizational engagement (Saks, 2006) and an employees' positive evaluation of their supervisor (McFarlin & Sweeney, 1992). According to a survey conducted in 32 countries by Walker Information, the factors that most influence employee commitment are fairness, trust as well as care and concern for employees (Frank, Finnegan, & Taylor, 2004). Fairness strives to make the employees' place of work better and has become more than just adequate pay, benefits and opportunities. It is suggested that if an organization treats its employees well, they will give back as much or more both physically and emotionally (Insightlink, 2005).

4.4 Health and Safety Standards in the Workplace

The respondents' perception for health and safety at workplace was measured by using a four-item index. The mean scores are presented in the table.

Table 4 suggests that the total mean scores for health and safety were relatively high, above the neutral central. In general, the respondents have a high positive view (64%) in relation to the health and safety standards provided by their employers. Seventy-three percent claimed that their company provides them with a comfortable working environment. It also seems that the employees are happy with the insurance coverage provided by their employers (86%). However, when the employees were asked whether their employer requires them to work reasonable hours 47% had a negative perception and 29% had a positive view. Therefore, even though the respondents have a strong favourable perception towards their work environment when it comes to the time spent at work most of them feel that they work too many hours. Undoubtedly an integral part of the company's efforts to be responsible towards their employees is ensuring that the health and safety standards in the workplace are met, the quality of the work environment is satisfactory, the

Table 4 Mean scores of health and safety standards in the workplace

Descriptive statistics		
	N	Mean
My organization employees with the latest health and safety standards	180	3.7167
My organization provides a comfortable working environment	180	3.9333
My organization provides insurance coverage to employees	180	4.0500
My organization requires employees to work a reasonable number to working hours	180	2.8333
Valid N (listwise)	180	

organisation provides its employees with satisfactory health insurance programs and ensures that the employees work reasonable hours. The banks studied give emphasis to the first three dimensions of internal CSR and appear to rather ignore the fourth dimension. The current literature reviewed did not suggest that this dimension is part of internal CSR perhaps because the employees' health and safety is a legal requirement of employers in European countries. However, if the activity is guided by the social responsibility principles then companies should be willing to go a step ahead of what is required by law and address their employees' concerns and desires when it comes to their health and safety in their workplace.

5 Conclusion

The chapter blends both the findings that emerged from a literature review in the area of CSR and an empirical study. The results of the empirical research study contribute to the literature on internal CSR by providing insights into how the employees that participated in the study perceive specific dimensions/elements of internal CSR used to implement the central idea of corporate social responsibility that is, addressing the needs and concerns of internal stakeholders beyond what is required by law. The study was conducted within twelve SMEs operating across different industries in the business sector in Cyprus. The primary objectives of the study were whether there is enough evidence to suggest that internal CSR is implemented in the SME sector in Cyprus and to what degree the four dimensions are perceived to be effectively implemented within the internal CSR context.

The study revealed that the corporations that were studied use various human resource management tools, some of which the literature reviewed showed them being an integral part of the internal CSR. The study produced some descriptive statistics that showed that the businesses studied although they practice the four dimensions they tend to prioritize employee training and development (mean score, 3.63), and health and safety in the workplace (mean score, 3.63), whilst organizational justice and fairness (mean score, 3.10), and work-life balance (mean score, 2.85) are rather neglected. This finding highlights a potential risk in that if the managers fail to identify and respond to their employees' needs and concerns, the employees will feel unhappy and dissatisfied and these feelings can impact negatively on their motivation and productivity. Obviously, there is a need to ensure that all the dimensions of internal CSR are effectively implemented and managed in order to avoid any frictions between the employees and the managers/owners. Businesses need to carry out periodic internal research studied in order to reveal the employees' feelings and concerns and use it as a basis for taking corrective action.

The study that provided the empirical data on which the discussion in the chapter was based could be extended in order to include a bigger number of SMEs in order to provide more representative results. Since the empirical evidence and the relevant literature on internal CSR are still at an embryonic stage there are ample opportunities for future research. For example, future research could address the antecedents or moderating variables of internal CSR, such as the credibility of CSR and the perceived importance of CSR by the employees. Future research studies could also aim in developing a multidimensional construct for internal CSR by using empirical data derived from SMEs across the EU.

Acknowledgement The author would like to thank Ms. Evgenia Hadjicosta for carrying out the empirical research used in this chapter as part of her MBA dissertation at the University of Nicosia. The thesis is entitled: Corporate Social Responsibility as a Management Tool: an investigation of CSR practices based on employees' perception in Cyprus organisations.

References

Aguilera, R., Rupp, D. E., Ganapathi, J., & Williams, C. A. (2006). Justice and social responsibility: A social exchange model. Paper presented at the Society for Industrial/Organizational Psychology Annual Meeting, Berlin.

Aguilera, R. V., Rupp, D. E., Williams, C. A., & Ganapathi, J. (2007). Putting the S back in corporate social responsibility: A multilevel theory of social change in organisations. *Academy of Management Review, 32*(3), 836–863.

Aguinis, H., & Kraiger, K. (2009). Benefits of training and development for individuals and teams, organisations and society. *The Annual Review of Psychology, 60*, 451–474.

Ali, I., Rehman, K. U., Ali, S. I., Yousaf, J., & Zia, M. (2010). CSR Influences employee commitment and organisational performance. *African Journal of Business Management, 4* (12), 2796–2801.

Armstrong, J. S., & Green, K. C. (2013). Effects of corporate social responsibility and irresponsibility policies. *Journal of Business Research, 66*, 1922–1927.

Barnett, M. L. (2007). Stakeholder influence capacity and the variability of financial returns to corporate social responsibility. *Academy of Management Review, 32*(3), 794–816.

Berthoin, A., & Sobczak, A. (2007). CSR in France. *Business and Society, 46*, 9–32.

Blowfield, M., & Murray, A. (2008). *Corporate Responsibility – A critical introduction*. Oxford, UK: Oxford University Press.

Brammer, S., Millington, A. & Rayton, B. (2005). *The contribution of corporate social responsibility to organizational commitment* (pp. 1–42). Working Paper Series, University of Bath. http://www.bath.ac.uk/management/research/pdf/2005-20.pdf

Brammer, S., Millington, A., & Rayton, B. (2007). The contribution of corporate social responsibility to organizational commitment. *International Journal of Human Resource Management, 18*(10), 1701–1719.

Calveras, A. (2013). External and internal corporate social responsibility: Complements through product quality. Evidence from the hotel industry. doi:http://www.webmeets.com/files/papers/earie/2013/339/CSR%20and%20product%20quality%20EARIE%202013.pdf

Carroll, A. B. (1979). A three-dimensional conceptual model of corporate social perfor-mance. *Academy of Management Review, 4*, 497–505.

Carroll, A. B. (1999). Corporate social responsibility – Evolution of a definitional construct. *Business and Society, 38*(3), 268–295.

Carroll, A. B., & Buchholtz, A. K. (2006). *Business and society. Ethics and stakeholder management*. Mason, OH: Thomson/South-Western.

Christensen, L. T., Morsing, M., & Thyssen, O. (2011). The polyphony of corporate social responsibility. Deconstructing accountability and transparency in the context of identity and hypocrisy. In G. Cheney, S. May, & D. Munshi (Eds.), *Handbook of communication ethics* (pp. 457–474). New York: Lawrence Erlbaum Publishers.

Collier, J., & Esteban, E. (2007). Corporate social responsibility and employee commitment. *Business Ethics, 16*(1), 99–33.

Collins, J. and Hussey, R. (2003). *Business research*. Palgrave, 9780333983256.

Colquitt, J. (2001). On the dimensionality of organizational justice: A construct validation of a measure. *Journal of Applied Psychology, 86*(3), 386–400.

Commission Green Paper. (2001). Promoting a European framework for corporate social responsibility. COM(2001)366 Final. http://www.eubusiness.com/topics/social/corporate-social-responsibility-in-the-eu/

European Commission. (2001). Promoting a European framework for corporate social responsibility. Green Paper. European Commission, Directorate-General for Employment and Social Affairs. Luxembourg. http://ec.europa.eu/education/lifelong-learningpolicy/doc/mobility/com329_en.pdf

Foster, D., & Jonker, J. (2005). Stakeholder relationships: The dialogue of engagement. *Corporate Governance, 5*(5), 51–57.

Frank, F., Finnegan, R., & Taylor, C. (2004). The race for talent: Retaining and engaging workers in the 21st century. *Human Resource Planning, 27*(3), 12–26.

Godfrey, P. C., & Hatch, N. W. (2007). Researching corporate social responsibility: An agenda for the 21st century. *Journal of Business Ethics, 98*, 70–98.

Hartel, C., Fujimoto, Y., Strybosch, V. E., & Fitzpatrick, K. (2007). *Human resource management: Transforming theory into innovative practice*. Frenchs Forest: Pearson Education.

Hawkins, D. (2006). *Corporate social responsibility: Balancing tomorrow's sustainability and today's profitability*. New York: Palgrave MacMillan. http://www.insightlink.com/employee_commitment.html

Insightlink Communications. (2005). Employers and employees – Making the marriage investors in people standard. http://www.investorsinpeople.co.uk/

Jenkins, R. (2001). *Corporate codes of conduct: Self-regulation in a global economy*. Technology, Business and Society. Programme Paper number 2 (April), United Nations Research Institute for Social Development (ISSN 1020-8216) (pp. 2–35).

Kim, T.-Y., Lin, X.-W., & Leung, K. (2013). A dynamic approach to fairness: Effects of temporal changes of fairness perceptions on job attitudes. *Journal of Business Psychology, 30*, 163.

Kramar, R. (2004, December 30). Corporate social responsibility: A challenge for human resources. *Human Resource Magazine*. Published in 2010. http://www.humanresources magazine.com

Kuhn, T., & Deetz, S. (2008). Critical theory and corporate social responsibility – Can/should we get beyond cynical reasoning? In Crane et al. (Eds.), *The Oxford handbook of corporate social responsibility* (Chapter 8, pp. 173–196). Oxford, UK: Oxford University Press.

Lee, M. (2004). National human resource development in the United Kingdom. *Advances in Developing Human Resources, 6*, 334–345.

Lee, C. H., & Bruvold, N. T. (2003). Creating value for employees: Investment in training and education. *International Journal of Human Resource Management, 14*(6), 981–1000.

Lee, E. M., Park, S., & Lee, H. J. (2013). Employee perception of CSR activities: Its antecedents and consequences. *Journal of Business Research, 66*, 1716–1724.

Maignan, I., & Ferrell, O. C. (2001). Antecedents and benefits of corporate citizenship: An investigation of French businesses. *Journal of Business Research, 51*(1), 37–51.

Maignan, I., Ferrell, O. C., & Hult, G. T. (1999). Corporate citizenship: Cultural antecedents and business benefits. *Journal of the Academy of Marketing Science, 27*, 455–469.

Martinuzzi, A., Gisch-Boie, S., & Wiman, A. (2010). Does corporate responsibility pay off? Exploring the links between CSR and competitiveness in Europe's industrial sectors, final raport to the Europena Commission. Directorate-General for Enterprise and Indutsry, Wien.

McFarlin, D. B., & Sweeney, P. D. (1992). Distributive and procedural justice as predictors of satisfaction with personal and organizational outcomes. *Academy of Management Journal, 35* (3), 626–637.

McPherson, M., & Reed, P. (2007). Work-life balance in the New Zealand context. In M. Waring & C. Fouche (Eds.), *Managing Mayhem: Work-life balance in New Zealand*. Wellington: Dunmore Publishing.

McWilliams, A., & Siegel, D. (2001). Corporate social responsibility: A theory of the firm perspective. *Academy of Management Review, 26*(1), 502–504.

McWilliams, A., Siegel, D., & Wright, P. M.. (2006). *Corporate social responsibility: International perspectives (PDF)* (Working Papers (0604)). Troy, NY: Department of Economics, Rensselaer Polytechnic Institute.

Meyer, J. P., Allen, N. J., & Gellatly, I. R. (1990). Affective andcontinuance commitment to the organization: Evaluation of measures andanalysis of concurrent and time-lagged relations. *Journal of Applied Psychology, 75*, 710–720.

Mohr, L. A., Webb, D. J., & Harris, K. E. (2001). Do consumers expect companies to be socially responsible? The impact of corporate social responsibility on buying behavior. *Journal of Consumer Affairs, 35*(1), 45–72.

Morison, E., & Thurnell, D. (2012). Employee preferences for work-life benefits in a large New Zealand construction company. *Construction Economics and Building, 12*(1), 12–25.

Mory, L., Wirtz, B. W., & Gottel, V. (2015). Factors of internal CSR and the effect on organisational commitment. *The Internatioal Journal of Human Resource Management., 27* (13), 1393–1425.

Mullerat, R. (2013, June). CSR: A European perspective. *European Union Centre of Excellence, 13*(6).

Papasolomou, I., Koundouros, H., & Kitchen, J. P. (2012). Developing a framework for successful symbiosis of corporate social responsibility, internal marketing and labour law in a European context. *Journal of Marketing Review, 12*(2), 109–123. Summer. ISSN: 1469-347X.

Papasolomou-Doukakis, I., Krambia-Kapardis, M., & Katsioloudes, M. (2005). Corporate social responsibility: The way forward? Maybe not! *European Business Review, 17*(3), 263–279.

Peterson, D. K. (2004). The relationship between perceptions of corporate citizenship and organizational commitment. *Business and Society, 43*(3), 269–319.

Polak-Weldon, R., Balogh, A., Bogdany, E., & Csizmadia, T. (2013). Internal CSR Models in Hungary, Active Citizenship by Knowledge Management and Innovation Management, Knowledge and Learning International Conference, 19–21 June, Zadar, pp. 755–763.

Ramus, C., & Steger, U. (2000). The role of supervisory support behaviors and environmental policy in employee‘s eco-initiatives at leading edge European companies. *Academy of Management Journal, 43*, 605–626.

Riordan, C. M., Gatewood, R. D., & Bill, J. B. (1997). Corporate image: Employee reactions and implications for managing corporate social performance. *Journal of Business Ethics, 16*, 401–412.

Roberts, J., & Markley, A. (2011). *CSR: New EU strategy threatens UK and European communication*. Washington, DC: The Heritage Foundation.

Rowley, T., & Berman, S. (2000). A brand new brand of corporate social performance. *Business and Society, 39*, 397–418.

Rupp, D. E., Gananpathy, J., Aguilera, R. V., & Williams, C. A. (2006). Employees' reactions to corporate social responsibility: An organizational justice framework. *Journal of Organizational Behaviour, 27*, 537–543.

Saks, A. (2006). Antecedents and consequences of employee engagement. *Journal of Managerial Psychology, 21*, 600–619.

Saunders, M. N. K., Lewis, P., & Thornhill, A. (2015). *Research methods for business students*. Essex: Pearson. ISBN: 978-1-292-01662-7.

Simmons, J. (2004). Managing in the post-managerialist era: Towards socially responsible corporate governance. *Management Decision, 32*(3/4), 601–611.

Sims, R. L., & Keon, T. L. (1997). Ethical work climate as a factor in the development of person-organisation fit. *Journal of Business Ethics, 16*(11), 1095–1105.

Smith, V., & Langford, P. (2009). Evaluating the impact of corporate social responsibility programs on consumers. *Journal of Management and Organization, 15*(1), 97.

Survey by Energy Dalen. (2005). *CSR in Sweden*, 12 May.

Turban, D. B., & Greening, D. W. (1996). Corporate social performance and organisational attractiveness to prospective employees. *Academy of Management Journal, 40*, 658–672.

Turker, D. (2008). How corporate social responsibility influences organisational commitment. *Journal of Business Ethics, 89*(2), 189–204.

Turker, D. (2009). Measuring corporate social responsibility: A scale development study. *Journal of Business Ethics, 85*, 411–427. doi:10.1007/s10551-008-9780-6.

Valentine, S., & Fleischman, G. (2008). Ethics programs, perceived corporate social responsibility and job satisfaction. *Journal of Business Ethics, 77*, 159–172.

Vives, A. (2006). Social and environmental responsibility in small and medium enterprises in Latin America. *The Journal of Corporate Citizenship, 21*, 39–50.

Vogel, D. J. (2005). *The market for virtue: The potential and limits of corporate social responsibility*. Washington, DC: Brookings Institution Press.

Weber, M. (2008). The business case for corporate social responsibility: A company-level measurement approach for CSR. *European Management Journal, 26*, 247–261.

Welford, R. (2004). Corporate social responsibility in Europe and Asia: Critical elements and best practive. *Journal of Corporate Citizenship, 13*, 31–47.

Welford, R. (2008). *Work life balance in Hong Kong: Survey results*. Retrieved October 31, 2012, from http://csr-asia.com/upload/WLB%202008%20Final.pdf

Wood, D. J., & Jones, R. E. (1995). Stakeholder mismatching: A theoretical problem in empirical research in corporate social performance. *International Journal of Organizational Analysis, 3* (3), 229–267.

www.worklifebalance.com. (2015). Accessed December 22, 2015.

Corporate Social Responsibility in the Romanian Public Sector

Radu Florin Ogarcă and Silvia Puiu

1 Introduction

The concept of Corporate Social Responsibility (CSR) is mostly known as a feature for companies in the private sector. However, the term has been extended to the public sector, more and more governments understanding the need for contributing to a sustainable development or at least facilitating it.

Our analysis will start from the definition offered by World Bank (2003): "Corporate Social Responsibility is the commitment of business to contribute to sustainable economic development, working with employees, their families, the local community and society at large to improve quality of life, in ways that are both good for business and good for development."

Fox, Ward, and Howard (2002) identified four key roles that the government may have related to CSR activities:

- Mandating refers to legislation, regulations. The government has a proactive role because the legislative background in a country has many implications in the CSR activity at a national level;
- Facilitating refers to the incentives that public authorities could offer for the private actors in order to encourage CSR projects, but also to a higher transparency regarding this area;
- Partnering implies that the public authorities should cooperate with private actors on different CSR projects, sharing human and material resources;
- Endorsing refers to the support offered by the government that should act as a model for the private sector, by implementing practices in accordance with CSR standards.

R.F Ogarcă • S. Puiu (✉)
Faculty of Economics and Business Administration, University of Craiova, Craiova, Romania
e-mail: silviapuiu@yahoo.com

S.O. Idowu et al. (eds.), *Corporate Social Responsibility in Times of Crisis*,
CSR, Sustainability, Ethics & Governance, DOI 10.1007/978-3-319-52839-7_6

Steurer (2010) mentions five reasons for which governments should be more involved in CSR activities: this could help them to meet policy objectives on a voluntary basis; CSR regulations can reduce the negative impact of some undesirable regulations; even if CSR extends beyond legal framework, governments should focus more on non-binding initiatives; CSR becomes an important feature of a partnership between public and private entities; the variety of stakeholders.

Corporate social responsibility in the public sector can be approached not only in terms of public policies, but also at the level of public management of specific public organizations. Even if the term CSR refers to a corporate area, the public institutions should also have a greater impact in the community and develop more CSR projects.

Di Bitetto, Chymis, and D'Anselmi (2015, p. 133) offer an explanation about the reasons for which CSR is more common for the private companies than for the public sector. They argue that companies had to become more oriented to community because of the competition. In many areas, the public institutions are still in a monopolistic situation. But societies evolve, citizens become more and more demanding, requiring from the governments to be more efficient and transparent.

In Romania, the concepts related to CSR are not so well known as in other European countries, because public authorities became more proactive only after 2000, aiming the adherence to the European Union (Anca, Aston, Stanciu, & Rusu, 2011, p. 19). At the beginning, the government's role was limited to the first one, that of mandating.

Mandl and Dorr (2007, p. 12) argue that "In Romania, the concept of social responsibility originates in the 1990s when many NGO's were founded with the assistance of international public or private institutions. The major involvement of both large and small companies in CSR activities started after 2000." The authors mention as causes the presence of multinationals and the adherence to EU.

The government implemented laws related to the environmental protection, trying to reduce the pollution and offered tax incentives for the companies becoming green, by acquiring advanced technologies, less pollutant. Regarding the social domain of CSR, Romania also implemented some laws and regulations for assuring equal chances on the labor market, reducing discrimination. The Law no. 57/1992 states that companies hiring people with disabilities will have some fiscal advantages. Other laws refer to volunteering, transparency and anti-corruption measures.

Public authorities in Romania are more focused on the first two traditional roles: mandating and facilitating. The government should encourage more the partnership between public and private entities. As partners, resources and knowledge are commonly shared and so, the synergy effect brings a lot of advantages for both sides. Endorsing and acting as a model for the private sector are other aspects that are not enough developed in Romania.

Romanian managers do not fully understand the benefits of CSR. Most of them see only the involved costs, but not the advantages for CSR projects (Korka, 2005). The resistance against changes is very high, especially in the public sector. Hristea (2011, p. 66) argues that the implementation of a CSR policy gives companies many social, economic and financial advantages.

Our opinion is that public institutions should become more open to corporate social responsibility, because this type of activities will bring a better image for citizens, partners and investors. CSR is linked to competitiveness and may be seen as an important part of a marketing strategy. Concepts like public marketing, competitiveness and corporate social responsibility in the public sector are not well known and understood, some of them being considered as a feature for the private sector.

In recognition of the importance of CSR activities, the Romanian Government established a national strategy for this field in 2010. The National Strategy on CSR covers the period between 2011 and 2016 and it is inspired from the Norwegian one. This strategy presents the key domains for promoting CSR in Romania, the main actors involved, the role played by the state and the private sector, the international directives regarding CSR and the economic consequences of CSR.

As any strategy, the National Strategy for promoting CSR as a national priority starts from a SWOT analysis presented in Table 1.

We can add to the mentioned weaknesses also the wrong perception that the costs involved are very high and not covered by the benefits brought by CSR

Table 1 SWOT analysis for CSR in Romania

Strengths	**Weaknesses**
The interest of the society and the relative novelty for the public sector The involvement of multinationals and big Romanian companies in CSR activities The increase of the CSR projects in the last years Consumers are more and more oriented on socially responsible companies	The concept of CSR is not well known and understood The lack of studies and research regarding CSR in Romania A lack of involvement of SMEs in CSR activities Lack of information regarding consumers' rights A high level of corruption The lack of a business culture built around moral values
Opportunities	**Threats**
The benefits for the community, companies, environment The importance of CSR at international level The development of SMEs and a higher openness on the European market A higher competitiveness of Romanian companies on the international markets The possibility to learn from countries where CSR is more developed Financial opportunities for initiating CSR projects Positive reactions of citizens regarding the involvement of the companies and the state in CSR activities A healthier environment	The lack of previous public policies regarding CSR The lack of a coherent and complete legal background Problems in implementation of the laws The gap between Romania and other European countries regarding CSR and the implemented directives on different areas The risk to be excluded from certain markets because of the lack of CSR Insufficient support from public authorities in order to promote CSR at a national level The lack of public resources needed for promoting CSR or initiating CSR activities

Source: National Strategy for Promoting CSR in Romania, 2011–2016, http://www.sgg.ro/nlegislativ/docs/2011/05/2x4bdfjnwskv89h17pzq.pdf

activities. Public institutions or SMEs, for example, do not understand how many advantages they can get on the long term. Another wrong perception is that in a crisis period, you do not have to invest in CSR. There are various types of CSR projects; some of them are not so costly (volunteering), but the benefits are important: competitive advantages, a better image and reputation in society, higher opportunities because the community is more vulnerable in a crisis period and therefore, the connection created between potential clients and the entity initiating CSR is stronger and on a long term. Another important weakness is also the resistance to change, many managers in the public sector being used with obsolete strategies.

Public authorities should become more open to CSR, because benefits are definitely higher than the involved costs on the long run, leading to a better positioning of our country on the European markets in terms of competitiveness and image.

The National Strategy on CSR established three main domains that require attention: creating the premises for developing more CSR activities both at national and local level; more promotional campaigns to encourage SMEs and other entities to invest in CSR projects; a higher involvement from public institutions and an increased openness to these types of activities.

The strategy mentions nine specific domains for corporate social responsibility: health, security and working conditions; environment; ethics; prevention and fight against corruption; civil protection; gender equality; restructuring process of the public sector; research and development; integration of disadvantaged groups.

The main consequences for public authorities are the legislative and the budgetary implications. The legislative initiatives should take into consideration the European Directives related to CSR activities and the support offered by public authorities has to be limited to the established public budget.

2 Barriers in Front of a Higher CSR Activity in the Romanian Public Sector

Some of the barriers are mentioned even in the National Strategy for promoting CSR. We could say that the weaknesses and the threats from the SWOT analysis are some of the most important barriers: the lack of knowledge and awareness regarding CSR and its importance for community; the lack of a proper legislation or a poor implementation of laws and European Directives; the lack of public resources; the lack of political will; the high level of corruption; the lack of a long term perspective; resistance to change; CSR in Romania is mostly done for PR reasons than ethical ones (European Communities, 2007, p. 75)

We appreciate that the lack of awareness and obsolete mentalities are the most important barriers, because if this problem is solved, the other will gradually fade away. According to the National Strategy on Promoting CSR (2011–2016), the document is in accordance with the following laws and regulations:

- The standard ISO 26000 regarding CSR;
- Council Resolution on the follow-up to the Green Paper on CSR (2002/C 86/03);
- Council Resolution 2003/C 39/02;
- Regulation no. 761/2001 of the European Parliament and of the Council regarding voluntary participation by organisations in a Community eco-management and audit scheme;
- Directive 2004/18/EC of the European Parliament and of the Council on the coordination of procedures for the award of public works contracts, public supply contracts and public service contracts;
- COM (2006) 136 implementing the partnership for growth and jobs: making Europe a pole of excellence on CSR;
- COM (2001) 366 Green Paper promoting a European framework for CSR;
- COM (2009) 400 Mainstreaming sustainable development into EU policies;
- European Council (2006) Review of the EU Sustainable Development Strategy (EU SDS)—Renewed Strategy, 10917/06, 26 June 2006;
- Lisbon Strategy;
- The Universal Declaration of Human Rights;
- The *OECD Convention* on Combating Bribery of Foreign Public Officials (1997).

Another important European Directive that is mandatory to be implemented for all member states in the EU is the Directive 2014/95/EU of the European Parliament and of the Council of 22 October 2014 amending Directive 2013/34/EU as regards disclosure of non-financial and diversity information by certain large undertakings and groups.

Romania will have to integrate this directive into the national legislation until 6th December 2016 and the companies should apply it starting with 2017. If we analyze the barriers, we can conclude that the problem is related to the implementation of laws.

The Directive 2014/95/EU states that public-interest entities "shall include in the management report a non-financial statement containing information to the extent necessary for an understanding of the undertaking's development, performance, position and impact of its activity, relating to, as a minimum, environmental, social and employee matters, respect for human rights, anti-corruption and bribery matters". The directive does not specify a certain standard for reporting, the states having the freedom to establish which standards will be applied.

Scott (2013) was explaining in an interview that the barriers in reporting CSR are very high for small entities because of a few reasons like limited resources or the fact that they do not have the expertise needed for preparing a proper report. And because they do not have the resources, they cannot hire a specialist or assure the trainings for their employees.

In 2015, The Romanian Center for European Policies published the results of a quantitative research on the Romanian companies that already publish a non-financial statement regarding the CSR activity. The barriers mentioned by the companies were: the difficulties in understanding the standards for reporting

(58%); the absence of an electronic system to easily collect the needed data (50%); the lack of a specific procedure (33%); the reluctance of the employees (29%); the reluctance of the managers to publish these data (17%).

If these are the barriers for companies that already apply the Directive 2014/95/EU, we may conclude that the barriers for the other entities that will have to apply it too from 2017 will be even higher. The same is the situation with the public institutions because the resources are limited and the social pressure is very high.

These barriers may be reduced if there is a strategy for increasing the awareness on the importance of CSR and having some standards in reporting it. If the managers and the employees understand the benefits of reporting in terms of competitiveness and reputation, the resistance to change will gradually decrease.

Education is a very important factor in this process too, because trainings on CSR, standards and legal background would assure an increase of the number of entities developing CSR projects in both the public and the private sector. The government should contribute to this process, facilitate and support entities involved in these activities through education, national campaigns for promoting CSR and also by being a model.

3 The Level of Knowledge About CSR in the Romanian Public Sector

3.1 Research Objectives

The objectives of this quantitative research were:

- To identify the level of knowledge and awareness regarding CSR in the public sector;
- To identify the obstacles that stay in front of a higher CSR activity in this sector;
- To find some solutions that could be applied in order to increase the number of CSR projects in the public sector (where the public authorities or public institutions are initiators or partners with private actors).

We started from two ***hypotheses*** in our research:

H1. : One of the most important barriers for a higher CSR activity is the lack of financial resources.

H2. : Employees working in the public sector are not well informed on CSR, the level of awareness and knowledge being low.

3.2 Methodology

The research was conducted on a number of 94 employees and managers working in the public sector between 14th of November 2015 and 21st of December 2015. The questionnaire was realized in Google Forms, this online research assuring the respondents of the confidentiality. A face to face questionnaire would have been difficult, people having a problem in recognizing that they are not familiar with some terms or legal aspects.

In order to have a more representative sample, the respondents were from different domains of the public sector: education, healthcare, army, police, justice, cultural institutions, public administration or other. The questionnaire was sent to 500 potential respondents; their e-mail addresses were gathered from the websites of the public institutions.

Surveygizmo (2015) published an article based on their experience, stating that the response rate for external surveys (like mine) is 10–15% on online surveys. Of course, there are different opinions regarding this subject. Other websites specialized on online surveys mention a 20%. The idea is that obstacles and limits of online surveys are higher, but the answers are more honest, precise and close to the truth than those based on paper, where respondents may be afraid to be sincere or they are in a hurry.

The response rate for our research is 18.8% (94/500*100) and we appreciate it as being average. The results of our study can offer a direction for the responsible people in the public sector, these being able to implement some measures in order to increase the awareness and the knowledge on CSR.

3.3 Results

The *respondent's profile* was revealing the following:

- According to their age, 35.1% of them were between 36 and 45 years old, followed by those between 26 and 35 years old (25.5%) and between 46 and 60 years old (22.3%);
- According to their graduated studies, 63.8% of them graduated master's degree, 28.7% bachelor's degree and 7.4% only high school;
- According to their position within the company (manager or employee), 81.9% were employees and 18.1% were managers.

Even if they answered in a proportion of 74.5% that their institutions are sometimes involved in CSR projects, the number of these types of activities is still very low, comparing with those initiated by private companies (mostly multinationals).

The respondents were asked about the barriers in front of a higher CSR activity in their public institutions. They had to check the first three barriers from the

Table 2 The barriers in front of a higher CSR activity in the public sector

Barriers	Percentage (%)
The lack of financial resources	61.6
The lack of a political will	27.4
The lack of motivation	46.6
The lack of information	35.6
The insufficient promotion of the National Strategy on CSR	35.6
The lack of a proper legislation to encourage CSR initiatives	41.1
Other reasons	5.5

following ones: the lack of financial resources, the lack of a political will, the lack of motivation for employees, the lack of information, the insufficient promotion of the National Strategy on CSR, the lack of a proper legislation to encourage CSR initiatives, other reasons.

The main three reasons for not having a higher CSR activity in the public sector were the lack of financial resources, the lack of motivation and the lack of a proper legislation as we see in Table 2.

The lack of financial resources as an important barrier was also mentioned by Scott (2013), especially for small entities. He was mostly referring to SMEs, but the situation is similar for public institutions. Even the National Strategy for promoting CSR states that the budgetary implications are one of the consequences, because the public budget is limited and the costs involved could put a pressure on it.

Of course, limited resources are not the only barrier; lack of motivation for employees and managers is another important aspect that should be taken into account. In our opinion, this could be diminished if there would be some training in the public institutions in order to make the employees understand the importance of CSR and the advantages that this could bring. The lack of a proper legislation is also seen as a barrier and sometimes even if the laws exist, they are not properly implemented, monitored or disseminated to the community.

These three represent the main barriers but if we look at the other percentages, we observe that the others are also important obstacles, being mentioned by 35.6% of the respondents (the lack of information and the insufficient promotion of the National Strategy on CSR, in equal proportions) and by 27.4% (the lack of a political will). We can say that the first hypothesis is validated, because more than 60% of the respondents checked the lack of financial resources as a main barrier in front of a higher CSR activity in the public sector.

For the following questions, we used Likert Scale and the results were calculated in Table 3. The respondents checked if they agree or disagree with these statements. The data showed a gap between what the respondents think and how things are in fact.

In this calculation, we used 5 for Total Agreement, 4 for Partial Agreement, 3 for Indecisive, 2 for Partial Disagreement and 1 for Total Disagreement. If we analyze

Table 3 Perceptions of employees in the public sector on CSR

Statements	Mean
CSR projects should be reduced or even cancelled during recession	1.83
CSR projects lead to a better image of the public institution	4.53
Benefits of CSR are higher than the involved costs	3.85
CSR is a good way to differentiate the public institution from other entities	3.77
There are no advantages of CSR so the efforts and the costs do not have any justification	1.88
If there were more fiscal facilities, the public institution would be more involved in CSR projects	3.86
Introducing courses of social responsibility in the educational system is a necessity	4.34
Public institutions should be a model for the other entities in an economy regarding the implementation of CSR projects	4.31
The public institution where I work is rather a secondary partner in a project with a private actor or an NGO than an initiator of a CSR project	3.36
The CSR of the public institution was an important factor when I made the decision of applying for a job there	3.19
A higher support from central administration to encourage public institutions to develop CSR projects is needed	4.52

numbers, we see that the respondents consider that CSR projects should not be reduced during recession, 77.6% of them having this opinion.

This answer should be the correct one but the reality shows that public institutions or SMEs decrease the amount of CSR in recession because of the costs involved. We can correlate this question with the following four. At all these five questions, respondents had a similar opinion. They understand that the benefits of CSR are higher than costs, 62.8% of them partially or totally agreeing with this statement.

One of the benefits is the image and reputation of the public institution or the capacity to differentiate itself from other entities. The correlation between CSR and a better image is agreed by 89.3% of the respondents and strengthened by those who appreciate CSR as a strategy for differentiation (64.9%). So, corporate social responsibility should be integrated in the marketing strategy of a public institution, because CSR leads to a better image and can increase the competitiveness of an entity through differentiation and the creation of a competitive advantage.

The fifth question is a reiteration of the first four statements and was aiming to verify the honesty and the attention of the respondents. They maintained their opinions regarding the advantages offered by CSR, fact revealed by the mean 1.38 (Disagreement).

Of course, this should be the reality, but it is not. So, we may conclude that employees understand the importance of CSR for a public institution and this is a first step in promoting CSR at a national level—to make people really see the benefits and not just the costs involved. But these aspects show the perceptions on how things should be. Even if the attitude of employees is the correct one, the financial barrier is a tangible limit and we have to admit that CSR is dependent on a budget. There are also some CSR activities that do not involve a lot of money but require the total implication of human resources in an institution (volunteering).

The following questions tried to bring light on what might be the causes for the low CSR in the public sector: few facilities from public authorities, lack of education regarding corporate social responsibility, lack of models, lack of various types of support for public institutions, lack of transparency on CSR and problems regarding CSR reporting, lack of communication on these issues and even lack of promotion on CSR.

Definitely, the National Strategy for CSR 2011–2016 (adopted in 2010) is an important step for Romania, but there are more steps needed in order to increase the level of CSR and the quality of the implemented projects, not just the quantity of them.

The statement "If there were more fiscal facilities, the public institution would be more involved in CSR projects" is agreed by 67% of the respondents, the mean of 3.86 also showing that tax incentives are considered one of the factors leading to an increase in the CSR activity. This is the role of facilitator that the state should have. These facilities should encourage donations, sponsorship and the employment of youngsters, elders, women, people with disabilities or other disadvantaged categories of society.

In 2015, ViitorPlus association and EY Romania realized a research on the fiscal facilities companies know and use. Some of the facilities proved that are not even known by the companies. The Law no. 448/2006 was aiming the insertion of people with disabilities on the labor market. According to this law, companies with more than 50 employees may buy products or services from entities based on the activity of these persons and so they are exempted from paying a specific tax to the state.

The insertion of persons with disabilities on the labor market in Romania comparing with other European countries is very low. The study from 2015 revealed that the percentage in Romania is 9.48%, meanwhile in the European Union was 45.8% since 2010, according to Academic Network of European Disability Experts.

The introduction of some courses on corporate social responsibility in the educational system (starting with elementary school) is a solution for increasing the awareness of citizens related to CSR. This solution was agreed by 85.1% of the respondents, the intensity of agreement being highlighted by the mean 4.34. We saw at the first questions that employees working in the public system understand the importance of CSR, but there are still a lot of gaps in knowledge related to the legislation, the facilities they could use, the reporting standards and so on.

Education in school should also be completed with trainings in public and private entities, in order to keep employees up to date regarding CSR problems and challenges. This is a good way for the government to receive a grounded feedback from the community and involve it in the debates and decisions. Having a good legislation requires a proper knowledge on the needs in the territory.

The respondents appreciated in a proportion of 84% that public institutions should be a model for the other entities in an economy when dealing with CSR projects. This would be the endorsing role of the state, when this acts like a model and supports the other entities wanting to be involved in CSR activities. This should be the ideal situation, but the reality in Romania is different. The most important entities developing CSR projects are the multinationals and large companies. Public institutions and SMEs represent only a small part of the amount of CSR in Romania. So, there is a long way from the present state of being to the ideal one in which the public authorities are the model to follow.

More than 50% of the respondents said that the public institution where they work is rather a secondary partner than an initiator of CSR projects. The percentage of those being indecisive was rather high (25.5%), showing that they are not fully aware of the activities in which their institution is involved, otherwise they would have checked "agree" or "disagree". This distribution of answers explains also the mean of 3.36.

"The CSR of the public institution was an important factor when I made the decision of applying for a job there" is another statement where the employees offered heterogeneous answers, the mean of 3.19 suggesting their hesitation. 42.5% of the respondents were taking into account the CSR in the institution when applying for a job, 28.7% were indecisive and 28.7% disagreed.

We appreciate that the percentages for agreement are too high, representing a tendency for an ideal situation. In fact, public institutions are not so active when it comes to CSR, so taking into account this factor when you want to get a job in the public sector is not a good idea, because the chances to find a public entity with an important activity of CSR are not so high. The unemployment rate and a general uncertainty on the labor market make it difficult to choose an employer considering its CSR involvement.

Our opinion is also strengthened by the answers given to another question: "Would you accept a reduction of the salary if this were the only solution for the public institution to get involved in CSR projects?" Only 19.1% of the respondents said "yes", the rest of 80.9% said "no". This is also a consequence of the general perception that public money is not properly spent in valuable CSR projects, corruption being a real problem for Romania.

The last question for which we used Likert Scale was referring to the higher support that the central authorities should offer to public institutions in order to encourage them to develop CSR projects. This endorsing role of the government is more complex, comprising the legislation part, the facilities offered, the marketing part, because CSR needs to be promoted in order to change some mentalities, to reduce the reluctance to change and to fully understand why CSR is important for a society. More than 90% of the respondents (92.6%) agreed that this support should increase, this being the question with the highest consensus.

The second set of questions was aiming to find out which is the level of knowledge for the employees in the public sector. These referred to the National Strategy on CSR, to CSR reporting, to the strategy of the institution regarding CSR, as we can see in Table 4. These questions have an average level of generality and the answers to them should be known by any employee.

If we take a look at the first question in Table 4, we see that 19.1% of the respondents said there is no National Strategy on CSR, even if this strategy exists

Table 4 The level of knowledge regarding CSR matters in the public sector

	Is there a National Strategy on CSR? (%)	Is there a report on CSR published on the website of your institution? (%)	Does your institution have a strategy on CSR? (%)
Yes	14.9	19.1	22.3
No	19.1	40.4	27.7
I do not know	66	40.4	50

since 2011. So, we could make the assumption that the percentage for "no" is in fact added to that for "I do not know". Only 14.9% of the respondents knew there is a National Strategy on CSR and this reflects a real and important problem: the lack of knowledge on CSR derived from a lack of promotion.

If the National Strategy was better promoted, the level of awareness and knowledge would be higher. It is ironical that the full name of the strategy is "National Strategy for Promoting Corporate Social Responsibility". The strategy aims the promotion of CSR but employees and managers do not know that it exists.

In order to maintain the confidentiality of the respondents, we do not know if the institutions in which some respondents work have a strategy or a CSR report on their websites. So, we cannot be sure that those who said "yes" were completely honest. There are few public institutions in Romania that publish a report on CSR, so we could appreciate that the proportion of those who checked "I do not know" is higher.

It is very difficult for the employees to talk about the institution where they work in non-favorable terms and this could be a reason for not being completely honest. But the general idea is obvious, most of the respondents answered "I do not know" at all the three questions related to the knowledge on CSR.

We can conclude that the second hypothesis "employees working in the public sector are not well informed on CSR, the level of awareness and knowledge being low" is also validated, even if the employees understand, at least at a declarative level, that CSR is important and offers a lot of advantages for the institution and for the community.

In the last part of our research, the respondents were asked about the importance of CSR for themselves and for the public institutions where they work (the last one being their perception). These two questions were formulated as scales from 1 to 5, where 1 was *not important at all* and 5 was *very important*. The mean and the standard deviation are shown in Table 5 and both show that employees perceive CSR as important.

The mean 3.98 and the proportion of those who said CSR is important and very important for them (75.5%) show us that people are willing to learn and there is a potential. The mean of 3.43 for how important is CSR for the public institution where they work revealed that the employees perceive its importance also within the institution.

Table 5 Descriptive statistics on the importance of CSR

	Mean	Standard deviation
How important is CSR for you?	3.98	1.037
How important is CSR for your institution, in your perception?	3.43	1.15

To perceive the importance of corporate social responsibility and to be well informed on these issues related to CSR (National Strategy, legislation, facilities, standards, European Directives, indicators, reporting) are two different aspects. Of course, the former one may be seen as a premise for the latter. People should be at least willing to change something and the change is possible only if they understand which are the most important advantages of CSR, the costs involved and that benefits are higher than the costs on the long run.

4 Conclusions

The main barriers in front of a higher CSR activity in the public sector (but also for SMEs) are the lack of financial resources, the lack of knowledge on CSR, the insufficient promotion of the National Strategy on CSR and the lack of a political will. Employees understand the importance of CSR, its benefits, advantages, but the financial component is very important too. For example, more than 80% of the respondents would not be willing to accept a reduction of their wages in order to support the institution in involving in CSR projects. But this is normal in the context of a society, where a lot of people are paid with the minimum wage (about 323 Euros).

The results can be correlated with the political regime existent in Romania before 1989, because one of the main causes for this state of being is the mentality and passivity of citizens. After 1990, people became more individualistic and the responsibility for others diminished. A comparison between the elders and the youngsters could be the subject for a future research, because there are probably some differences between these categories. Youngsters tend to have a modern view on CSR issues, not being so influenced by the communist regime before the 90s.

Tax incentives and other fiscal facilities are considered one of the factors leading to an increase in the CSR activity. We could correlate this factor with the most important barrier that is related to limited resources. Public institutions are rather secondary partners than initiators of CSR projects. Even the National Strategy on CSR mentions the budgetary implications.

Even if the National Strategy exists since 2011, there are a lot of employees that are not aware of its existence and therefore we identify a problem—the lack of transparency and the insufficient promotion of this strategy. Employees working in the public sector are not well informed on CSR, the level of awareness and knowledge being low; they are confused about the term, the strategy on CSR, what reporting means or how a reporting should be done. There are important differences between a CSR reporting and a report on CSR activities. The reporting is a more complex process, implies some indicators, some standards for reporting and the report is just the final step that should be public for the stakeholders.

There is no perspective on a long term regarding corporate social responsibility, this being done mostly for PR reasons, for a temporary change in terms of image. Multinationals and large companies are more involved in CSR projects because they have the financial resources and the expertise required. The problem is with public institutions and small or medium enterprises.

CSR in Romania cannot be a model of doing things, but other countries in EU or future EU states can learn from our country that the lack of a CSR strategy is not the only problem. The real obstacle in front of a higher CSR implementation is the poor communication of these issues, the lack of transparency, the passivity of citizens and a low public marketing on CSR.

After a documentary and also an empirical research, we could offer some recommendations regarding CSR in the public sector and how it can be increased:

- Harmonize the national legislation with the new European Directives on CSR;
- Assure a higher dissemination of the legislation and a continuous monitoring and evaluation of how laws are implemented;
- Increase the number of partnerships with the private sector;
- Offer a higher support for the entities developing CSR projects, not just tax incentives but endorsing them in the process of CSR reporting and using standards;
- Initiating CSR and have a more proactive attitude. There are also some forms of CSR that do not involve a lot of money, so the problem with limited resources would be covered (volunteering);
- A permanent dialogue with the private actors, NGOs, community and other public institutions;
- Introduce more courses on CSR in schools to develop a culture more oriented to social responsibility and reduce reluctance to change;
- Organize more trainings on CSR at the level of public institutions in order to increase the level of awareness and knowledge on these issues;
- Organize national campaigns for promoting CSR, the National Strategy on CSR and the new European Directives like the Directive 2014/95/EU.

Corporate social responsibility in the public sector of Romania is at its first steps and cannot be compared with the projects developed by large companies. We appreciate the potential is high and the National Strategy on CSR should be renewed and also improved. The central authorities should learn from the mistakes and make a new SWOT analysis in order to reevaluate the new situation. CSR could also be integrated in a marketing strategy at different levels (national, regional or local), offering better chances for increasing the competitiveness and the image of the public sector.

References

Academic Network of European Disability Experts (2010). *Annual activity report*. Retrieveed from http://sid.usal.es/idocs/F8/FDO26327/ANED_2010_Activity_Report.pdf

Anca, C., Aston, J., Stanciu, E., & Rusu, D. (2011). *Corporate social responsibility in Romania. Situation analysis and current practices review*. Retrieved from undp.ro/libraries/projects/CSR/deliverables/Analiza%20Situatiei%20RSC%20in%20Romania%20ENG.pdf

Di Bitetto, M., Chymis, A., & D'Anselmi, P. (Eds.). (2015). *Public management as corporate social responsibility: the economic bottom line of government*. Switzerland: Springer. Retrieved from https://books.google.ro/books?id=caQxBQAAQBAJ

European Communities (2007). *Corporate social responsibility—national public policies in the European Union*. Retrieved from www.spolecenskaodpovednostfirem.cz/wp-content/uploads/2013/09/csr_en.pdf

Fox, T., Ward, H., & Howard, B. (2002). *Public sector roles in strengthening corporate social responsibility: A baseline study*. Retrieved from http://pubs.iied.org/pdfs/16017IIED.pdf

Hristea, A. M. (2011). Corporate social responsibility—between desideratum and reality. *Theoretical and Applied Economics*, *10*(563), 57–74. Retrieved from store.ectap.ro/articole/650.pdf

Korka, M. (2005). Corporate social responsibility in Romania: from theory to practice. *Transition Studies Review*, *12*(1), 47–57. doi:10.1007/s11300-005-0034-3.

Law no. 57/1992. Retrieved from http://legislatie.resurse-pentru-democratie.org/legea/57-1992.php

Law no. 448/2006. Retrieved from www.equalrightstrust.org/ertdocumentbank/LEGE%20448%20engleza.pdf

Mandl, I., & Dorr, A. (2007). *CSR and competitiveness European SMEs' good practice. Consolidated European report*. Vienna:Austrian Institute for SME Research. Retrieved from www.ifm-bonn.org/assets/documents/CSREuropa.pdf

Romanian Government (2010). *National strategy on promoting CSR 2011-2016*. Retrieved frome http://www.sgg.ro/nlegislativ/docs/2011/05/2x4bdfjnwskv89h17pzq.pdf

Scott, P. (2013). *Interview about tendencies in CSR reporting*. Retrieved from www.responsabilitatesociala.ro/editoriale/interviu-paul-scott-despre-tendintele-in-raportarea-de-csr.html

Steurer, R. (2010). *The role of governments in corporate social responsibility: Characterising public policies on CSR in Europe*. *Policy Sciences*, *43*(1), 49–72. Retrieved from www.wiso.boku.ac.at/fileadmin/data/H03000/H73000/H73200/InFER_Discussion_Papers/InFER_DP_10_2_The_Role_of_Governments_in_Coporate_Social_Responsibility_Characterising_Public_Policies_on_CSR_in__Europe.pdf.

Surveygizmo (2015). *Survey response rate*. Retrieved from www.surveygizmo.com/survey-blog/survey-response-rates/

The European Parliament and the Council of the European Union (2014). Directive 2014/95/EU. *Official Journal of the European Union*. Retrieeved from http://eur-lex.europa.eu/legal-content/EN/TXT/?uri=uriserv:OJ.L_.2014.330.01.0001.01.ENG

The Romanian Center for European Policies (2015). *Research on non-financial reporting*. Retrieved from www.crpe.ro/wp-content/uploads/2015/06/Raport-cercetare-raportare-non-financiara-FINAL.pdf

ViitorPlus Association and EY Romania (2015). *Less known and used fiscal facilities*. Retrieved from www.eyromania.ro/sites/default/files/attachments/Raport_Omenia%20ajuta%20compania_EY%20Romania%20ViitorPlus_FINAL.pdf

World Bank (2003). *Public policy for corporate social responsibility*. Retrieved from http://info.worldbank.org/etools/docs/library/57434/publicpolicy_econference.pdf

References

Part II
Corporate Social Responsibility in Times of Crisis: North America

United States of America

Corporate Social Responsibility Policy in the United States of America

Mark Anthony Camilleri

1 Introduction

The US markets for labor and capital are fairly unregulated as there are low levels of welfare state provision. Consequently, many social issues, such as education, healthcare or community investment have traditionally been at the core of corporate social responsibility (CSR) in the American context. CSR initiatives and communicating activities within the areas of philanthropy, stewardship, volunteerism and environmental affairs are not treated as a regulatory compliance issue in the United States of America (USA or U.S.). Therefore, CSR in the USA is often characterized by voluntary societal engagements by businesses as they are not obliged to undertake social and environmental responsibility practices. Such laudable behaviors are also referred to as corporate citizenship (Fifka, 2013; Matten & Crane, 2005). Social Responsibility and corporate citizenship encompass responsible behaviors that go beyond financial reporting requirements. These behaviors are particularly evidenced in cause-related marketing, stewardship initiatives, philanthropic and charitable contributions (Porter & Kramer, 2002; Varadarajan & Menon, 1988). In fact, US companies donate ten times as much as their British counterparts (Brammer & Pavelin, 2005). Notwithstanding, at this point in time, the United States is currently consuming some 207% of its ecological capacity (Worldwatch, 2015) and the average U.S. citizen uses 11 times as many resources as the average Chinese, and 32 times as much as the average Kenyan (Worldwatch, 2015). Moreover, the United States was a net importer of 67 non-fuel minerals and metals out of the 92 tracked by the U.S. Geological Survey (2010). Nonetheless, the American policy makers handle the issues that are related to global warming or the use of genetically modified organisms in food production, quite differently than

M.A. Camilleri (✉)
University of Malta, Msida, Malta
e-mail: mark.a.camilleri@um.edu.mt

S.O. Idowu et al. (eds.), *Corporate Social Responsibility in Times of Crisis*,
CSR, Sustainability, Ethics & Governance, DOI 10.1007/978-3-319-52839-7_7

their counterparts (Doh & Guay, 2006). In other parts of the world, the provisions of healthcare or issues pertaining to the climate change have traditionally been considered in the realms of government's responsibilities. Corporate responsibilities for social and environmental issues seem to have become the object of codified and mandatory regulation in certain jurisdictions (Camilleri, 2015). Therefore, it may appear that the larger firms rather than small and medium sized enterprises (SMEs) are the leading actors and drivers of CSR engagement and sustainable behaviors.

This conceptual paper reports that responsible practices are increasingly being embedded into core business functions and corporate strategic decisions. In this light, it sheds light on major US institutional frameworks and principles that have been purposely developed to foster a climate for social and environmental responsibility engagement. Policies and voluntary instruments include formal accreditation systems and soft laws that stimulate businesses and large organizations to implement and report their CSR-related activities. Several agencies of the US Government are currently employing CSR programs that are intended to provide guidance on corporate citizenship and human rights; labor and supply chains; anticorruption; energy and the environment; as well as health and social welfare among other issues. Firstly, this chapter introduces its readers to the notion of corporate citizenship as it draws comparisons with other CSR concepts. Secondly, it makes reference to empirical studies that have explored how organizations were engaging in economic, legal, ethical and discretionary behaviours toward stakeholders. Thirdly, this contribution also reports on the American corporate citizenship policies that are intended to support major areas of responsible corporate conduct in the realms of social responsibility and environmental sustainability. The author provides an interesting discussion on how policies and principles are raising awareness of the CSR agenda in the U.S. context. Finally, this chapter implies that appropriate policy frameworks and initiatives necessitate considerable discretionary investments as they pursue the sustainable path.

2 Corporate Citizenship

Corporate citizenship offers ways of thinking and behaving responsibly (Matten & Crane, 2005). It has potential to unlock significant benefits to both business and society (Carroll & Shabana, 2010). Sound environmental practices could be linked to improvements in economic performance and productivity, operational efficiencies, higher quality, innovation and competitiveness (Porter & Kramer, 2011). Therefore, corporate citizenship (through social responsibility and environmental sustainability) can be strategic in its intent and purposes (Basu & Palazzo, 2008; Burke & Logsdon, 1996). An integration of these different perspectives has led to the definition of corporate citizenship. The conceptual grounds to better understand the nature of corporate citizenship can be found in the bodies of literature on corporate social responsibility (e.g., Carroll, 1979), corporate social responsiveness (e.g., Clarkson, 1995), corporate social performance (e.g., Albinger & Freeman, 2000; Wartick & Cochran, 1985; Wood, 1991), and stakeholder engagement (Strand & Freeman, 2013). Carroll (1979) attempted to synthesise the fundamental principle of social responsibility. He explained the rationale behind social responsibility initiatives and went on to describe the corporate responses to social issues. Businesses always had a commitment towards society as they are obliged to engage in economic, legal, ethical and discretionary (philanthropic) activities (Carroll, 1979).

For instance, CSR's economic responsibilities include the obligations for businesses to maintain economic growth, and to meet consumption needs. The economic component of CSR represents the fundamental social responsibility of businesses. Many firms produce goods and services and sell them at fair prices. This will in turn allow the business entity to make a legitimate profit and to pursue growth. Legal responsibilities imply that businesses must fulfil their economic mission within the extant framework of regulations and legal parameters. The legal component recognises the obligation of the enterprise to obey laws. However, it could prove harder to define and interpret the ethical responsibilities of businesses. This component is often referred to as a "grey area", as it "involves behaviours and activities that are not embodied in law but still entail performance expected of business by society's members" (Carroll, 1979: 30). Ethical responsibilities require that businesses abide by moral rules that define appropriate behaviours within a particular society. Another category of corporate responsibility is related to discretionary, voluntary or philanthropic issues. Corporate philanthropy is a direct contribution by a corporation to a charity or cause, most often in the form of cash grants, donations and/or in-kind services' Kotler and Lee (2005: 144). This category of social responsibility is totally dictated at the "discretion" of the organisation as there are no laws or codified expectations guiding the corporations' activities (Rasche, De Bakker, & Moon, 2013). "Discretionary responsibilities include those business activities that are not mandated, not required by law, and not expected of businesses in an ethical sense" (Carroll, 1979: 500). Practically, some examples where organisations meet their discretionary responsibilities include; when they provide day-care centres for working mothers, by committing to philanthropic donations, or by creating pleasant work place aesthetics. Carroll (1991) describes these four distinct categories of activity by illustrating a "Pyramid of Corporate Social Responsibility". He maintained that his conceptualisation of the pyramid depicts the obligations of the business. Eventually, Schwartz and Carroll (2003) suggested an alternative approach that is based on three core domains (economic, legal and ethical responsibilities). The authors produced a Venn diagram with three overlapping domains; which were later transformed to seven CSR categories. This development was consistent with the relentless call on the part of the business community for the business case of CSR. Kotler and Lee (2005) demonstrated how a CSR approach had established a new way of doing business that led to the creation of value (Porter & Kramer, 2011; Wheeler, Colbert, & Freeman, 2003) with a respectful and proactive attitude towards stakeholders (Strand & Freeman, 2013).

Corporate citizenship continues to receive specific attention, particularly by those facilities that are operating outside their own domestic markets. At the same time, multinational corporations (MNCs) have been (and still are) under increasing pressure to exhibit "good corporate citizenship" in every country or market from where they run their business. MNCs have always been more closely monitored and scrutinised than the home country firms. No doubt this will continue to be the case in the foreseeable future.

2.1 Measuring Corporate Citizenship

Several empirical studies have explored the respondents' attitudes and perceptions on corporate social responsibilities. Very often, the measurement of corporate citizenship could have involved quantitative analyses on organisational commitment toward responsible organisational behaviours (Aupperle, Carroll, & Hatfield, 1985; Maignan, Ferrell, & Hult, 1999). Therefore, their survey responses could not have revealed and explained actual corporate citizenship practices. Other research could have focused on investigations of managerial perceptions of corporate citizenship rather than focusing on corporate behaviours (e.g., Basu & Palazzo, 2008; Singhapakdi, Kraft, Vitell, & Rallapalli, 1995). A number of similar studies have gauged corporate citizenship by adopting Fortune's reputation index (Fryxell & Wang, 1994; Griffin & Mahon, 1997; Stanwick & Stanwick, 1998), the KLD index (Fombrun, 1998; Griffin & Mahon, 1997) or Van Riel and Fombrun's (2007) Reptrak. Such measures require executives to assess the extent to which their company behaves responsibly toward the environment and the community (Fryxell & Wang, 1994). Despite their wide usage in past research, the appropriateness of these indices remains doubtful. For instance, Fortune's reputation index failed to account for the multi-dimensionality of the corporate citizenship construct and is suspected to be more significant of management quality than of corporate citizenship (Waddock & Graves, 1997). Fortune's past index suffered from the fact that its items were not based on theoretical arguments as they did not appropriately represent the economic, legal, ethical, and discretionary dimensions of the corporate citizenship construct. Hunt, Wood, and Chonko's (1989) investigated broad based perceptions on (a) the extent to which employees perceive that managers are acting ethically in their organisations (b) the extent to which employees perceive that their managers are concerned about the issues of ethics in their organisations and (c) the extent to which employees perceive that ethical (or unethical) behaviour is rewarded (or punished) in their organisation. Other authors, including Webb, Mohr, and Harris (2008) also explored the philanthropic values that were related socially responsible consumption.

Pinkston and Carroll (1994) identified four dimensions of corporate citizenship, including; orientations, stakeholders, issues, and decision-making autonomy. They argued that by observing orientations, one may better understand the inclinations or posturing behaviours of organisations with respect to corporate citizenship. The stakeholder dimension should better define to whom the organisation feels responsible as it could identify where the corporate citizenship issues or social concerns are originating. The aspect of decision-making autonomy was believed to illuminate the perceived importance of corporate citizenship as one that determines at what organisational level corporate citizenship decisions are actually made. In a similar vein, Griffin and Mahon (1997) combined four estimates of corporate citizenship: the Fortune reputation index, the KLD index, the Toxic Release Inventory (TRI), and the rankings provided in the Directory of Corporate Philanthropy. They admitted that their four measures do not necessarily track one another. Such findings suggest that these indicators may not be representative of the same underlying construct and their items may not be sufficient to provide an overall understanding of corporate citizenship.

Singh and Del Bosque (2008) adopted a multi-dimensional perspective on three domains, including; commercial responsibility, ethical responsibility and social responsibility. Firstly, they proposed that commercial responsibility of businesses relates to their continuous development of high quality products and truthful marketing communications of their products' attributes and features among customers. Secondly, they maintained that ethical responsibility is concerned with businesses fulfilling their obligations toward their shareholders, suppliers, distributors and other agents with whom they make their dealings. Singh and Del Bosque (2008) argued that ethical responsibility involves the respect for the human rights and norms that are defined in the law when carrying out business activities. They hinted that respecting ethical principles in business relationships has more priority over achieving superior economic performance. Their other domain, the social responsibility is concerned about laudable behaviours. The authors suggest that businesses could allocate part of their budget to the natural environment, philanthropy, or toward social works that favoured the most vulnerable in society. This perspective supports the development of financing social and/or cultural activities and is also concerned with improving societal well-being.

3 Social Responsibility Policies

The governments are usually considered as the main drivers of CSR policy. However, there are other actors within society, such as civil organizations and industry (Camilleri, 2015). It is within this context that a relationship framework has been suggested by Mendoza (1996) and Midttun (2005). It seems that at the time, there was a need for a deeper understanding of the governments' role and function in promoting CSR. Societal governance is intrinsically based on a set of increasingly complex and interdependent relationships (Albareda, Lozano, & Ysa, 2007). There are different expectations and perceptions within each stakeholder relationship, which have to be addressed in order to develop an appropriate CSR policy, as featured in Fig. 1 of chapter "Corporate Social Responsibility in Challenging Times: A consideration of how Small and Medium Scale Enterprises

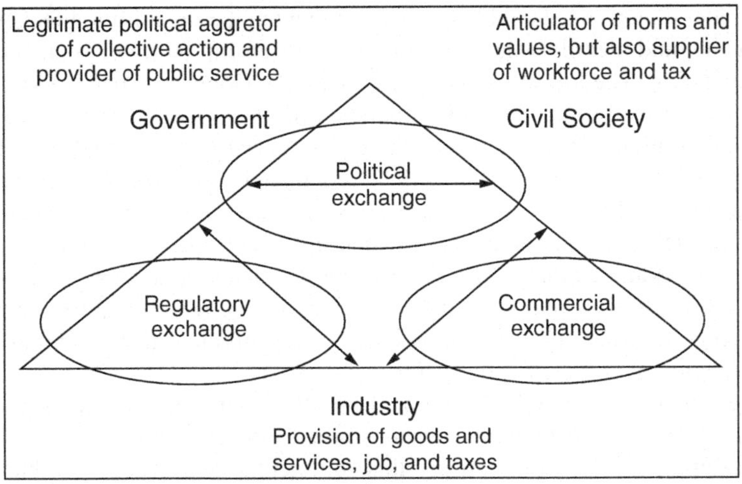

Fig. 1 Actors and exchange arenas (Camilleri, 2015, Adapted from Albareda et al., 2007)

attempt to deal with CSR Challenges in Nigeria" (Camilleri, 2015). Essentially, this relational approach is based on the idea that recent changes and patterns affecting the economic and political structure may transform the roles and capacities of various social agents (Albareda, Lozano, Tencati, Midttun, & Perrini, 2008). These exchange relationships among different actors and drivers that are shaping CSR policy and communications are depicted in Fig. 1.

This exchange arena is exemplified in the U.S. government's comprehensive approach to providing support and guidance on areas of corporate conduct and sustainable behaviours. The U.S. secretary of state's agenda is to ensure effective coordination and partnerships with individual bureaus and offices in order to harness global economic tools that advance U.S. foreign policy goals on responsible initiatives. For example, the U.S. Bureau of Economic and Business Affairs (EB) leads a corporate social responsibility team. Its primary purpose is to promote responsible business practices and fostering sustainable development whilst building economic security. This team provides guidance to American companies and their stakeholders to engage in corporate citizenship. EB's CSR team supports major areas of responsible corporate conduct, including: 'good corporate citizenship', 'human rights', 'labour and supply chains', 'anticorruption', 'health and social welfare', 'contribution to the growth and development of the local economy', 'innovation, employment and industrial relations', 'environmental protection', 'natural resources governance' including the Kimberley Process, 'transparency', 'trade and supply chain management', 'intellectual property' and the 'women's economic empowerment' among other issues. Most of EB's corporate policies are drawn from the Organisation for Economic Co-operation and Development (OECD) 'Guidelines for Multinational Enterprises' and from U.S. national contact point for the guidelines (as explained hereunder).

3.1 *Good Corporate Citizenship and Human Rights*

In 1998, the Bureau of Democracy, Human Rights and Labour (DRL) set up a Human Rights and Democracy Fund (HRDF) to fulfil the bureau's mandate of monitoring and promoting human rights and democracy in the global context. The HRDF fund was designed to act as the department's "venture capital" fund for democracy and human rights issues, including; the promotion of democratic principles and personal liberties. Such programs enabled the U.S., "to minimize human rights abuses, to support democracy activists worldwide, to open political space in struggling or nascent democracies and authoritarian regimes, and to bring positive transnational change". DRL's important efforts have brought positive change as its funding of HRDF has grown from $7.82 million in 1998 to over $207 million in 2010 (HRDF, 2015).

In parallel, an 'Office to Monitor and Combat Trafficking in Persons (TIP) works with business leaders to prevent and stop human trafficking. TIP does this by advancing the Luxor Guidelines, which focus on corporate policy, strategic planning, public awareness, supply chain tracing, government advocacy and transparency to reduce forced labour in supply chains. In 2015, TIP Office awarded over $18 million in grants and cooperative agreements to combat human trafficking. This office continues to fund an emergency global assistance project that provides services on a case-by-case basis for individuals that have been identified as trafficked persons (TIP, 2015).

Currently, many NGOs and international organisations are working in tandem as they support 27 projects that address prosecution, protection and prevention of sex and labour trafficking in different places around the globe (TIP, 2015). On the 28th October, 2015, the Partnership for Freedom in collaboration with the Department of State and four other federal agencies launched "*Rethink=Supply Chains: The Tech Challenge to Fight Labour Trafficking*", an innovation challenge that calls for technological solutions that identify and address labour trafficking in global supply chains for goods and services. The Partnership for Freedom has awarded $500,000 in prizes and services that are aimed to spur innovative solutions to end human trafficking, and to support victims of human trafficking in the United States.

3.2 *Labour and Supply Chains*

America made human trafficking illegal in 2000, after which it started to publish annual assessments of other countries' efforts to tackle it. But it has only slowly turned up the heat on offenders within its borders. Australia and the UK have recently passed light-touch laws requiring transparency in supply chains. This legislation required manufacturers and retailers that earn global revenues above the $100 million threshold to list their efforts on how they are eradicating modern slavery and human trafficking from their supply chains. For the time being, a firm can comply by simply reporting that it is doing nothing. But it seems that few corporations are willing to admit such a statement that will surely affect their CSR credentials. Hence, it seems that this issue is forcing its way on to managers' to-do lists. Moreover, the ILO has launched a fair-recruitment protocol which it hopes will be ratified by national governments. The ILO's intention is to cut out agents. In this light, TIP has partnered with Slavery Footprint to provide online tools to initiate marketplace action and ongoing dialogues between individual consumers and producers about modern slavery practices in supply chains (TIP, 2015). Similarly, DRL continues to promote labour rights throughout the supply chain as it enforces labour law and provides due diligence. DRL has also strengthened legal advocacy that expanded livelihood opportunities for many individuals, as it advanced multi-stakeholder approaches. EB, in cooperation with DRL and other stakeholders, has coordinated the U.S. Department of State's participation in the Kimberley Process to stem the flow of conflict diamonds and to address their traceability across supply chains.

3.3 Anti-corruption

The corruption undermines sound public financial management and accountability at all institutional levels: It deters foreign investment in many countries, it stifles economic growth and sustainable development, it distorts prices, and undermines legal and judicial systems (INL, 2006). The high-level, large-scale corruption by public officials is also referred to as kleptocracy. It can have a devastating effect on democracy, the rule of law, and economic development. Those who contribute to such corruption by paying or promising to pay bribes or by giving other undue advantages to foreign public officials will undermine good governance and alter fair competition. The U.S. has long led by example in its enduring fight against corruption. Through its Foreign Corrupt Practices Act (FCPA) in 1977, the U.S. became the first country to criminally penalise its nationals and companies that bribe foreign public officials in commercial transactions. In fact, the United States denies safe haven to egregiously corrupt officials and other public figures as specified in the Presidential Proclamation 7750 (of January 2004). Moreover, the United Nations Convention Against Corruption (UNCAC) has also provided a framework for international cooperation against corruption, including preventative and enforcement measures. The U.S. government has participated in drafting U.N. legislative guide materials prior to its implementation and enforcement (INL, 2006). The USA is also member of the OECD's Anti-Bribery Convention where EB represents the U.S. Department of State within the OECD Working Group on Bribery in International Business Transactions.

3.4 Health and Social Welfare

In the United States, public education was not considered as a social welfare activity, probably because it is taken for granted, since its inception 125 years ago. On the other hand, public health and vocational rehabilitation are not included within the Social Security Act, but are present in separate Federal laws. However, medical care and cash benefits have always been provided under the workmen's compensation laws. These laws cover work-injuries and members of the armed forces and their dependents, and veterans who are entitled to medical care at public expense.

Interestingly, landmark reform on the Patient Protection and Affordable Care Act (PPACA), and the Health Care and Education Reconciliation Act (HCERA) of 2010 (H.R. 4872) was passed and enacted through two federal statutes. *PPACA was* signed in March 23, 2010. This act which is also known as 'Obamacare', provided the phased introduction over 4 years of a comprehensive system of mandated health insurance with reforms that were designed to eliminate "some of the worst practices of the insurance companies", including pre-existing condition screening and premium loadings, policy cancellations on technicalities when illness seems imminent, annual and lifetime coverage caps, among other issues. It also sets a minimum ratio of direct health care spending to premium income; and creates price competition that was bolstered by the creation of three standard insurance coverage levels to enable like-for-like comparisons by consumers; and a web-based health insurance exchange where consumers can compare prices and purchase plans (PPACA, 2010). This system preserves private insurance and private health care providers and provides more subsidies to enable the poor to buy insurance. Notwithstanding, the Health Care and Education Reconciliation Act of 2010 (H.R. 4872), which amended PPACA (that was passed a week earlier), was enacted by the 111th United States Congress and became law on March 30, 2010 (Reuters, 2010). This latter act (H.R. 3221) also incorporated the Student Aid and Fiscal Responsibility Act (SAFRA) expanded federal Pell Grants to a maximum of $5500 in 2010 and tied grant increases to annual increases in the Consumer Price Index, plus 1%. Therefore, SAFRA ended the practice of federal subsidization of private loans. This has translated to cutting the federal deficit by $87 billion over a period of 10 years.

4 Policies for Environmental Sustainability

4.1 Energy and the Environment

Historically, the United States prides itself of a long tradition of environmental leadership, that dates back to President Teddy Roosevelt. As a matter of fact, in the 1960s and 1970s the U.S. established a series of progressive laws and institutions. For example The National Environmental Policy Act (NEPA) of 1969 committed the United States to sustainability, declaring it a national policy "to create and maintain conditions under which humans and nature can exist in productive harmony that permit fulfilling the social, economic and other requirements of present and future generations" (NEPA, 1969).

The formulation of the Environmental Protection Agency's (EPA) policies and instruments have anticipated Brundtland's concept of "sustainable development" and his idea that generates clean prosperity today whilst preserving resources and ecological functions for use by future generations. Arguably, policies on social and environmental development are expected to reinforce responsible practices on resource management, energy efficiency and measures that mitigate climate change. In this regard, EPA has developed a variety of methods, tools and guidance programmes that are aimed at supporting the application of environmental sustainability. Moreover, the Bureau of Energy Resources (ENR) advances U.S. interests with regards to secure, reliable and ever-cleaner sources of energy. ENR promotes good governance and transparency in the energy-sector as it supports the Extractive Industries Transparency Initiative (EITI). Countries implementing the EITI disclose information on tax payments, licences, contracts, production and other key elements that revolve around resource extraction. This information is disclosed in an annual EITI Report. This transparent report allows citizens to see for themselves how their country manages its natural resources and it also specifies the revenue that they generate. The EITI Standard contains a set of requirements that countries, including the U.S., need to meet in order to qualify as an EITI Candidate or EITI Compliant country (EITI, 2015).

5 Discussion and Conclusions

Arguably, the social and environmental responsibility is the only way forward for all nations, particularly for big economies like China, the U.S., Russia and India. These countries are the largest producers of emissions and greenhouse gases in the world. This article shed light on the US governmental institutions and agencies' credentials on socially and environmentally responsible policies. It described in detail relevant instruments including relevant legislation and executive orders that were intended to unlock corporate citizenship among business and industry. At the same time, it reported how many commentators including academia are suggesting that the United States is lagging behind many other countries, in developing more sustainable economic processes and energy infrastructure. Environmental lobbyists argue that in the past three decades, average temperatures in the continental U.S. rose five times as much, than in a century-long period. A new report from the Worldwatch Institute, entitled; "Creating Sustainable Prosperity in the United States: The Need for Innovation and Leadership" called for a broad range of policy innovations in the areas of renewable and non-renewable resource use, waste and pollution, and population. This NGO purports that U.S. leaders have not implemented adequate and sufficient reforms on social and environmental responsibility. Arguably, at the moment many businesses are still characterised by their unsustainable practices such as linear flows of materials, heavy dependence on fossil fuels, disregard for renewable resources, and resource use. According to Columbia University's Environmental Sustainability Index (ESI), the US has merely scored 38 out of 100 in "global stewardship" and 27 out of 100 in "reducing stresses".

These results suggest the US's poor performance in mitigating air and water pollution and ecosystem stresses is the outcome of the country's minimal responsibility and sensitivity toward global environmental institutions (and international treaties). Notwithstanding, in a recent survey among seventeen countries by National Geographic, the American consumers ranked among the last in their green consumption habits (Greendex, 2012). Moreover, Chen and Bouvain (2009) reported that the percentage of U.S. companies that were members of the Global Compact was much lower than in the other countries. This finding could indicate that certain aspects of the Compact may not be acceptable to the U.S. corporations. Maybe, the relatively low environmental credentials among U.S. businesses and individual citizens transcends from the political arena. Although, the U.S. regularly attends to the annual conferences of the parties (COPs) that are organized by to the United Nations Framework—Convention on Climate Change (UNFCCC), yet consecutive governments, since Clinton's administration did not transpose Kyoto's protocol. One of the strengths of the Kyoto treaty was the establishment of an international emissions trading system, where countries can earn credits toward their emission target; by investing in emission clean-ups outside their own country.

This review paper reported there are a number of corporate citizenship and social responsibility policies that are still evolving in the US context. Arguably, national institutional structures are creating both challenging opportunities and threats for businesses. US corporations are already operating in various contexts where they could be mandated by law to abide by national legislation and regulation. Notwithstanding, there are different CSR communications and stakeholders' evaluations of given firms across countries. Despite the growing commitment to corporate citizenship, past research did not sufficiently link this notion with CSR policy. This contribution has reported how different U.S. institutions, including bureaus, agencies and other stakeholders are pushing forward the social responsibility, environmental sustainability as well as the responsible corporate governance agenda. The US CSR policies and instruments are generally (i) based on sound theoretical arguments (ii) tackle the economic, legal, ethical, and discretionary dimensions. However, these regulatory tools could contain disclosure guidelines and reporting mechanisms for the monitoring and controlling of corporate responsible behaviours in the U.S.

The U.S. Government to trigger companies to invest in more efficient technologies by subsidising cleaner production and circular economies. Alternatively, businesses can be penalised when they do not conform to regulatory requirements on responsible behaviours (e.g. reducing environmental impact) (Moratis & Slaa, 2016). For instance, with carbon pricing, governments cannot interfere with management decisions. The businesses themselves ought to decide on effective ways on how they cut their emissions. Carbon markets are there and are expanding (e.g. The EU's Emissions Trading Scheme—ETS). There are many lessons to be learned from the countries' that have resorted to ETS to curb their pollution on the environment. Perhaps, one of the challenges for policymakers is the monitoring and controlling of carbon markets. Indeed, it is in the businesses' interest to anticipate the reinforcement of extant regulatory instruments or any mandatory compliance procedures to new legislation. The firms' proactive corporate citizenship behaviours will inevitably lead them to a sustainable competitive advantage, particularly at times when the market is not working well.

6 Implications and Recommendations

Although there have been many contributions on corporate citizenship practices (Fifka, 2013; Maignan et al., 1999; Matten & Crane, 2005; Pinkston & Carroll, 1994), there is still considerable potential for research that focuses on regulatory policy, in this regard. Future research could measure the comparability of policy frameworks for corporate citizenship in the US with other states. Notwithstanding, CSR policies, procedures, and activities necessitate considerable discretionary investments, in terms of time and resources by policy makers, civil authorities, businesses and non-governmental organisations. The underlying question is to establish whether both companies and non-for profit organisations perceive a business or a political case for corporate citizenship, as their potential to create value for themselves and for society as they pursue the sustainable path.

The increased quality of life has brought unsustainable **consumption** behaviors among customers. Notwithstanding, increased productivity levels are rapidly depleting the world's natural resources. This research has indicated that on paper there are several policies frameworks and initiatives that are pushing forward the corporate citizenship agenda in the U.S. However, the proof is in the pudding. Debatably, the U.S. government and its agencies should ensure that the true ecological cost of environmental degradation and climate change is felt in the market. In this light, there is scope in promoting circular economies that are characterised by resource efficiencies through recycling, reducing and reusing. Moreover, organisations should be urged to find alternative ways for sustainable **energy** generation, **energy** and water **conservation**, environmental protection and greener transportation systems.

Corporate citizenship policies should be promoting socially-responsible investing (SRI), responsible supply chain management and the responsible procurement of sustainable products. Fiscal policies and tools could encourage consumers to purchase sustainable, eco-labelled products, standardized items and 'fair-trade' goods.

References

Albareda, L., Lozano, J. M., & Ysa, T. (2007). Public policies on corporate social responsibility: The role of governments in Europe. *Journal of Business Ethics, 74*(4), 391–407.

Albareda, L., Lozano, J. M., Tencati, A., Midttun, A., & Perrini, F. (2008). The changing role of governments in corporate social responsibility: Drivers and responses. *Business Ethics: A European Review, 17*(4), 347–363.

Albinger, H. S., & Freeman, S. J. (2000). Corporate social performance and attractiveness as an employer to different job seeking populations. *Journal of Business Ethics, 28*(3), 243–253.

Aupperle, K. E., Carroll, A. B., & Hatfield, J. D. (1985). An empirical examination of the relationship between corporate social responsibility and profitability. *Academy of Management Journal, 28*(2), 446–463.

Basu, K., & Palazzo, G. (2008). Corporate social responsibility: A process model of sensemaking. *Academy of Management Review, 33*(1), 122–136.

Brammer, S., & Pavelin, S. (2005). Corporate community contributions in the United Kingdom and the United States. *Journal of Business Ethics, 56*(1), 15–26.

Burke, L., & Logsdon, J. M. (1996). How corporate social responsibility pays off. *Long Range Planning, 29*(4), 495–502.

Camilleri, M. A. (2015). Valuing stakeholder engagement and sustainability reporting. *Corporate Reputation Review, 18*(3), 210–222.

Carroll, A. B. (1979). A three-dimensional conceptual model of corporate performance. *Academy of Management Review, 4*(4), 497–505.

Carroll, A. B. (1991). The pyramid of corporate social responsibility: Toward the moral management of organizational stakeholders. *Business Horizons, 34*(4), 39–48.

Carroll, A. B., & Shabana, K. M. (2010). The business case for corporate social responsibility: A review of concepts, research and practice. *International Journal of Management Reviews, 12*(1), 85–105.

Chen, S., & Bouvain, P. (2009). Is corporate responsibility converging? A comparison of corporate responsibility reporting in the USA, UK, Australia, and Germany. *Journal of Business Ethics, 87*(1), 299–317.

Clarkson, M. E. (1995). A stakeholder framework for analyzing and evaluating corporate social performance. *Academy of Management Review, 20*(1), 92–117.

Doh, J. P., & Guay, T. R. (2006). Corporate social responsibility, public policy, and NGO activism in Europe and the United States: An Institutional-Stakeholder perspective. *Journal of Management Studies, 43*(1), 47–73.

EITI. (2015). *Extractive industries transparencies initiative*. Retrieved November 20, 2015, from https://eiti.org/eiti

Fifka, M. S. (2013). Corporate citizenship in Germany and the United States—Differing perceptions and practices in transatlantic comparison. *Business Ethics: A European Review, 22*(4), 341–356.

Fombrun, C. J. (1998). Indices of corporate reputation: An analysis of media rankings and social monitors' ratings. *Corporate Reputation Review, 1*(4), 327–340.

Fryxell, G. E., & Wang, J. (1994). The *fortune* corporate 'reputation' index: Reputation for what? *Journal of Management, 20*(1), 1–14.

Greendex. (2012). *Consumer choice and the environment. A worldwide tracking survey*. Retrieved November 25, 2015, from http://images.nationalgeographic.com/wpf/media-content/file/NGS_2012_Final_Global_report_Jul10-cb1341951434.pdf

Griffin, J. J., & Mahon, J. F. (1997). The corporate social performance and corporate financial performance debate twenty-five years of incomparable research. *Business and Society, 36*(1), 5–31.

HRDF. 2015. *DRL programs: Human rights and democracy fund*. Retrieved November 17, 2015, from http://www.state.gov/j/drl/p/

Hunt, S. D., Wood, V. R., & Chonko, L. B. (1989). Corporate ethical values and organizational commitment in marketing. *The Journal of Marketing, 53*(3), 79–90.

INL. (2006). *U.S. strategy to internationalize efforts against kleptocracy: Combating high-level public corruption*. Denying Safe Haven, and Recovering Assets. Retrieved November 20, 2015, from http://2001-2009.state.gov/p/inl/rls/fs/70365.htm

Kotler, P., & Lee, N. (2005). *Corporate social responsibility. Doing the most good for your company and your cause*. Hoboken, NJ: Wiley.

Maignan, I., Ferrell, O. C., & Hult, G. T. M. (1999). Corporate citizenship: Cultural antecedents and business benefits. *Journal of the Academy of Marketing Science, 27*(4), 455–469.

Matten, D., & Crane, A. (2005). Corporate citizenship: Toward an extended theoretical conceptualization. *Academy of Management Review, 30*(1), 166–179.

Mendoza, X. (1996). *Las transformaciones del sector público en las sociedades avanzadas. Del estado del bienestar al estado relacional*. Papers de Formació 23 Diputació de Barcelona.

Midttun, A. (2005). Policy making and the role of government. Realigning business, government and civil society. Emerging embedded relational governance beyond the (neo)liberal and welfare state models. *Corporate Governance: International Journal of Business in Society, 5* (3), 159–174.

Moratis, L., & Slaa, A. M. (2016). Sticks in a bundle are unbreakable: The creation of a Kenyan CSR knowledge centre and business network. In *Key initiatives in corporate social responsibility* (pp. 399–417). Cham: Springer International Publishing.

NEPA. (1969). *Summary of the national environmental policy act*. Law and Regulations, US Environmental Protection Agency. Retrieved November 16, 2015, from http://www.epa.gov/laws-regulations/summary-national-environmental-policy-act

Pinkston, T. S., & Carroll, A. B. (1994). Corporate citizenship perspectives and foreign direct investment in the US. *Journal of Business Ethics, 13*(3), 157–169.

Porter, M. E., & Kramer, M. R. (2002). The competitive advantage of corporate philanthropy. *Harvard Business Review, 80*(12), 56–68.

Porter, M. E., & Kramer, M. R. (2011). Creating shared value: How to reinvent capitalism—And unleash a wave of innovation and growth. *Harvard Business Review, 89*(January/February), 62–77.

PPACA. (2010). *The patient protection and affordable care act: An overview of its potential impact on state health programs*. Retrieved October 2, 2015, from http://www.lao.ca.gov/reports/2010/hlth/fed_healthcare/fed_healthcare_051310.aspx

Rasche, A., De Bakker, F. G., & Moon, J. (2013). Complete and partial organizing for corporate social responsibility. *Journal of Business Ethics, 115*(4), 651–663.

Reuters. (2010). *Factbox-U.S. healthcare bill would provide immediate benefits*. Retrieved October 31, 2015, from http://www.reuters.com/article/usa-healthcare-rollout-idUSN1914020220100322#vlGrBoWEAWZE7UER.97

Schwartz, M. S., & Carroll, A. B. (2003). Corporate social responsibility: A three-domain approach. *Business Ethics Quarterly, 13*(4), 503–530.

Singh, J., & Del Bosque, I. R. (2008). Understanding corporate social responsibility and product perceptions in consumer markets: A cross-cultural evaluation. *Journal of Business Ethics, 80* (3), 597–611.

Singhapakdi, A., Kraft, K. L., Vitell, S. J., & Rallapalli, K. C. (1995). The perceived importance of ethics and social responsibility on organizational effectiveness: A survey of marketers. *Journal of the Academy of Marketing Science, 23*(1), 49–56.

Stanwick, P. A., & Stanwick, S. D. (1998). The relationship between corporate social performance, and organizational size, financial performance, and environmental performance: An empirical examination. *Journal of Business Ethics, 17*(2), 195–204.

Strand, R., & Freeman, R. E. (2013). Scandinavian cooperative advantage: The theory and practice of stakeholder engagement in Scandinavia. *Journal of Business Ethics, 127*(1), 65–85.

TIP. (2015). *Office to monitor and combat trafficking in persons*. Retrieved November 14, 2015, from http://www.state.gov/j/tip/

U.S. Geological Survey. (2010). *Mineral commodity summaries. U.S. department of the interior.* Retrieved November 10, 2015, from http://minerals.usgs.gov/minerals/pubs/mcs/2010/mcs2010.pdf

Van Riel, C. B., & Fombrun, C. J. (2007). *Essentials of corporate communication: Implementing practices for effective reputation management.* London: Routledge.

Varadarajan, P. R., & Menon, A. (1988). Cause-related marketing: A co-alignment of marketing strategy and corporate philanthropy. *The Journal of Marketing, 52*(3), 58–74.

Waddock, S. A., & Graves, S. B. (1997). The corporate social performance-financial performance link. *Strategic Management Journal, 18*(4), 303–319.

Wartick, S. L., & Cochran, P. L. (1985). The evolution of the corporate social performance model. *Academy of Management Review, 10*(4), 758–769.

Webb, D. J., Mohr, L. A., & Harris, K. E. (2008). A re-examination of socially responsible consumption and its measurement. *Journal of Business Research, 61*(2), 91–98.

Wheeler, D., Colbert, B., & Freeman, R. E. (2003). Focusing on value: Reconciling corporate social responsibility, stakeholder theory and sustainability in a network world. *Journal of General Management, 28*(3), 1–28.

Wood, D. J. (1991). Corporate social performance revisited. *Academy of Management Review, 16* (4), 691–718.

WorldWatch. (2015). *U.S. must commit to sustainability to overcome mounting economic and ecological strains.* Retrieved November 25, 2015, from http://www.worldwatch.org/us-must-commit-sustainability-overcome-mounting-economic=-and-ecological-strains

Part III
Corporate Social Responsibility in Times of Crisis: Africa

Nigeria
Kenya
Ghana I
Zimbabwe
Ghana II
Ghana III

Corporate Social Responsibility in Challenging Times: A Consideration of How Small and Medium Scale Enterprises Attempt to Deal with CSR Challenges in Nigeria

Adebimpe Lincoln

1 Introduction

Corporate Social Responsibility (CSR) is often considered as a firm's obligation for its economic, legal, ethical and philanthropic activities (Carroll, 1991). This view was echoed by McGuire (1963) when he asserts "corporations have economic, legal obligation and responsibilities to society at large which extends beyond these obligations (p. 144). Various authors state that CSR involves compliance with the law, philanthropy, and responsibility in the marketplace to various stakeholder groups who can affect or are affected by the achievement of the organisations objectives (Freeman, 1984 pp 46). Stakeholders are broadly defined to include investors, consumers, suppliers, employees, the local community and the environment. The theory posits that mechanisms should be adopted to ensure that management safeguards the rights and interests of stakeholders and that management are accountable for the generation and distribution of wealth in firms (Aguilera, Filatotchev, & Gregory Jackson, 2008). While stakeholder theory conceptualises ways that managers might balance their responsibilities to shareholders and legitimate stakeholders, the reconciliation of such competing 'stakes' can be difficult (Carney, Gedajlovic, & Sujit, 2011).

Various academics and researchers have put forward divergent classifications in a bid to gain deeper insight into stakeholder identification and management. For example, Freeman (1994) posits three main stakeholder taxonomies. The first is the stakeholder enabling principle, which is that firms are managed in the interests of stakeholders, such as employees, financiers, customers, employees and communities. The second is the principle of director responsibility. This means that firms have the legal authority and moral responsibility to use reasonable judgment to

A. Lincoln (✉)
University of Liverpool, Liverpool, L69 3BX, UK
e-mail: adebimpe.lincoln@online.liverpool.ac.uk

S.O. Idowu et al. (eds.), *Corporate Social Responsibility in Times of Crisis*, CSR, Sustainability, Ethics & Governance, DOI 10.1007/978-3-319-52839-7_8

identify and distribute rights and obligations among stakeholder groups. The third is the principle of stakeholder recourse, which means that stakeholders are provided with a platform to bring actions against directors who fail to discharge the duty of care owed to the firm and stakeholders (Freeman, 1994). Stakeholder theory seeks to strike a balance between maintaining the economic and social goals of the firm and its stakeholders, while also encouraging the efficient use of resources, accountability in the use of power and stewardship by those in control of the firm (Okeahalam & Akinboade, 2003; Lincoln, Adedoyin, & Croad, 2015). Therefore, maintaining harmonious relationships with stakeholders should be of high strategic importance to the firm as this determines its success (Donaldson & Preston, 1995).

Various scandals such as that seen in Enron and Worldcom in the United states and Pamalat in Italy, and the most recent financial crisis in 2009 have brought to the fore international deliberations on the firms responsibilities to their stakeholders, the efficacy of CSR and existing corporate governance standards and practices (Mallin, 2015). In a bid to ensure good governance and bolster socially responsible practices, various countries have devised strategies to ensure transparency and accountability of corporate practices not only as a means of inculcating investor confidence but also as a way of further strengthening their legal systems, democratic governance and ensuring local communities are accounted for in CSR practices. There is general consensus amid academics, researchers and government agencies worldwide that such systems helps to provide a degree of confidence, integrity, stability, investment and sustained economic growth, which is necessary for the proper functioning of a market economy. In-spite of the growing recognition of the significance of CSR, much of the scholarly discussion in this area has focused exclusively on Multinational Corporations (MNC's) and other large corporations (Coppa & Siramesh, 2013). The concept remains elusive within the African continent (Muthuri, 2012) consequently there are unanswered questions on CSR within the African milieu. It is argued that the conclusions drawn from studies which have their focus primarily in the 'Developed World' do not necessarily apply to the situation in Nigeria which has peculiarities that distinguish it from the outside world generally and from Western countries where many MNC's and large corporations originate (Lincoln, Fields, & Adedoyin, 2013). This study seeks to explore the dimensions of CSR within the SMEcontext in Nigeria and factors, which hinder SME active engagement in CSR, related activities. This study is of significant importance given the lack of adequate investigation on CSR practices of Nigerian SME's within their local communities. Consequently, this research contributes to pragmatic theoretical development on SME approach and engagement in CSR activities in Nigeria. This study seeks to examine CSR concepts from a Nigerian perspective with specific focus on Nigerian SME's. In particular the study seeks to provide answers to the following questions:

- What challenges do SME's experience when undertaking CSR related activities
- What are SME owner/managers perceptions on how to overcome the challenges?

- Are there similarities or differences in challenges experienced or perception in overcoming CSR related challenges

Following the introductory section, the paper progresses with a review of SME Corporate Social Responsibility approaches adopted and challenges encountered. In addition, the review provides useful information on Small and Medium Scale Enterprises and Corporate Social Responsibility and Sustainability in Nigeria. Following the literature review the paper provides the methodological consideration informing the research. The findings of the study are presented in four main parts. The first part of the findings contains a demographic profile of the Male and Female Entrepreneurs who took part in the study. The second part provides the perception of SME owner/manager on CSR Practices in Nigeria. The third part of the finding provides an analysis of the challenges experienced by SME Owner/managers when undertaking CSR. The final part of the analysis focuses on the SME Owner/managers perception on strategies to overcome CSR related challenges.

2 Small and Medium Scale Enterprises Corporate Social Responsibility Approach and Challenges

SME approach to CSR differs significantly from the approach adopted by larger firms. This is due in part to the nature of the SME firm and associated characteristics such as size, sector and the SME owner/managers orientation towards profit. For example Sweeney (2007) notes that in small firms' ownership and management are not separated as such control remains in the hands of the owner. It stands to reasoning thus that SME's acceptance of CSR is largely a factor of the personal attitudes of the owner/manager. Quinn (1997) asserts that the SME owner/managers personal values, ethical as well as moral convictions to an extent impact CSR strategies adopted in light of the lack of separation between ownership and control of the business. Murillo and Lozano (2006) and Fassin (2008) also suggest that CSR advancement in SME's is linked to the owner/manager and their value system. Accordingly, Spence (2004) posit that SME entrenchment within their communities and their close nexus to their customers and local community creates a situation whereby the SME owner/manager feels morally obligated to engage in a socially responsible manner.

Murillo and Lozano (2006) and Russo and Perrini (2010) the social capital theoretical perspective closely aligns with SME's approach to CSR. SME entrenchment in the local socio-economicenvironment, their interaction with clients and employees in the local communities make them key contenders where CSR is concerned since a good reputation is essential to achieving competitive advantage (Perrini, Russo, Tencati, & Vurro, 2007; Murillo & Lozano, 2006; Lee, 2008). SME's are also more aware of local practices, associated risks and emerging social issues within their local communities than large corporations. Furthermore, SME owner/managers due to the informal hierarchical structure of the firm are more suited to fostering effective networking, interpersonal relationships and cooperation based on trust of stakeholders and legitimacy (Murillo & Lozano, 2006; Lepoutre & Heene, 2006; Fassin, 2008; Russo & Perrini, 2010). Consequently SME's are more likely to conduct their business in a socially responsible manner, respond to social and community needs and engage in activities, which take into consideration key stakeholder interests Giovanna et al. (2012).

In-spite of the associated benefits, which can be reaped from SME engagement in CSR, evidence shows that many SME's are unaware of CSR initiatives (Jenkins, 2006). A review of the literature highlights that CSR and social community activities undertaken by SME's tend to be fragmented and informal in nature (Maitland, 2002; Perrini et al., 2007; Jamali, Safieddine, & Rabbath, 2009). Lepoutre and Heene (2006) suggest that SME's seldom engage in formal CSR strategies. SME's are also said to encounter greater barriers and challenges in their quest to engage in CSR activities. SME's are prevented from active CSR engagement due to various factors ranging from their size, lack of time and support, lack of financial resources and expertise on CSR issues (Lepoutre & Heene, 2006; Ortiz & Kuhne, 2008; Blomback & Wigren, 2009). For example Hsu and Cheng (2012) state that many SME's consider that they lack requisite skills and guidance in CSR, this is because some of the guidance developed on a global level or sector specific level are not suitable for their purpose. Murillo and Lozano (2006) state that SME's lack knowledge and know-how in relation to CSR issues which act as a barrier to their ability to effectively engage in CSR. Furthermore SME's make little or no attempt to communicate their involvement in CSR related activities or publish annual reports, which discloses their CSR engagements (Murillo & Lozano, 2006; Fassin, 2008; Giovanna, De Massis, & Lucio, 2012). Fassin (2008) however assert that SME's actively engage in socially responsible practices although they do not formally publish or report their activities.

3 Small and Medium Scale Enterprises in Nigeria

CSR is important to all firms large and small. Accordingly, businesses have the responsibility of aligning diverging interests of stakeholders by effectively adopting appropriate strategies. The CSR literature has traditionally focused on large firms and MNC's; consequently, very little empirical research is available on CSR in small firms. Giovanna et al. (2012) state that relatively little is known CSR practices of Small and Medium Sized Enterprises (SME's). Evidence suggests that recent developments and studies on CSR in Africa in general and in particular Nigeria have been mainly promulgated by MNC's as such there is a dearth of academic literature on CSR amongst SME's in Nigeria (Amaeshi, Adi, Ogbechie, & Amao, 2006; Lincoln et al., 2015). This is surprising in light of the number of SME's in various countries around the world. SME's are an important force for economic growth and development in both developed and developing countries. SME's are said to account for more than 50 per cent of employment and about 90 per cent of all businesses worldwide. According to the OECD (2015) the SME sector contributes to 46% of the Nigerian GDP and employs more than 30 million individuals. Furthermore, these firms are key engines of job creation and economic growth in developing countries, particularly following the recent 2007–2008 global financial crises.

Entrepreneurship is a dynamic field particularly as it relates to the process of change creation and vision. The entrepreneurs' vision to recognise opportunities and willingness to take calculated risks are critical factors needed for success (Kuratko & Hodgetts, 2007). Consequently, the importance of the Small and Medium Scale Enterprise (SME) and the appreciation of their role cannot be overemphasised. In Nigeria, entrepreneurship is often the only option for those seeking an income to overcome poverty and unemployment. The unemployment rate in Nigeria is a staggering 23.90 per cent (National Bureau of Statistics, 2014). In light of recent economic downturns there is now recognition that the ability to increase global wealth and wider communitydevelopment is dependent to an extent on successful entrepreneurship in transition economies (Lincoln et al., 2015). The importance of the SME sector to these economies has also been recognised (OECD, 2015). The prominence of the sector is matched by the appreciation of its unique role as one of the most vital contributors to people's income and to economic development and innovation. SMEs are perceived to be the seedbed for indigenous entrepreneurship and are known to generate many small investments, thus encouraging entrepreneurship is a key policy instrument for promoting economic growth and employment creation including social and environmental development in Africa (Woldie et al. 2008). SMEs in Africa represent a large proportion of businesses contributing immensely to the mobilisation of resources, employment and poverty alleviation. Consequently, these firms are seen as dominant forces for economic development and industrialisation in the African region. Due to their small and flexible character, SMEs are able to withstand adverse economic environments and survive where many large enterprises fail.

The Nigerian small and medium enterprise sector is however characterised by low productivity, inability to compete with imports, lack of diversification, unfavourable businessenvironment, infrastructural deficiencies, corruption, low access and high cost of finance and weak institutions (OECD, 2015). For example an OECD Report (2015) states, according to a federal survey of Nigerian micro, small and medium enterprises undertaken in 2010, weak infrastructure, lack of access to finance and inconsistent government policies are the main challenges faced by SME's in Nigeria. The Small and Medium Enterprise Development Agency of Nigerian (SMEDAN) established in 2003 is responsible for devising strategies at federal level to promote, monitor and co-ordinate the development of the SME sector in Nigeria (SMEDAN, 2007). According to the OECD (2015) SEMDAN performs the following activities:

- Information dissemination on markets, inputs and suppliers
- Business development services to help SME's build capacities to run their enterprise sustainably and profitably

- Promotion of clustering and providing access to common facilities for small businesses, such as layouts, incubators and industrial parks
- SME policy design and advocacy, voicing the needs and challenges of SME's to government
- Facilitation of access to finance for SME's

On a state level, each State is also tasked with devising strategies to cater for SME development within the State. Various policy supports have been devised to support SME's in Lagos State such as the Lagos State Development Plan 2012–2015. According to the OECD (2015) in recognition of the need to develop the SME sector, the Lagos State Government has identified the following priority actions namely; developing an SME/informal sector plan; including gender issues in SME/informal sector development with programmes targeted at women entrepreneurs; establishing business incubator centre; assisting SMEs with access to new markets through procurement processes; promoting business linkages between the informal and formal sectors; promoting social responsibility, health and environmental health education among SME's and informal economic enterprises; strengthening micro-credit institutions that service the SME and informal sector; holding regular consultations with SMEs to understand their needs and determine ways of meeting those needs. Other initiatives include streamlining existing regulations on small business to facilitate business creation, survival of new and existing enterprises and increased development of entrepreneurs and establishing information and monitoring systems to keep track of development and evolution within the SME sector.

4 Corporate Social Responsibility and Sustainability in Nigeria

Many countries have witnessed radical change in the relationship between businesses and the society in which they operate. Consequently, counties around the globe are adopting tougher stance on businesses in relation to governance and ensuring that they embrace socially responsible practices within their strategic framework. There is a need to ensure that CSR rules and regulations adopted supports SME development rather than undermining their efforts (UNIDI, 2002). While there are valid arguments for legislative intervention, such as increased compliance and enforceability, increased public trust, promoting transparency and accountability and ensuring uniformed practices across all spheres and promoting fairness and certainty (Gatto, 2002), one cannot discount arguments against legislating which includes high cost and possibility that such legislative provisions would be reactive rather than proactive (Gatto, 2002; Ihugba, 2012). Furthermore, Ihugba (2012) argues that there is possibility of drafting complexities, which may arise from tokenism and ingenuous attempts to pacify lobbyists.

There are various legislative provisions in Nigeria, which regulate business activities and social responsible practices. For example the Nigerian National Environmental standards and Regulations Enforcement Agency Act (2007) contains provisions, which cater for CSR. Section 20 of the Act grants the Agency power to make regulations setting specifications and standards to protect and enhance the quality of Nigeria's air resources, so as to promote public health or welfare and natural development and productive capacity of the nations' human, animal, marine or plant life. Furthermore, Section 25 contains powers to implement Regulations for the purpose of protecting public health and promotion of sound environmental sanitation. Section 27 contains provisions regarding the discharge of hazardous substances and related offences and allows the Agency to impose fines and prison terms to those who contravene the provisions. The Agency is also empowered under Section 30 to enter premises, examine any article, and seize any items believed to contravene any provision of the Act. Furthermore, the Nigerian Criminal Code Act (1990) contains provisions on socially responsible practices. Sections 243 contain provisions with regards the sale of food or drinks. Section 245 deals with fouling water and sections 247 and 248 contain provisions with regards noxious acts and sale of matches made with white phosphorus and use of white phosphorous in manufacturing matches. In addition, the Harmful Waste (special Criminal Provisions) Act (1988) contains relevant provisions and powers with regards carrying, depositing and dumping of harmful waste on land, territorial waters and matters under Section 1. Section 7 focuses on acts committed by body corporate and provides for negligence on the part of a director, manager, and secretary in contravening provisions. Section 10 contains enforcement powers to search, seize and arrest those found to have contravened the provisions contained under the Act. In-spite of the provisions available, law enforcement mechanisms in Nigeria are weak and ineffective. Ineffective institutions and corrupt practices among Nigerian law enforcement agencies further compound the situation.

The Nigerian government in a bid to reform CSR practices put forward a Corporate Social Responsibility Bill pending before the National Assembly since 2008. The purpose of the Corporate Social Responsibility Bill is to bring into effect an Act to establish the Corporate Social Responsibility Commission. Part 3 of the Bill states that the Commission will be tasked with ensuring that companies are accountable not only to employees and their trade unions, but to investors, consumers, host communities and the wider environment and develop environmental guidelines that need to be met by corporations doing business in Nigeria. In addition, the Commission is vested with powers to carry out a variety of function, which include creating a standard for social responsibility of corporate organizations consistent with international standards, serving notices of social responsibility requests to organizations and implementing social and environmental regulations. The Commission is also tasked with corporate organizations classification, ranking them according to organizational size and magnitude of investment, which determines the nature of corporate social responsibility expected of them. Publish annual reports on social and environmental impacts of company's direct activities on communities, develop policies to encourage corporate organizations to undertake communityengagements as part of corporate social responsibility and ensure that companies sponsor cultural and educational activities that offer added value to Nigeria's socio-political and technological development.

The CSR Bill is not without its critics, for example Mordi, Opeyemi, Tonbara, and Ojo (2012) while recognising that the Bill is a welcome development in helping to firmly establish corporate ethics among Nigerian firms, also highlights some of the deficiencies inherent in the Bill. According to Mordi et al. (2012) the Bill fails to provide for extensive enlightenment of the Nigerian society on the importance and benefits of CSR as well as identifying the minimum constitutional duties that are imposed on the Nigerian government. Furthermore the Bill on the face of it has the potential to add to the high cost and complexities of doing business in Nigeria. Part 5 of the Bill grants the Commission power to temporarily shut down and suspend operations of an organization, corporation or a company for a minimum of 15 of 30 working days as a penalty for non-compliance with statutory requirement. The Commission is vested with the power where an organization fails to comply with the statutory requirement of the Act to impose a fine of 2 % of its gross annual profit and on subsequent violation to impose a fine of not less than 3.5 % of its annual gross profit in addition to compliance with statutory corporate social responsibility within the given period. Firms operating in Nigeria face many obstacles and constraints ranging from poor infrastructure such as roads, electricity supply, lack of security, weak enforcement of contracts and high costs of finance Amaeshi et al. (2006), consequently the CSR contributory charge proposed by the Bill could act as a disincentive to investment in Nigeria in light the already high cost of doing business (Mordi et al., 2012). This line of reasoning finds support in research work carried out by UNIDO (2002) who contend that if CSR demands are protectionist, culturally inappropriate or unreasonably bureaucratic, the net effect will be to undermine people's livelihoods.

The Lagos State Climate Change Policy 2012–2014 is another key piece of policy in enhancing sustainability on a State level (OECD, 2015). The policy recognises the adverse effects of climate change on individuals, business and infrastructure. The Policy emphasises the importance of mainstreaming sustainability initiative into the development plans, taking a sectoral approach which focus on key socio-economic sectors more likely to be adversely affected by climate change and/or with a greater mitigation potential. These include but not limited to, energy, water, agriculture and food security, biodiversity and ecosystem services (wetlands, coastal and marine ecosystems, forests, and wildlife), human health, land use and soil, tourism, industry, human settlements, transport and other infrastructure. The policy recognises the need to put in place measures geared towards adapting and enhancing resilience to the adverse impacts of climate change as well as contributing to global mitigation efforts (Lagos State Climate Change Policy, 2012). There is consensus that various challenges exist in preventing effective adoption of CSR and sustainability initiative in Nigeria. For example the Lagos State Climate Change Policy 2012–2014 recognises that challenges exist in developing countries.

According to the policy documents, the following are some of the challenges identified:

1. Inadequate financial resources
2. Inadequate technical capacity
3. Bureaucratic processes
4. High transaction costs
5. Limited required datasets and lack of project monitoring and evaluation
6. Weak institutional capacities
7. Lack of legal and regulatory frameworks

The policy also recognises the following specific challenges plague the SMEs sector in particular:

1. Increasing cost of energy exacerbated by inadequate grid power supply, forcing many businesses to close shop;
2. Low investment in research and infrastructure to stimulate and support the growth of low carbon SMEs.
3. Information and synergy gap between existing research institutions and entrepreneurs;
4. Prohibitively high costs of low carbon technologies for a majority of SMEs.

Accordingly, the policy proposes the following sector specific objective in order to stimulate and support the growth and sustainability of SMEs through research and investment in green energy, infrastructure and green technology, and access to new market and finance opportunities including carbon finance. Consequently the Policy identifies the need to:

- Support the development of low-cost green initiatives and technologies that can be adopted by the SMEs sub-sector and provide incentives for their adoption;
- Promote sustainable waste management practices such as recycling and re-use including waste-energy opportunities that can among other benefits, attract carbon finance;
- Promote green energy technologies such as solar, biomass and wind power in the SMEs sector; and
- Support the co-location of firms in an industrial park that will draw its power from a central generation system and reduce the emissions associated with multiple single-user generated power, as is the present situation.

5 Research Methodology

The study adopts a mixed method approach involving the use of multiple data collection techniques in collecting data (Creswell, Clark, Gutmann, & Hanson, 2003). Questionnaire survey and semi-structured interviews were used in order to collect data for the study. The initial part of the data collection involved the use of a face-to-face questionnaire survey with 371 SME owner/managers in Lagos Nigeria. Bryman and Bell (2011) define a survey as *"a cross-sectional design in relation to which data collected predominantly by questionnaire or by structured interview in order to collect a body of quantitative or quantifiable data in connection with two or more variables, which are then examined to detect patterns of association"* (p.54). The questionnaire uses a combination of closed-ended and Likert scale questions. Closed-ended questions were adopted in relation to the demographic characteristics of the entrepreneurs and their business.

A five point Likert scale was used to capture data in relation to challenges male and female SME owner/managers experience when undertaking CSR related activities, strategies adopted in overcoming these challenges. The advantage of these types of questions rests on the fact that they are easier and relatively quicker to administer. Questions can be answered within a short time, and the respondents can perform with greater reliability the task of answering the questions (Oppenhiem, 1992). The questionnaire is divided into three main sections. The first section contains demographic information of the entrepreneurs. The second section sought to measure the entrepreneur's awareness of CSR initiatives. In particular the questions in this section were designed to assess the entrepreneur's adoption of CSR related practices and readiness to adopt CSR initiatives. The final section of the questionnaire sought to explore challenges of CSR among Nigerian entrepreneurs and their perception on strategies to overcome challenges faced.

As a result of a lack of established data sets and records from which meaningful random sampling could be drawn, convenience sampling was used to identify the entrepreneurs who participated in the study. Convenience sampling involves the use of readily available participants to participate in the study (MacNealy, 1999). Malhotra (2010) notes the basic principle of sampling is that by selecting some of the elements in a population, a researcher can draw his/her own conclusion about the entire population. Consequently, the test of sample design is how well the sample structure represents the characteristics of the population it purports to be a representative of (Cooper & Schindler, 2011), bias inevitably occurs where the number of subjects is too small. Babbie (2010) states that the larger a sample the more representative of the population it becomes, and so the more reliable and valid the results based on it will become. This view is supported by Saunders, Lewis, and Thornhill (2009) who note that the larger the sample size, the lower the likely error in generalising to the population. The decision was therefore made to adopt a larger sample size in order to minimise the potential bias, which can often arise through convenience sampling. Consequently, a sample size of 400 was considered as an adequate and workable sample. Indeed several studies have worked with remarkably fewer samples and have achieved satisfactory results. From previous research done in Nigeria, the highest concentrations of SMEs are thought to be in Lagos, the sample target was set in Lagos state as Lagos is the commercial nerve centre of the country and is the most populated and diverse state in the country. In order to ensure that the sample is representative of the general population, the decision was made to focus the study in areas with a large cluster of small and medium scale enterprises. The sample includes SME firms in different industries, different trading companies and companies in different geographical locations throughout Lagos in order to ensure diversity. In order to ensure validity and reliability, no leading questions were posed to the female entrepreneurs. Furthermore, the questionnaire questions are based on stable sources of literature and sound theoretical underpinning derived from the SME and CSR literature. As Gill and Johnson (2010) rightly puts it, in the case of questionnaire, reliability derives from the clear formulation of questions, which facilitates understanding and hence correct answering. From the 400-questionnaire survey carried out with the female entrepreneurs in Nigeria, a total of 371 were fully completed. Data obtained from the questionnaire was analysed using the Statistical Package for Social Sciences (SPSS 22.0 for Mac). Descriptive statistics such as percentages were used to analyse the information obtained from the SME owner/managers.

The second stage of the data collection involved the use of semi-structured interviews with 64 SME owner/managers who took part in the survey. This sample size is of an acceptable limit for an exploratory study of this kind (Burns, 2000; Jamali, 2009). The owner/managers selected for the semi-structured interviews were chosen for their level of awareness of CSR concept and understanding of CSR challenges and strategies to mitigate the challenges. The rationale for adopting semi-structured interviews rests on the premise that the study being exploratory in nature seeks to provide a narrative of real life occurrences and events from the respondents own perspective (Burns, 2000). The semi-structured interviews provided the researcher with ample opportunity to follow up ideas from the face-to-face questionnaire survey carried out with the SME owner/managers, probe responses and investigate motives and feelings on some of the questions listed on the questionnaire (Bell, 1993). Information obtained from the semi-structured interviews was transcribed in order to make it more presentable and bolster the results obtained from the questionnaire survey.

6 Analysis of Findings

The analysis of findings is divided into two parts. The first part of the finding provides descriptive statistics in form of percentages used to provide a general description of the demographic characteristics of the SME owner/managers obtained from the face-to-face questionnaire survey. The data presented in the second part of the analysis contains findings obtained from both the face-to-face questionnaire survey and the semi-structure interviews carried out with the SME owner/managers.

6.1 Demographic Profile of the Male and Female Entrepreneurs

Findings in relation to gender shows that 51.2 per cent of the SME owner/managers were male, 48.8 per cent were female. The finding shows that majority of the SME owner/managers were in the 31–40 age category representing 53.6 per cent. Findings in relation to academic qualifications show that many of the SME owner/managers representing 46.4 per cent stated that they had obtained a diploma, 16.7 per cent stated that they had obtained a bachelor's degree, 27.2 per cent stated that they had obtained a postgraduate/professional qualification. The finding also shows that 26.1 per cent of the SME owner/managers stated that they had skilled manual experience and a further 27.2 per cent states that they had professional experience and 21.0 per cent stated that they had prior SME experience. 25.6 per cent percent stated that they were unemployed before starting their business. Majority of the SME owner/managers stated that they were motivated to set-up their business because of financial motives representing 48.5 per cent. 35.6 per cent stated that they were motivated to start their business as a result of their desire to be independent representing.

The finding in relation to SME owner/managers firm characteristics shows that many of the firms operated as sole traders representing 61.7 per cent, 32.3 per cent of the SME owner/managers operated as companies. Majority of the SME owner/managers operated in the service sector representing 29.9 per cent. 22.4 per cent of SME owner/managers stated that they operated in commerce and retail sector. 20.8 per cent stated that they operated in the agriculture sector and 16.4 per cent stated that they operated in manufacturing. The finding in relation to firm size shows that the majority of the SME owner/managers were small and employ between 1 and 5 employees representing 41.0 per cent. 27.2 per cent employed 11–20 and 21.8 per cent employ more than 21 employees. 48.8 per cent of the SME owner/managers stated that their firms were between 11 and 20 years of age and 30.5 per cent stated that their firms were between 1 and 5 years of age.

6.2 Perception of SME Owner/Manager on CSR Practices in Nigeria

Majority of the owner/managers stated that socially responsible activities implemented within their firm were done on an informal basis. According to Jenkins (2004) SMEs are likely to have an informal, trusting, and personally engaging approach to CSR. The finding is supported by research work carried out by various authors who found that CSR activities of SME's tend to be informal in nature and often times fragmented (Maitland, 2002; Perrini et al., 2007; Jamali et al., 2009). The owner/managers who took part in the study stated that the informal and often times piece meal nature of corporate social responsibility in

Nigeria makes it less likely that businesses will embrace the concept. When asked their perception on CSR activities they engage in, 88.4 per cent of male owner/managers and 60.2 per cent of female owner/managers stated that their firms engage in activities related to environmental protection such as waste management, pollution prevention, water use efficiency, energy use optimization, recycling, refuse collection and disposal. 33.7 per cent of the male owner/managers compared to 9.9 per cent of the female owner/managers stated that they adopted policy statement, which measure and evaluate CSR related activities. Some of the SME owner/managers stated:

> "There is no formal approach to CSR among many Nigerian SME's. We do what we can but many of the practices are informal...people need more awareness of the concept in order for it to take root...there are challenges in doing business in Nigeria, many entrepreneurs are faced with more dire problems... We try however way we can to help support our local community" (Female SME owner/manager)

> "More formal approach is required with government institutions and NGO's taking a more proactive step in setting out formal rules and educating Nigerians on the concept" (Male SME owner/manager)

> "There are no rules or enforcement really, many SME's act in a responsible manner because they want to do so, it is voluntary effort on their part. I do not think strict enforcement will work in a place like Nigeria. What we need is more awareness of what is need and also substantial support from the government" (Female SME owner/manager)

When asked about their perception of CSR, some of the owner/managers who took part in the interview stated:

> "Corporate social responsibility involves active engagement within your community and ensuring that you carry out your business in a moral and ethical manner... your business practices need to add value to Nigeria as a whole" (Female SME owner/manager)

> "...It includes how I interact with my employees, customers, community and the environment" (Female SME owner/manager)

The findings show that the owner/managers agree with the statement that their ethical, moral and personal values and their religious and cultural beliefs were primary drivers of their CSR engagement. 70.2 per cent of the female owner/managers and 55.8 per cent of the male owner/managers stated that religious beliefs impact on their engagement in socially responsible practices. 60.2 per cent of the female owner/managers and 55.3 per cent of the male owner/managers stated that Nigerian cultural traditions impact their engagement with CSR, while 70.2 per cent of female owner/managers and 66.8 per cent strongly agreed that societal values impact on their firm's engagement in socially responsible practices. More male owner/managers (56.9 per cent) compared to female owner/managers (50.2 per cent) strongly agree that the state of the Nigerian economy makes CSR engagement challenging. In addition more female owner/managers (40.3 per cent) compared to male owner/managers (22.1 per cent) stated that they strongly agree that the legal system impact on their engagement in socially responsible practices.

The finding supports the notion that countries such as Nigeria have unique differences from Western countries due to their culture, religion and political economy, as such CSR initiatives and practices would to an extent be mirrored by socio-cultural norms thus CSR practices may be quite different from those practiced by their Western counterparts. The result finds support in previous research, which suggests that SME actively engage in philanthropic activities for example through religion/church (Giovanna et al., 2012). The result is not surprising, for example the finding in relation to religion as a driver of CSR is to be expected as many Nigerians openly profess some type of religion and religious associations are very dominant in many aspects of people's lives. Amaeshi et al. (2006) suggest that CSR as philanthropy in the Nigerian context could be connected to religious influences. In light of the importance attached to gifts and sacrifices within certain religion, such practice could be easily transposed into business understanding and practices. Furthermore the high responses received from the SME owner/managers in relation to ethical, moral and personal values as key drivers of CSR find support in the seminal article by Quinn (1997). Some of the SME owner/managers involved in the study stated:

> "Social responsibility deals with morality and charity, businesses do not function in isolation... Businesses need to act in an ethical and moral way towards their customers and local community" (Female SME owner/manager)

> "Nigerians are very charitable, one can say that Nigerians engage in some form of socially responsible practice. They also actively engage in the local community through church groups and so on" (Male SME owner/manager)

> "In Nigeria, social responsibility usually involves charitable contributions and also sponsorship through education or training in order to help contribute to people's lives and the community at large" (Female SME owner/manager)

7 Challenges Experienced by SME Owner/Managers When Undertaking CSR

The finding obtained from the male and female SME owner/managers shows that various challenges were identified when undertaking CSR related activities. 66.8 per cent of male SME owner/managers compared to 59.6 per cent of female SME owner/managers stated that lack of time was a challenge to effectively engage with CSR. 49.7 per cent of the female owner/manager stated that lack of know-how and expertise was a major challenge compared to 44.8 per cent of male owner/managers. Furthermore, more female owner/managers representing 70.2 per cent stated that lack of efficient government and NGO institutions was a major challenge compared to 55.8 per cent of male owner/managers. The perception echoed by the SME owner/managers is supported by past research which asserts that SMEs lack the resources needed for CSR and are prevented from active CSR engagement due to factors ranging from their size, lack of financial resources and expertise on CSR issues. Various writers note that SMEs by their nature experience specific barriers and challenges. Factors such as time, human and financial constraints act as a barrier preventing them from undertaking CSR related activities (Lepoutre & Heene, 2006; Ortiz & Kuhne, 2008; Blomback & Wigren, 2009). In-spite of the above mentioned challenges some SMEs are found to be more environmentally responsive compared to large businesses (Keinert, 2008).

More female owner/managers (49.7 per cent) compared to male owner/managers (22.6 per cent) stated that lack of understanding of CSR related issues and their benefits possess as a major challenge in undertaking CSR related activities. There was consensus among the owner/managers that lack of financial resources was a major challenge representing 79.6 per cent of female owner/managers and 77.9 per cent of male owner/managers. Challenges such as accessing financial, know-how, technical and technological resources hinder progression. The Lagos State Climate Change Policy, 2014 recognises some of the challenges identified. The Policy cites the following as potential challenges affecting CSR initiative among SMEs, inadequate financial resources, technical capacity, and bureaucratic processes, high transaction costs, weak institutional capacities and lack of legal and regulatory frameworks. In light of the aforementioned, it is not surprising that the Lagos State Climate Change Policy, (2014) proposes SME sector specific objective in order to stimulate and support the growth and sustainability of SMEs through research and investment in green energy, infrastructure and green technology, and access to new market and finance opportunities.

8 SME Owner/Managers Perception on Strategies to Overcome CSR Related Challenges?

The SME owner/managers were asked their perception on how to overcome the challenges faced. Many of the owner/managers stated that the provision of financial initiatives, training and educating SMEs, networking and the need for more concerted efforts from government institutions and NGOs would help in alleviating some of the challenges. Some of the owner/managers involved in the interview stated:

> "Networking is key to ensuring CSR principles are promoted and shared in Nigeria. Like-minded business people can use this opportunity to share best practice. Besides, those who are new to the concept can be enlightened by those who already incorporate it into their business practices" (Male SME owner/manager)

> "Financial assistance would help ensure many of the key CSR practices are embraced by more Nigerians. The Nigerian government is weak where corporate social responsibility is concerned. There is a clear lack of initiatives, lack of understanding of the concept and a lack of resources to help in making CSR a key area within business and society" (Male SME owner/manager)

"There are financial implications and this is a major challenge. Many SME's are not in a position to meet some of the standards. Nigerian government needs to work towards providing incentives for businesses to act in a more responsible manner" (Female SME owner/manager)

"The cost of being socially responsible is high in an environment like Nigeria. Many businesses in the first instance have to provide their own resources such as water, electricity and so on. Assistance is needed in order to ensure that these businesses embrace socially and environmentally friendly practices. . .many of the businesses will not survive, the cost of compliance is too high" (Female SME owner/manager)

"Most people are not aware of the CSR concept, we need to promote CSR through education and training in Nigeria. Awareness of CSR is lacking, people are not really aware of corporate social responsibility, the government can work towards promoting it by putting in place workshops and training facilities to educate SME owner/managers about CSR and put measures in place to monitor it and also provide financial assistance for those businesses who want to take part" (Female SME owner/manager)

"There is more to CSR and public awareness through more education and training will go a long way in ensuring people and businesses are more informed" (Male SME owner/manager)

"SMEs need to be willing to dedicate their time and effort to promote CSR practices. Many SME's struggle to cope with the myriad challenges often encountered when doing businesses in Nigeria and so it leaves little time to focus on CSR. If they are able to devote time then they may be able to actively engage in CSR practices" (Female SME owner/manager)

"The state of infrastructure in the country makes it more challenging to engage with more meaningful forms of social responsibility practice. If the state of infrastructure were better then there would be an argument for ensuring that business practices are carried out in a socially responsible way. SME's are left to struggle alone. The added cost of being socially responsible cripples many businesses and this is why they do not fully engage" (Male SME owner/manager)

"In order to overcome the challenges, resources needs of businesses must be met. SME's need resources in order to enable them embrace socially responsible practices, financial incentives will go a long way in aiding compliance. In addition, provisions of useful information will help in promoting knowledge and awareness of CSR. We also need institutions who know what they are talking about and can offer assistance" (Male SME owner/manager)

"The legal regulatory environment needs to be more effective in promoting CSR related activities. . .the legal regulatory environment must also be effective in ensuring compliance. Nigeria has more pressing issues, which require attention. . .unfortunately CSR is at the bottom stage in Nigeria" (Male SME owner/manager)

"Clearer understanding of CSR and trainingfacilities are needed to educate people and promote wide spread understanding among Nigerians. SME's cannot be left to do this alone. The cost of doing business in Nigeria is astronomical. This will only add to the already high cost, if some form of incentive is provided to help reduce cost then this must be considered by the government and Ngo's" (Male SME owner/manager)

> "Corporate social responsibility needs to be introduced in schools and universities so that it is ingrained in young Nigerians and they can easily identify what practices are socially responsible and which are not" (Female SME owner/managers)

The finding shows that the SME owner/managers (66.8 per cent male) and (50.3 per cent female) engaged in training and development activities within the firm. 77.9 per cent of the male owner/managers and 30.4 per cent female owner/managers also stated that they support social activities through employee involvement in community activities. The SME owner/managers agreed that adopting effective CSR practices is more challenging in a country like Nigeria. They also stated that there are specific constraints in the Nigerian institutionalcontext that affect the evolution of socially responsible practices where SME's are concerned. Some of the SME owner/managers stated:

> "SME's are essential in this as they are the ones who operate in the communities and are closer to the customers and those living within the local communities. Nigeria has many small businesses operating in these communities. It is in everyone's interest if SME's are provided with the necessary resources and tools to be able to promote responsible practices" (Female SME owner/manager)

> "We are more familiar with the environments and the communities. Many of us and our employees live here, we are from the local community, and so it is in our own best interest to act responsibly" (Female SME owner/manager)

> "SME's are powerless to do much unless they are provided with necessary resources, many of the businesses are trying to survive. They need help otherwise it is futile" (Female SME owner/manager)

The SME owner/managers who took part in the interviews were asked their perception on how SME's in Nigeria could contribute to the promotion of CSR in their own business practices regardless of some of the challenges experienced. The SME's felt that more needs to be done to ensure a higher percentage of businesses adopt CSR policy statements, which measure and evaluate CSR. Some of the SME owner/managers stated:

> "Businesses need to promote more uptake of CSR practices from senior personnel in order for the concept to trickle down to the employees" (Female SME owner/manager)

> "Our business includes sustainability goals into the personal objectives of our senior employees to ensure everyone is accountable for successfully driving change and this is reflected in our business practices" (Male SME owner/manager)

> "Customers react positively where sustainability programs are part of the firms objective, this can improve the efficiency of a business and bring about much needed repeat custom. Nigeria is working its way towards achieving a system where ethics drives business. We are not there yet" (Male SME owner/manager)

> "In order to bring about change, every one in the business will need to get involved. Many of our employees are in the front-line interacting with customers, suppliers and the community. There is a greater chance of success where employees engaged and are provided with skills through training" (Male SME owner/manager)

9 Conclusion

The need to ensure businesses act in a socially responsible manner and make sure that sustainability, social and environment concerns are at the core of a firms' strategic agenda is now widely recognized. The paper presented findings obtained from questionnaire surveys and semi-structured interviews carried out with male and female SME owner/managers in Nigeria. The findings show similarities and differences in male and female owner/manger response on CSR practices and challenges in their CSR activities. The study identifies the need for an effective CSR policy in Nigeria, which clearly outlines essential social responsibility practice and strategies. More guideline is needed where CSR is concerned so as to provide a benchmark on which current practices can be measured.

Various reports have confirmed that engaging in CSR has positive ramifications on business performance and society as a whole. Consequently, concerted effort is required from the Nigerian government, NGO's and businesses both large and small in ensuring societal and environmental concerns and sustainability issues are at the forefront of company and socio-economicstrategy. The difficulties, which often arise within the Nigerian setting is the lack of co-ordinated attempt between Federal and State programmes. This study shows that existing policies, programmes, actions and measures in Nigeria are insufficient in addressing CSR and sustainability issues. Weak technical capacity and lack of appropriate institutional framework and governance instruments particularly at the national level are additional challenges. It is essential that SME support mechanism devised both at State and Federal level should complement active efforts to establish a sound environment for SME's to engage with CSR related activities.

There is a need to ensure more support from government institutions in promoting the importance and benefits associated with CSR. Current assistance available is insufficient and more wide reaching initiatives are needed. There is a need for stronger emphasis on the SME and informal sector development and growth. There is also a need to address skill gaps and ensure close linkages promoted through wider government efforts in capacity building and investments in education and training. Training and facilitation can be invaluable in encouraging linkages between SME's, larger companies and local communities. In addition there is a need to continue to engage with SME owner/managers in order to better appreciate the challenges faced when engaging in CSR.

This study has several limitations first; the study focus is on Lagos State and may not be representative of other States in Nigeria. Future research may focus on comparison across different industrial sectors and States in Nigeria. The study adopts a qualitative approach to data analysis in order to provide SME owner/managers perception of CSR issues in the Nigerian context. Further research would benefit from adopting quantitative data analysis technique to identify relationships between the variables.

References

Aguilera, R. V., Filatotchev, I., & Gregory Jackson, H. G. (2008). An organisational approach to comparative corporate governance: Costs, contingencies and complementarities. *Organisation Science, 19*(3), 475–492.

Amaeshi, K., Adi, B., Ogbechie, C., & Amao, O. (2006). *Corporate social responsibility in Nigeria: Western mimicry or indigenous influence?* Available at SSRN: http://ssrn.com/abstract=896500 or http://dx.doi.org/10.2139/ssrn.896500.

Babbie, E. (2010). *The basics of social research* (5th ed.). Belmont, CA: Cengage Learning, Wadsworth Publishing Company.

Bell, J. (1993). *Doing your research project*. Buckingham: Open University Press.

Blomback, A., & Wigren, C. (2009). Challenging the importance of size as determinants for CSR activities. *Management of Environmental Quality: An International Journal, 20*(3), 255–270.

Bryman, A., & Bell, E. (2011). *Business research methods* (3rd ed.). Oxford: Oxford University Press.

Burns, R. B. (2000). *Introduction to research methods* (4th ed.). French Forest, Australia: Pearson Education.

Carney, M., Gedajlovic, E., & Sujit, S. (2011). Corporate governance and stakeholder conflict. *Journal of Management and Governance, 15*, 483–507.

Carroll, A. B. (1991). The pyramid of corporate social responsibility: Toward the moral management of organizational stakeholders. *Business Horizons, 34*, 39–48.

Cooper, D. R., & Schindler, P. S. (2011). *Business research methods* (11th ed.). Boston: Irwin Publishers.

Coppa, M., & Siramesh, K. (2013). Corporate Social Responsibility among SME's in Italy. *Public Relations Review, 39*, 30–39.

Creswell, J., Clark, V., Gutmann, M., & Hanson, W. (2003). Advanced mixed methods research designs. In A. Tashakkori & C. Teddlie (Eds.), *Handbook of mixed methods in social and behavioral research*. Thousand Oaks, CA: Sage.

Donaldson, T., & Preston, L. E. (1995). The stakeholder theory of the corporation: Concepts, evidence and implications. *Academy of Management Review, 20*(1), 65–91.

Fassin, Y. (2008). SMEs and the fallacy of formalising CSR. *Business Ethics: A European Review, 17*(4), 364–378.

Freeman, R. E. (1984). *Strategic management: A stakeholder approach*. Boston: Pitman.

Freeman, R. E. (1994). The politics of stakeholder theory: Some future directions. *Business Ethics Quarterly, 4*(4), 409–421.

Gatto, A. C. (2002). *The European Union and Corporate Social Responsibility: Can the E U contribute to the accountability of multinational enterprise*? Institute of International Law, Working Paper No. 32, September. Retrieved from https://www.law.kuleuven.be/iir/nl/onderzoek/wp/WP32e.pdf (February 14, 2011)

Gill, J., & Johnson, P. (2010). *Research methods for manager* (4th ed.). London: Paul Chapman.

Giovanna, C., De Massis, A., & Lucio, C. (2012). Corporate Social Responsibility: A survey among SME's in Bergamo. *Procedia- Social and Behavioral Sciences, 62*, 325–341.

Harmful Waste (special Criminal Provisions) Act. (1988). Available at: http://www.placng.org/new/laws/H1.pdf (Last accessed 06/04/2015).

Hsu, J., & Cheng, M. (2012). What prompts small and medium enterprises to engage in Corporate Social Responsibility? A study from Taiwan. *Corporate Social Responsibility and Environmental Management, 19*(5), 288–305.

Ihugba, B. (2012). Compulsory regulation of CSR: A case study of Nigeria. *Journal of Politics and Law; Canadian Center of Science and Education, 5*(2), 68–81.

Jamali, D. (2009). Constraints and opportunities facing women entrepreneurs in developing countries. *Gender Management Journal, 24*(4), 232–251.

Jamali, D., Safieddine, A., & Rabbath, M. (2009). Corporate governance and corporate social responsibility: Synergies and inter-relationships. *Corporate Governance: An International Review, 16*(5), 443–459.

Jenkins, H. (2004). A critique of conventional CSR theory: An SME perspective. *Journal of General Management, 29*(4), 37–57.

Jenkins, H. (2006). Small business champions for Corporate Social Responsibility. *Journal of Business Ethics, 67*, 241–256.

Keinert, C. (2008). *Corporate Social Responsibility as an International Strategy*. Heidelberg: Physica-Verlag.

Kuratko, D. F., & Hodgetts, R. M. (2007). *Entrepreneurship: Theory and process, practice* (7th ed.). Mason, OH: Thompson South Western Publishing.

Lagos State Climate Change Policy. (2012). Available at: http://developmentdiaries.com/wp-content/uploads/2013/03/Lagos-State-Climate-Change-Policy-First-Draft_27.03.2012.pdf (Accessed 28/05/2015).

Lee, M. (2008). A review of the theories of corporate social responsibility: Its evolutionary path and the road ahead. *International Journal of Management Reviews, 10*, 53–73.

Lepoutre, J., & Heene, A. (2006). Investigating the impact of firm size on small business social responsibility: A critical review. *Journal of Business Ethics, 67*, 257–273.

Lincoln, A., Fields, R., & Adedoyin, O. (2013). Chairman/CEO duality and its effects for the management of corporate governance in Guayana. *International Journal of Business and Management, 8*(20), 36–47.

Lincoln, A., Adedoyin, O., & Croad, J. (2015). Fostering Corporate Social Responsibility among Nigerian small and medium scale enterprises. In S. O. Idowu (Ed.), *Corporate Social Responsibility: A global dimension of key initiatives implemented by corporate entities*. Germany: Springer.

MacNealy, M. (1999). *Strategies for empirical research in writing*. New York: Longman.

Maitland, A. (2002, May 16). Affordable responsibility: Small suppliers. *Financial Times, 14*.

Malhotra, N. K. (2010). *Marketing research: An applied orientation* (6th ed.). Indiana, USA: Prentice Hall.

Mallin, C. A. (2015). *Corporate governance* (5th ed.). Oxford: Oxford University Press.

McGuire, J. W. (1963). *Business and society*. New York: McGraw-Hill.

Mordi, C., Opeyemi, I., Tonbara, M., & Ojo, S. (2012). Corporate Social Responsibility and the legal regulation in Nigeria. *Economic Insights—Trends and Challenges, LXIV*(1), 1–8.

Murillo, D., & Lozano, J. (2006). SMEs and CSR: An approach to CSR in their own words. *Journal of Business Ethics, 67*, 227–240.

Muthuri, J. (2012). Corporate Social Responsibility in Africa definition, issues and processes. In T. Lituchy, B. J. Punnett, & B. Puplampu (Eds.), *Management in Africa macro and micro perspectives*. New York: Taylor & Francis Ltd.

National Bureau of Statistic. (2014). *Nigerian unemployment rate*. Available at: http://www.tradingeconomics.com/nigeria/unemployment-rate

Nigerian Criminal Code Act. (1990). Available at: http://www.nigeria-law.org/Criminal%20Code%20Act-Tables.htm (Last accessed 18/05/2015)

Nigerian National Environmental Standards and Regulations Enforcement Agency Act. (2007). Available at: http://faolex.fao.org/docs/pdf/nig120569.pdf (Last accessed 01/03/2015).

OECD Investment Policy Reviews OECD Investment Policy Reviews: Nigeria. (2015). OECD Publishing. Available at: https://books.google.co.uk/books?id=A_BzCQAAQBAJ&pg=PA248&lpg=PA248&dq=Supervisory+Intervention+Guideline+%282011%29+Nigeria&source=bl&ots=oo_3dSIWPG&sig=pVrAinBnKTam4hNLW334daSbJR4&hl=en&sa=X&ved=0ahUKEwiT3dDts8rLAhUI0RQKHRpHCa8Q6AEIPDAF#v=onepage&q=Supervisory%20Intervention%20Guideline%20%282011%29%20Nigeria&f=false (Accessed November 2015).

Okeahalam, C. C., & Akinboade, O. A. (2003). *A review of corporate governance in Africa: Literature, issues and challenges*. Paper presented for the Global Corporate Governance Forum, 1–34.

Oppenhiem, A. N. (1992). *Questionnaire design and attitude measurement*. London: Heinemann Publishers.

Ortiz, A., & Kuhne, S. (2008). Implementing responsible business behavior from a Strategic Management Perspective: Developing a framework for Austrian SME's. *Journal of Business Ethics, 82*, 463–475.

Perrini, F., Russo, A., Tencati, A., & Vurro, C. (2007, July). *Going beyond a long-lasting debate: What is behind the relationship between corporate social and financial performance*? Working paper, Bocconi University, Milano.

Quinn, J. J. (1997). Personal ethics and business ethics: The ethical attitudes of owner/managers of small business. *Journal of Business Ethics, 16*(2), 119–127.

Russo, A., & Perrini, F. (2010). Investigating stakeholder theory and social capital: CSR in large firms and SMEs. *Journal of Business Ethics, 91*, 207–221.

Saunders, M., Lewis, P., & Thornhill, A. (2009). *Research methods for business students* (5th ed.). London: Pitman.

SMEDAN. (2007). *National policy on micro*. Abuja: Small and Medium Enterprises.

Spence, L. (2004). Small firm accountability and integrity. In G. Brenkert (Ed.), *Corporate integrity and accountability* (pp. 115–128). London: Sage.

Sweeney, L. (2007). Corporate social responsibility in Ireland: Barriers and opportunities experienced by SMEs when undertaking CSR. *Corporate Governance, 7*(4), 516–523.

United Nations Industrial Development Organization (UNIDO). (2002). *Corporate Social Responsibility: Implications for small and medium enterprises in developing countries*. Available from http://www.unido.org/fileadmin/import/userfiles/puffk/corporatesocialresponsibility.pdf (Last accessed 28/04/2015).

Woldie, A., Leighton, P., & Adesua, A. (2008). Factors influencing small and medium enterprises (SME's) in Nigeria: An exploratory study of owner/manager and firm characteristics. *Journal of Banks and Bank System, 3*.

Corporate Social Responsibility in Kenya: Blessing, Curse or Necessary Evil?

Thomas Kimeli Cheruiyot and Daniel Kipkirong Tarus

1 Introduction

The concept of Corporate Social Responsibility (CSR) has been characterized by contestations and operational lacunas. This makes the concept unclear, fuzzy and subject to a multiplicity of interpretations (Idemudia, 2014; Okoye, 2009). Furthermore, CSR is understood differently in diverse contexts, which results in its underdevelopment (Visser, 2008). This has led to confusions and difficulties in efforts to implement the CSR agenda.

Despite the increasing interest in CSR research in different contexts and situations, the role of CSR in developing countries, and in particular during challenging times, has been neglected. Such challenges include the following: financial crisis, climate change, political upheavals, extreme poverty, and conflict. Indeed, most studies on CSR in challenging times have focused mainly on financial crisis. It has been reported that attention to CSR has increased in the recent past following serious financial scandals and the collapse of large corporations in Europe and the United States of America (USA).

Since these challenging environments are more prominent in certain contexts (e.g. developing countries) than others, the need for CSR studies in such contexts is critical. The issues facing developing countries have been worsened by challenges surrounding political, social, economic, legal and environmental issues in some developing countries. Thus, CSR calls for a range of strategies, depending upon different contexts and situations. For instance, while in times of economic crisis organizations could focus on cost reduction strategies, other challenging circumstances may call for alternative strategies. Inevitably, most organizations reduce their expenditure on CSR during times of financial crunch. However, some

T.K. Cheruiyot (✉) • D.K. Tarus
School of Business and Economics, Moi University, Eldoret, Kenya
e-mail: cherutho@gmail.com; cherutho@mu.ac.ke

S.O. Idowu et al. (eds.), *Corporate Social Responsibility in Times of Crisis*,
CSR, Sustainability, Ethics & Governance, DOI 10.1007/978-3-319-52839-7_9

organizations have managed to pursue CSR activities in spite of financial difficulties, and for good reasons. For instance, it is argued that pursuing environmental CSR initiatives in times of crisis and recession helps firms to cut costs and increase their CSR budgets without changing their prices (Quelch & Jocz, 2009).

Furthermore, CSR is founded on the premise that organizations interact with economic, social, cultural, environmental and legal systems because business affects and is affected by wider societal interests. The definition of CSR may vary from context to context, due to the differing focus, approach and expectations of stakeholders. CSR is generally used to describe the complex and multi-faceted relationship between business and society, and the accountability of business and business activity in terms of its social and environmental impact (Tarus, 2015). Thus, in developing countries, CSR is perceived as the formal and informal ways in which businesses contribute to improving the governance, social, ethical, labor and environmental conditions of the countries in which they operate, while remaining sensitive to prevailing religious, historical and cultural contexts (Visser, 2005).

CSR's breadth and flexibility has led to an operational and conceptual lacuna, although CSR is becoming increasing prominent in academic circles. Accordingly, no single universal definition of CSR is agreed upon in Kenya. Every practitioner and scholar perceives CSR differently and yet claims to know what CSR is or ought to be. However, most of what is perceived to be CSR in Kenya is actually not so. This scenario leads to serious local abuses of CSR. For instance, most CSR initiatives involve cash, voucher, food or other material donations, but these are purely philanthropic kneejerk responses that are not based on long term social perspectives., CSR is thus organizations' long-term commitment to social, economic, legal and environmental rights and responsible outcomes for the sustainability of humanity. Arising from the aforementioned, the purpose of this chapter is to succintly espouse the state of CSR in Kenya in such times, and ultimately critically examine whether it is a blessing, curse or necessary evil.

The chapter is organized as follows: we begin by discussing the Kenyan context, and what constitutes CSR in such a context; next, we examine the social, economic, environmental and politicalchallenges faced by corporate organizations operating in Kenya in pursuit of CSR and inherent status of CSR as either a blessing, curse or necessary evil; then, we examine the methods used by the Government of Kenya, SME's, International Organizations and NGO's to mitigate the identified challenges; finally, we derive some theoretical, practical and research implications from the results of the study.

2 The Kenyan Context

Kenya is an East African country that gained independence in 1963. It is a regional hub for trade and finance, with a population of approximately 46.2 million and an area of approximately 582.646 km^2. Kenya's Gross Domestic Product (GDP) is approximately US$22.78 billion, with an annual growth of 5.8% (World Bank,

2016), and a recently declared low middle-income status. It has a fairly good infrastructural facilities and services sector. Agriculture makes up 19% of GDP, industry 18%, and services 62.6%. The country is the world's third largest exporter of tea. Other important foreign exchange earners include coffee, horticulture, tourism, and mining. Thus, Kenya has unique economic, political, social, cultural, and regulatory conditions relevant to CSR. After independence in 1963, the government promptly declared war on poverty, disease, and illiteracy. Since then, Kenyans live two decades longer, fertility and infant mortality have been cut by half, and school attendance has more than doubled (World Bank, 2016). GDP per capita has increased eightfold, and the financial sector is now the third largest in sub-Saharan Africa, after South Africa and Nigeria.

However, despite this positive outlook, the country's socio-economic prospects face serious challenges as the country enters an '*adolescence age*' characterized by crisis and challenges including unprecedented political intolerance and insecurity, persistent poverty and high unemployment, rapid urbanization, persistent gender inequality, climate change, corruption, social decay and environmental degradation. Kenya's human development indicators have remained relatively low. The country has a vibrant multi-party democracy and active civil society groups.

Thus, a great deal of reflection is needed on how to transform the lives of the majority of Kenyans. Nearly four in ten Kenyans still live in poverty, many of them youths, women, and elderly people. Maternal mortality is among the highest in Africa, despite a workforce that is more educated than the African average. Although GDP growth is relatively solid, and Kenya is classified as lower middle-income country, this growth has yet to reach the takeoff level necessary to transform Kenya into a modern market economy and to change the lives of ordinary Kenyans (World Bank, 2016).

While Kenya's poverty level has declined, inequalities and regional disparities in wealth distribution remain high. Better macro-economic conditions in the past decade have helped to improve the welfare of Kenyans, but the poor remain susceptible to drought and climate induced vulnerability. Progress has been made across other dimensions of social development, including school attendance and the enrollment ratio between boys and girls. The educationtransition from one level to another has improved due to increased public and private investment in the education sector, child mortality has improved, and immunization coverage has increased. Access to electricity, clean drinking water and better sanitation has increased steadily, but coverage remains low.

On governance, Kenya continues to rank poorly in the Transparency International Corruption Perception index. Recent political reforms have strengthened Kenya's governance record, though it remains mixed. Kenya still remains vulnerable to external and domestic risks, and it continues to underperform relative to its potential and vision 2030 target, as well as compared with its peers in the East African Community and in Sub-Saharan Africa. Furthermore, there is emerging fiscal pressure resulting from the implementation of devolution, a key plank in addressing inequalities in the country that is at the heart of the new constitution as a key vehicle for addressing spatial inequalities (World Bank, 2016). Devolution promises to transform Kenya, through accountable and transparent institutions, inclusive growth, and the equitable delivery of public goods and services. But the sharing of power and resources between national and county governments continues to pose a major challenge.

Kenya's prospects for success are promising: the country has the largest economy in East Africa, with a significant potential to further capitalize on regional markets and to strengthen its position as the region's economic powerhouse. With over 40% of the population under the age of 15, Kenya stands to benefit from a significant demographic dividend as this population enters the workforce, provided that sufficient jobs can be found in time. The Kenya Vision 2030 development plan (vision, 2007) identifies three key pillars: economic, political, and social. Moreover, the recent discovery of oil, gas, and coal represents a great opportunity for Kenya's overall development, which may propel it further towards middle-income country status in the medium term, given the impetus provided by the Government's vision 2030 social and economic transformation framework.

3 CSR in Kenya

While CSR is a contested concept (Moon, 2002), its operationalization, implementation and application is context specific (Cheruiyot & Tarus, 2016). Mainstream CSR studies in Kenya, though minimal compared with similar research conducted in South Africa and Nigeria, have focused on institutional analysis (Muthuri & Gilbert, 2011), sectors prone to social irresponsibility, such as employees in the horticultural sector and tourist hotels (Cheruiyot & Maru, 2012; Opondo, 2006), human rights (Cheruiyot & Maru, 2014), CSR in Kenya reflections and implications (Cheruiyot & Tarus, 2016), and the general impact of CSR engagement on firm performance (Tarus, 2015).

Thus, CSR research in Kenya, as in other developing countries, is relatively underdeveloped, relies heavily on case studies, and focuses prominently on multinational companies (Cheruiyot & Tarus, 2016; Muthuri, Matten, & Moon, 2009). More succinctly, CSR in Kenya has been defined as the long-term commitment of organizations to social, economic, legal and environmental rights and responsible outcomes for the sustainability of humanity (Cheruiyot & Tarus, 2016). Furthermore, CSR in Kenya is characterized by voluntary actions of a philanthropic nature and is weakly regulated. In the Kenyan political-legal and socio-cultural nexus, four approaches to CSR have been identified: the political, ethical (Garriga & Melé, 2004), altruistic and philanthropic CSR (Cheruiyot & Tarus, 2016).

In our view, CSR in Kenya, like elsewhere in Africa and the developing world, is critical and controversial, and is subject to varied interpretations, conceptualizations and operationalization. This state of confusion can be attributed to its relative underdevelopment in theoretical terms and its inappropriate contextual application. There are several perspectives or motives that account for the development of CSR in Kenya. These contextual perspective include, but not limited to, the various developmental stages a country undergoes. In Kenya, one of the greatest challenges is that the country has entered an unprecedented stage of development. This is explored in the next section.

4 Is Kenya Facing Adolescence Age Crisis?

In this chapter, we argue that Kenya is facing challenging times, after 50 years as a republic. This era is akin to a human life cycle stage called adolescence. Thus, we have typified this Kenyan era as "Kenya at adolescence-age." Like the human developmental stage, it is crisis ridden. As a nation, the crisis is characterized by, among other things, unprecedented political intolerance and insecurity, economic difficulties and greed amid economic growth, persistent poverty and high unemployment, rapid urbanization, persistent gender inequality, climate change, corruption, social decay, and environmental degeneration. These factors arise in the midst of resilience, optimism, an enlightened population, new opportunities, and promising prospects in the country. Inherent in these factors is whether or not CSR is a blessing, curse or necessary evil. The following are the political, economic, social, legal and environmental challenges that characterize the country's developmental stage and ultimately characterize CSR either as blessing, curse or necessary evil.

Political Challenges
Political CSR stresses the power of companies in society over other actors and their corresponding responsibility in the political arena. As such, it articulates the presence of firms in relation to other political actors, such as civil society, the political elite, community leaders, and the public. Large local and multinational companies exert influence on the country's political environment by providing corporate donations and employment opportunities. Thus, examining political CSR at the macro level, as a framework for organizations to engage with the public and the state for sustainability is necessary, but faces challenges that may impede achievement of the overall goal. Hereinafter, some political challenges faced by corporate organizations operating in Kenya in pursuit of CSR are described.

First and foremost is weak *democracy and poor politicalgovernance*. Kenya is largely an open society, with democratic systems and good political governance structures, which at times is treated by some political actors as dispensable, especially in the pursuit of political power. This was more pronounced previously, before the advent of political pluralism in the country. Weak democracy and poor political governance fails to provide an enabling environment for business to implement social programs. Organizations face this challenge as they engage with political actors in search of support for their CSR programs.

Furthermore, a *high-level of corruption* impedes CSR implementation. In Kenya, corruption remains high. For instance, during the period 1996–2015, the Kenya corruption index stood at 22.16 points out of 100, as reported by Transparency International (where 0 is very corrupt and 100 very clean). Kenya has continued to score poorly in the global corruption index over the years. The country's state of corruption is important. First, corruption creates an impetus for the misplacement of social priorities, as actors corruptly influence CSR projects. Secondly, it creates an enabling environment for unethical practices by organizations. Organizations are therefore faced with the vice of corruption during the implementation of their social programs at the community level. The impact of corruption is worsened in cases where CSR programs are implemented by other agencies on behalf of donors.

Furthermore, there is a lack of corporate *transparencyand accountability*. Many corporations fail to disclose pertinent information to stakeholders, despite the requirement for full disclosure in accordance with the Capital Markets Authority (CMA) and corporate governance guidelines. This lack of accountability is seldom reported, even though the state of affairs is symptomatic of corporate failure. Furthermore, funding institutions are faced with this challenge as they monitor and evaluate CSR projects.

Moreover, there is heightened *politicalactivism*. Kenyan society is a relatively enlightened political society, with heightened political activism surrounding electoral cycles. CSR programs, while well intentioned and potentially supported at the community level when at their best, are likely when at their worst to face unwarranted political criticism, social cynicism, and resistance.

In addition, there is *constrained unionism*. While unionism is sanctioned by the Kenyan constitution and thus legally guaranteed, it is tacitly discouraged by some corporate actors, and responsible unionism is uncommon. Similarly, *critical advocacy* is another challenge. Civil society, while highly represented and vocal in Kenya, is unduly critical of organizations. This acts as a driver to CSR strategies and policies. Further, *enlightened Consumerism* presents a challenge. Customers are largely well informed and have a wide range of available choices. However, knowledge distribution is asymmetrical, and favors wealthy over poor customers. Moreover, traditionalCSR practices in Kenya are derived from Western perspectives on CSR and ethical consumerism.

Finally, *governance* is a critical issue. Kenya's aggregate governance performance is trailing the in the world average index, based on the following World Governance indicators: (i) Voice and Accountability, (ii) Political Stability and Absence of Violence, (iii) Government Effectiveness, (iv) Regulatory Quality, (v) Rule of Law, and (vi) Control of Corruption. Although the implementation of the constitution is a major milestone in uplifting Kenya's governance performance, there is a need for an integrated approach to governance. For example, corruption is a vice that for a long time has been related to government. However, the current trend is for the private sector to contribute to it through bribery. Improvement of Kenya's governance system has thus far been instigated by various actors, including, but not limited to the following: pressure groups, the media, NGOs, corporate entities and legislative mechanisms. There have been cases where corporate entities have played the role of whistleblowers and have taken a stand on matters of corruption, calling on government to account for specific corruption activities.

LegalChallenges

This category of challenges relates to laws and regulations that constraint organizations and the citizenry from engaging and benefiting from CSR. Legal challenges faced by corporate organizations operating in Kenya in pursuit of CSR include non-existent or weak legislation on CSR, which makes CSR a purely voluntary affair. Where legislation exists, it is poorly enforced. Kenya's constitution promulgated in 2010 with national and county Governments coexisting creates an opportunity for responsible Government. It creates a system of checks and balances in public positions. However, it is marred by poor coordination and overlaps between the two levels of Government. This creates difficulties in the execution of a CSR agenda. Furthermore, CSR practices are generally adopted from Western perspectives on CSR, and hence most legal provisions are characterized by legal lacunas. There is also lack of tax-exempt provisions that would motivate individuals and organizations to invest in CSR.

The Constitution of Kenya recognizes and protects human rights. Article 19 (2) provides that the protection of human rights is meant to preserve the dignity of individuals and communities and to promote social justice and the realization of the potential of all human beings. To achieve this, the Constitution has a raft of articles designed to uplift and protect human rights, including the following: right to life (Article 26), equality and freedom from discrimination (Article 27), human dignity (Article 28), freedom and security (Article 29), right against slavery, servitude and forced labor (Article 30), right to privacy (Article 31), and right to freedom of conscience, religion, belief and opinion (Article 32).

Other Articles covering human rights include freedom of expression (Article 33), freedom of media (Article 34), right of access to information held by the state and other persons required for the exercise or protection of any right or fundamental freedom (Article 35), freedom of association, which includes the right to form, join or participate in the activities of the association of any kind (Article 36), the right to assemble, demonstrate, picket and present petition to public authorities peacefully (Article 37), political rights, which includes the right to participate in the activities of political parties (Article 38), freedom of movement and residence (Article 39), and the right to acquire or own property (Article 40).

Furthermore, the Constitution has created Commissions charged with the responsibility of protecting and promoting human rights. Article 59 (1) states that the Kenya National Human Rights and Equality Commission shall be established to promote respect for human rights and to develop a culture of upholding human rights, promoting gender equality, monitoring and investigating complaints about human right abuses, and acting as a principal organ of the state in ensuring compliance and obligations under treaties and conventions relating to human rights. The Constitution mandates the Commission to receive complaints on human right abuses and take remedial action.

Social Challenges

The first social challenge is *Educationand Training*. This is a crucial aspect of CSR in Kenya, and certain companies invest significant resources in uplifting the educational standards in the country. Few recent examples of such firms are as follows: Safaricom, a Telecommunication company, has built a school that accommodates up to 1000 needy students. The secondary school dubbed M-Pesa Academy is a Ksh. 3 Billion ($ 30 Million) project that targets needy but bright students. It is built and owned by M-Pesa Foundation, a charity organization established by Safaricom in 2010. In the same vein, Equity Bank is pursuing the provision of education and mentorship for bright students. Equity Group Foundation and The MasterCard Foundation, with additional funding support from USAID, UKAid, KfW, Equity Bank, and individuals and institutions under the Wings to Fly program, has been implementing a comprehensive secondary scholarship support programs for top performing yet needy (orphaned or vulnerable) students in the sub-counties across Kenya. In providing this opportunity to Kenyan children who may have otherwise gone unnoticed, the Wings to Fly program is in line with Kenya's vision 2030 to transform Kenya into a middle-income economy led by well-educated and trained citizens.

The objective of the scholarship program is to provide financial support and leadership empowerment tools to such students. The scholars under this program receive support for all their needs (tuition, accommodation, books, uniform, transport to and from school, shopping and pocket money) while attending secondary school for 4 years. The program, which includes 2000 students this year, will bring the total number of scholarships given under the Wings to Fly program to a total of 10,377 in 5 years.

The Wings to Fly Program is a public-private partnership between the Equity Group Foundation, Mastercard Foundation, UKaid and USAID. The program provides full scholarships that include tuition, room and board to enable academically gifted but economically disadvantaged and vulnerable children to pursue secondary and tertiary education. USAID's contribution of $17.5 million will provide for comprehensive scholarships, leadershiptraining, and mentoring to at least 2166 Wings to Fly scholars over a 5-year period. Overall, the initiative will support 7300 young Kenyans with access to a full secondary education by 2019. The selected young people also receive leadership development, career guidance, and mentorship to help them realize their potential. Other banks that have invested in education include the Co-operative bank and the Kenya commercial bank.

Secondly, *Healthand HIV/Aids Challenge*: The HIV and AIDS pandemic is a global crisis, and constitutes one of the most formidable challenges to development and social progress. It is eroding decades of development gains, undermining economies, threatening security and destabilizing societies. The HIV/AIDS pandemic poses a serious challenge to development in Kenya. An estimated 1.6 million people, or roughly four per cent of the population, are living with HIV in Kenya. Every year an average of 89,000 adults and about 11,000 children in Kenya are infected with HIV. About 11,000 Aids-related deaths, a quarter of total deaths, occur among children (Otieno, 2015). The effects of HIV/AIDS is long-lasting, ranging from increased poverty arising from high spending on Medicare, increased absenteeism among employees, high labor turnover, reduced worker productivity, disruption of investment in child education, and reliance on philanthropy for survival. These consequences are severe for a country in which everybody is expected to be productive.

In recognition of the impact of HIV/AIDS in Kenya and its effect on individual business enterprises, the corporate sector, alongside government and other development agencies, have initiated a number of programs to mitigate HIV/AIDS. For example, the government has developed an HIV/AIDs Policy to guide organizations in the management of HIV/AIDS at workplace. Similarly, corporate entities have initiated programs to manage the spread of HIV/AIDS, such as non-discrimination of affected staff, taking appropriate infection control precaution at workplace, and providing care and support, including offering counseling services for infected and affected staff.

Furthermore, there are efforts concerning conflict and peace building. Although Kenya is considered to be peaceful, unlike her neighbors, closer scrutiny reveals an unprecedented wave of internal and cross-border conflicts. These conflicts, which mainly manifest themselves as political, economic, environmental conflicts, conflicts over natural resources, land and tribal clashes and, recently, terrorism, are sending signals that all is not as rosy in Kenya as the outside world has been erroneously led to believe. Pastoralists in northern Kenya have borne much of the brunt of internal conflicts, and considerable efforts have been directed at addressing their specific conflicts by a number of stakeholders that include government, religious organizations, community-based organizations and corporate entities. For example, Kengen Company has worked in partnership with other stakeholders to bring peace to some of the regions. These include Turkwel and Moroto in Eastern Uganda, where the company has supported trans-boundary initiatives.

CSR in Kenya often takes the form of investments in community projects, such as supporting or fully funding educational initiatives and health projects. By investing in the community, the corporation improves its own support system, enhances the loyalty of its workforce, and signals to consumers that it is part of the social fabric. An example of such an initiative is the Equity Bank's "Wings to fly project", which provides scholarships for bright but needy Kenyan students.

While social challenges exist everywhere, they are prominent in developing countries, thus giving impetus to CSR initiatives in such contexts. However, the social challenges are quite diverse, ranging from conflict, poverty, and even unemployment. Attention has been drawn to social problems such as conflicts, war, and insecurity in Africa, and their implication for CSR.

Some of the social challenges faced by corporate organizations operating in Kenya in pursuit of CSR include, among others: the prevailing social deprivation and destitution, where the enormity of the social challenges makes the contribution of CSR merely nominal, social discrimination, illiteracy and low levels of education, which make the uptake of some social initiatives difficult. Furthermore, disease and ill health in communities, particularly among those that migrate in pursuit of pasture and water, make it difficult to provide social services to the dispersed population. Such factors as social mobility, cultural barriers, ethnicity and social disparities between the rich and poor provide enormous challenge for organizations seeking to develop CSR strategies.

While research into CSR in times of war and insecurity is rare, particularly in Africa, these are the times when CSR is most needed, to alleviate human suffering in such contexts. Organizations have taken responsibility in countries characterized by poverty, conflict and limited statehood (Kolk & Lenfant, 2013).

Economic Challenges

African economies often face downturns, in comparison to developed economies. Accordingly, business organizations are faced with serious viability and survival challenges during such times. While this is so, communities and publics where firms operate are faced with higher expectations and demands, due to high levels of poverty and deprivation.

Kenya is classed among the world's low middle-income countries. Indeed, poverty in Kenya is a paradox, for the simple reason that although the economy seems to have expanded in the last several years, a large part of the population remains in poverty trap. For instance, in 2013, Kenya was ranked 145th among 186 countries in terms of the Human Development Index established by the United Nations Development Program. Overall poverty levels increased from 48.8% in 2007 to 50.8% in 2008, before declining marginally to 49.8% in 2012 (KIPPRA, 2013).

The fight against poverty is a top priority in Kenya's developmentpolicy agenda. The government's commitment to the realization of MDGs, the elimination of hunger and poverty, and the achievement of inclusive and equitable growth, is contained in various policy documents, such as the Medium Term Plan and Vision 2030. However, a concerted effort is required from all stakeholders, including the corporate sector. It is common in Kenya to find business enterprises implementing poverty reduction and community development programs within their host communities in order to uplift their lives.

Some additional economicchallenges faced by corporate organizations operating in Kenya in pursuit of CSR include, among others: transitory economic depression, leading to low household income; increasing levels of poverty, which increase pressure and expectation; financial crisis; high unemployment among women and youths; high levels of debt at the national level; and, finally, high interest rates and inflationary pressure that create a turbulent economic environment. These are huge challenges to the development of an economically viable CSR plan.

Most studies on CSR in challenging times have focused mainly on financial crisis. Some of the severe consequences of financial downturn are stock index failure, the collapse of financial institutions, higher unemployment, poverty, and governments needing rescue packages to bail out their financial systems (Adamu, 2009). Companies are compelled by financial circumstances to restrict their expenses, including reneging on their corporate social responsibilities as these generate costs (Fernández-Feijóo Souto, 2009; Orlitzky, Schmidt, & Rynes, 2003). Njoroge (2009) concludes that CSR initiatives can be delayed or cancelled because of financial crises. However, the global financial crisis provides great opportunities for companies such as corporate brands, employee satisfaction, economic performance, and increased productivity.

Times of financial crisis are likely to be characterized by uncertain business environments. During such challenging times, both organizations and the relevant parties in society seek to avoid the effects of crisis through remedial actions, such as cutting costs by laying off workers, postponing investments, reducing budgets for the following year, and consuming less. However, it is often suggested that it is a mistake for organizations to reduce their CSR projects.

Environmental Challenges

Environment represents the totality of our surrounding, such as plants, animals, microorganisms, and socio-economic and cultural factors. It includes the physical factors in human surroundings such as land, water, forests, atmosphere, sound, smell, taste, animals and plants, and the social factors of aesthetics, in both the natural and built environment. Environment is an essential feature of Kenya's developmentpolicy, because of the realization that the survival and socio-economic wellbeing of Kenyans is intertwined with the environment.

Having realized the critical role of the environment in Kenya's development, the government has established institutions, policies, laws, regulations and guidelines to manage the environment. For instance, the government enacted The Environment Management & Coordination Act which established the National Environment Management Authority (NEMA). The key mandate of NEMA is to coordinate environmental management activities undertaken by lead agencies and to integrate environmental considerations into development policies, plans programs and projects, examine land use patterns, and carry out surveys which assist in the proper management and conservation of environment. Further, the authority advises government on legislative and other measures for the management of the environment, including the interpretation of relevant international conventions, treaties, and agreements. The authority also coordinates research, surveys and disseminates information about environment management, among other functions. The Environment Management & Coordination Act clearly stipulates those activities that affect the environment, and provides mitigation measures required in order to protect the environment.

The government has also developed several policies to address environmental issues, such as a Wetlands Policy, Wildlife Policy, Forestry Policy, National Energy & Petroleum Policy, among others. Therefore, the Kenyan government has put mechanisms in place to ensure environmental protection and conservation in order to realize its strategic objectives.

Environmental issues are of critical importance to developing countries, in view of climate change, environmental pollution, and degradation. Organizations are faced with decisions concerning the environmental sustainability of their activities. Some of the environmental challenges faced by corporate organizations operating in Kenya include, among others; environmental pollution common in many urban and peri-urban settlements in Kenya, sustainability of productive activities and avoidance of waste, environmental degradation, and the increasingly damaging effect of climate change at the national and global level. The impact of climate change appears particularly critical for the country. Other challenges include environmental health and the increased risk of exposure to life threatening diseases.

5 Motivators of CSR Engagement

Cultural Values Kenyan culture is driven heavily by cultural values anchored in philanthropy and communal ties. In this regard, CSR is a product of Kenya's cultural, social, economic and politicaldevelopment over the centuries. A critical examination of Kenya's cultural, social, economic and political environment reveals this, and demonstrates the deep seated norms, traditions and values that promote social responsibility or irresponsibility among the citizenry. Perhaps more significantly, CSR in Kenya has its roots in a national, social, cultural, and political heritage. Firstly, at the birth of the republic in 1963, the government pledged to fight poverty, illiteracy, and disease. This has significantly shaped the Kenyan agenda for the past 50 years. In addition, the national anthem espouses the tenets of justice, peace, liberty, unity, and service. The cultural context is an important factor in defining CSR in any country (Muthuri & Gilbert, 2011). CSR in Kenya has its roots partly in the cultural and communal context, and more specifically in indigenous and religious concepts.

First, the '*Harambee*' spirit; a rallying call at independence by the founding president embodies and reflects the strong ancient value of mutual assistance, joint effort, social responsibility, and community self-reliance. It is guided by the principle of collective good rather than individual gain (Winston & Ryan, 2008), by helping others in times of need through donations (Tarus, 2015). The *harambee* spirit entails the provision of support to underprivileged members of the community, by fundraising for school fees and medical bills and the construction of schools, hospitals, and churches, among other community projects (Mbithi & Rasmusson, 1977; Musau, 2014). Although the *harambee* concept was intended to help the poor members of the society, it is common to find wealthy individuals using it to meet pressing obligations. Therefore, philanthropy is part of the Kenyan value system, and thus organizations that do not participate in corporate giving are perceived negatively by the community (Tarus, 2015).

Secondly, making *donations to the poor in society* is central to most of the religions with significant adherents among the Kenyan population, such as Christianity, Hinduism, Buddhism, African traditional religions and affiliated religious organizations. For instance, *Zekat* (or Zakat), a concept that means charity or alms to the poor, is one of the five pillars of Islam, and is practiced by a significant Kenyan Muslim population, who comprise approximately 10% of the country's population. The giving of alms is the practice of charitable giving by Muslims, based on accumulated wealth. It is an obligatory practice for all who are able to do so, and it is considered a personal responsibility for Muslims to ease the economic hardship of other Muslims and to eliminate inequality among the followers of Islam.

The Legal and Regulatory Environment There are several constitutional, legal and regulatory provisions that require CSR engagement in Kenya. First, the Constitution of Kenya explicitly underlines several CSR provisions. In particular, Chapter 2 Article 10 of the constitution clearly outlines the national values and principles of governance, including national unity, the sharing and devolution of power, the rule of law, democracy and the participation of the people, human dignity, equity, social justice, inclusiveness, equality, human rights, non-discrimination, and the protection of the marginalized. It also emphasizes good governance, integrity, transparency and accountability, and sustainable development. Chapter 4 Article 19 outlines the bill of rights for citizens. Further, Chapter 5 of the constitution of Kenya stipulates environmental protection and, consistently with this provision, the National Environment Policy was developed to address environmental concerns. The policy provides a broad range of measures and actions designed to respond to key environmental issues and challenges. It provides a framework for an integrated approach to the planning and sustainable management of natural resources in Kenya, recognizes vulnerable ecosystems, and prescribes various policy measures to promote mainstream sound environmental management practices, and recommends institutional and governance measures to anchor the achievement of these desired goals.

Market Access Kenya is integrated into the world economy, as evidenced by its annual exports and imports. For instance, the vast majority of exports are in tea, horticulture and coffee. In this regard, Kenyan exports are expected to comply with a raft of ever-growing sets of international rules and industry specific standards. Some of the set standards include the social and environmental conditions of production, such as labor standards and pollution. These form the basis of corporate codes of conduct that the industry has developed to assure key stakeholders of their commitment to corporate citizenship. Certain codes are becoming minimum market entry requirements: specifically, the Ethical Trading Initiative (ETI) in the UK. These codes may present barriers to market entry, especially for small producers, as are most producers in Kenya. Furthermore, there is a growing interest across the world in the adoption of ISO 14001 for export oriented companies.

Stakeholder Activism Several stakeholder groups have emerged in Kenya to put pressure on organizations to be socially responsible. Some of the key pressure groups include development agencies, trade unions, international non-governmental organizations (NGOs), business associations, religious groups, and the media.

Catastrophic Events Catastrophic events are likely to elicit CSR responses from corporate entities, especially those of a philanthropic nature. For example, the terrorist attack at Garissa University College (GUC) where up to 148 people were killed, was one of the most tragic events in the world at the time. This saw a number of diplomatic missions in Kenya and across the globe pulling out all the stops to aid the surviving students by offering scholarships, and medical and other support.

Similarly, the terrorist attack at Westgate Mall, Nairobi, resulted in massive donations and contributions from corporate entities to the affected families.

Community Expectations Community expectations put enormous pressure on organizations to be socially responsible in Kenya. According to Visser (2008), the failure by the government to provide basic public services compels citizens to look to the private sector as an alternative solution to their socio-economicchallenges. A typical case in Kenya was when the Maasai community demonstrated against the Magadi Soda Company in the year 2000. The local community staged a demonstration over the failure of the company to provide employment opportunities (Muthuri et al., 2009). The Maasai community has traditionally relied on livestock rearing as its main source of livelihood, but with persistent drought and the absence of government intervention, the Maasai turned to the Magadi Soda Company for increased social provision and employment. The company had employed few local people, owing to low skill levels and illiteracy among the Maasai community. As an offensive mechanism, the Maasai challenged the company's land lease rights of approximately 225,000 acres, claiming it as their ancestral land (Muthuri et al., 2009). These demonstrations triggered increased philanthropic giving by the company.

Government The Government has been shaping the CSR agenda through the formation of regulatory agencies to oversee CSR and the enactment of legislations and specific codes. The government has established several organs to oversee CSR implementation, such as the National Environmental Management Authority (NEMA), the Kenya Bureau of Standards (KEBS), the Competition Authority of Kenya, the Kenya Consumer Protection Advisory Committee (Kecopac), and the Capital Market Authority and Kenya Flower Council. Furthermore, the government has established legal and regulatory frameworks and codes of conduct to support the CSR agenda, such as National Environmental Policy, Kenya Flower Council Code of Practice for local exporters, Consumer Protection Act No. 46 of 2012, Competition Act, No. 12 of 2010, and Occupational Safety and Health Act, 2007, among others.

In addition, the government has also set up constitutional commissions to oversee ethical concerns such as an Ethics and Anti-Corruption Commission, a Kenya National Human Rights and Equality Commission (KNHREC), a Commission on Administrative Justice, and a National Gender and Equality Commission. The development of these laws and legislative organs is a deliberate attempt to institutionalize CSR in Kenya. However, the enforcement of and compliance with CSR regulations remains a challenge.

Multinational Corporations Although Kenya has a strong heritage that shapes its CSR profile, her large cosmopolitan foreign constituency, such as multinational organizations, partly shape the CSR profile. For example, Bamburi Cement Company attaches particular importance to the development and well-being of employees and local communities in the places where it operates. Their primary focus is on health and safety, education, and environmental conservation. The company also sponsors sports activities such as rugby. Other multinationals, such as East African Breweries, also invest in CSR programs. The company introduced a sorghum-based beverage—a drought local crop grown in dryland—as a way of empowering communities in such areas to have gainful agriculture and hence to address and improve food insecurity. These initiatives by multinational companies are being replicated by local firms because of increased expectations across society.

The Media The media plays an important role in creating awareness of social irresponsibility and empowering stakeholders with information on issues relevant to CSR initiatives, in areas such as health, education, security, environment, and society in general (Kensall, 2001). While the accuracy of information and the level of reporting is generally weak, Kenya's media industry has played a significant role in identifying social irresponsibility among corporate entities. The freedom of media which is envisaged in the constitution of Kenya has been at the forefront in highlighting the negative impact of entities in various sectors of the economy.

For instance, the lead incident in Owino Uhuru slums in Mombasa in 2014 is a good example of the role played by the media in exposing social irresponsibility. Indeed, the demand for lead batteries by the vehicle, telecommunications, and solar industries around the world has led to the lead battery recycling industries in Kenya, which have grown, while environmental controls have lagged behind. This was highlighted by the case of the Metal Refinery Export Processing Zone, a factory which manually extracts lead metal from used car batteries located in close proximity to the Owino Uhuru Slum. Residents experienced severe cases of lead poisoning with serious health consequences such as loss of neurological function, brain damage, decreased blood production, male infertility, increased blood pressure and even death. The effect of lead poisoning is more severe among small children.

As a result of these environmental hazards, some affected families approached the media with information about the health crisis in Owino Uhuru slum. The media flagged the hazard through investigative reporting, for more than 1 week, which involved interviewing the affected families, the owners of the enterprise, the public health officials, health experts and political class. Within just a few days, the business was shut and the owners left the country to avert arrest. The enterprise has since closed.

Civil Society and Advocacy Groups Africa's long history of social and economic deprivation arising from acts of slavery, colonialism, economic and trade inequalities, resource exploitation and degradation by multinationals calls for an urgent and clear focus on CSR as a social and political empowerment strategy. Indeed, Kenya has a vibrant civil society and advocacy groups, as enshrined in the constitution. There are hundreds of such organizations operating in Kenya, such as Consumers Federation of Kenya (Cofek), Central Organization of Trade Unions (COTU), Centre for Corporate Governance, and Community-based organizations (NGOs/CBOs). In addition, international organizations such as UN Global Compact, UNDP and UNIDO have played a key role in institutionalizing the CSR concept in Kenya.

6 CSR in Small and Medium Enterprises

Small and Medium sized Enterprises (SMEs) have been recognized as the engine that drives development and industrialization. SMEs have been variously defined, depending on the study context. However, in Kenya, consistent with the sessional Paper No. 2 of 2005, an SME is an enterprise with less than one hundred employees, and with an annual turnover of less than four million shillings (GOK, 2005). The role of SMEs in driving development and industrialization in Kenya cannot be overemphasized. The 1999 Baseline Survey conducted by the Central Bureau of Statistics (CBS) shows that there are over 1.3 Million Micro and Small-Scale enterprises employing a total of approximately 2.3 million workers, which accounts for 26 % of the total workforce. It contributes an estimated 13.8% of the country's Gross Domestic Product (GDP).

SMEs are present in almost all sectors of the economy, such as manufacturing and the service industry. In pursuit of their objectives, SMEs also contribute significantly to environmental degradation and pollution, due to factors such as the type and age of technologies used, awareness of the impact of their operations, and other industry-specific characteristics. In particular, the manufacturing SMEs consume substantial resources and also produce a significant share of pollution and resource depletion. For example, the constant logging of forests in Kenya is attributed to SMEs engaged in wood products.

7 Mitigation Strategies

7.1 Governments Mitigation Strategies

The government has formulated several legislations and policies to address CSR. For instance, corporate governance, which is a critical issue in managing organizations, is guided by the Capital Markets Act (CMA Cap. 485A) for public listed companies in Kenya (GoK, Gazette Notice No. 3362, 2002). These guidelines were developed in response to the growing importance of governance issues in Kenya, in order to promote growth in domestic and regional capital markets, to recognize of the role of good governance in corporate performance, capital formation and maximization of shareholders value, and to protect investors' rights, including the separation of power between three arms of governments and the balance of power between National & County Government.

7.2 International Organizations and NGO's Challenges Mitigation Strategies

Traditionally, many NGOs engaged in consumer or shareholder activism (Dahan, Doh, Oetzel, & Yaziji, 2010), and use adversarial strategies of litigation against corporate malpractice. They also conduct critical research and fostering public education (Utting, 2005). Other actors include the UN (specifically, UN Global Compact), the African Union, NEPAD, and the African Peer Review Mechanism. Others include international NGOs on human rights, corruption, governance and environmental impacts, the World Bank and IFC, indigenous approaches, and supply chainregulations (e.g. export market regulations). Some roles of these entities include funding opportunities, empowerment programs, training programs, and the provision of basic services.

7.3 Mitigation Strategies by SMEs

The role of SMEs in Kenya is enormous, from job creation, entrepreneurshipdevelopment, and contribution to GDP. As SMEs make their contribution to the national development, it is expected that their activities will have a negative impact not only on the environment but on other areas such as quality jobs and products that meet the required customer standards. For SMEs to be socially responsible, the following mitigation strategies are required:

- Capacity development and the promotion of environmentaleducation and awareness among SMEs
- Promotion of efficient and cleaner production technologies that meet socially responsible standards
- Promotion of employee rights among SMEs, in terms of the provision of conducive work environment and fair remuneration practices, among others

8 Is CSR a Blessing, a Curse or Necessary Evil?

The status of CSR espoused in earlier sections of this paper point to three possible scenarios. These are:

CSR as a Blessing
Most studies on CSR have examined its beneficial effects in developed countries. However, few such studies have been carried out in developing countries, particularly in Kenya, especially in challenging times. Thus, the beneficial effects of CSR have been pursued in various ways by different actors in Kenya.

As discussed earlier, persistence of political, economic, legal, social and environmental challenges calls for CSR agenda to mitigate these.

Traditionally, CSR has promoted the corporate or public image of organizations. Indeed, organizations engaged with CSR activities benefit through the enhancement of their corporate status and image. Indeed, CSR has helped organizations to differentiate themselves from others (Jamali & Mirshak, 2007). In addition, the empowerment of deprived communities, e.g. Safaricom supports economic empowerment projects such as income generation, food security, employment creation, and skills transfer through vocational training.

Moreover, organizations engage in CSR to gain social acceptance and legitimacy. This is akin to a license to operate in the community. Furthermore, CSR is used to bridge the widening divide between rich and poor. This is critical in Kenya, which remains very unequal and where the gap is widening. CSR is also pursued to meet the high social expectations of businesses. In Kenya, some businesses are very large relative to the national economy, which brings an increased level of social expectation.

In a normative sense, business and society are intricately interlinked, and businesses are expected to be good corporate citizens. In another sense, CSR is the right thing to do, and therefore organizations are ethically and morally obliged to undertake CSR.

CSR as a Curse

This perspective is completely opposed to the idea promoted by some organizations that CSR is an unnecessary digression. Such organizations believe that by performing their functions as private enterprises, they are already engaging in CSR, and that the pursuit of CSR would hurt shareholder interests. Others argue that CSR is a costly enterprise without direct benefit to the organization, and that it suffers from free riding problem. Thus amid several political, legal, economic, social and environmental challenges espoused earlier CSR is perceived as a curse.

CSR Is a Necessary Evil

This perspective emanates from actors who are reluctant to pursue CSR, but do so only to meet the minimum possible standard. They would rather not pursue CSR, and would not if it were voluntary. In Kenya, this is the more likely approach to CSR. Since it is mostly voluntary, many organizations shun CSR. However, a few do pursue CSR, as a philanthropic gesture to the communities.

9 Conclusions and Implications

The concept of CSR in challenging times is receiving interest from a range of actors—but it is a multifaceted concept and is far from clear. In Kenya, despite several challenges of a political, social, economic, legal and environmental nature, the lack of an institutional framework has led to haphazard responses to CSR from various stakeholders and actors. Indeed, deepening CSR to respond to crisis situations calls for a regulatory framework. This is because the practice of CSR in a situation of crisis or a challenging era is amenable to a variety of initiatives, approaches, and multiple actors. Thus, in Kenya's adolescence era, CSR is in a state of flux. There is a need to develop management education by focusing on CSR and ethics in local contexts, in order to bolster, deepen and broaden the CSR agenda, and to develop policies and strategies that are specific to local challenges and situations.

Theoretical, Practical and Research Implications

The theoretical implication of the paper is implicitly apparent. While definition of CSR and its operationalization and conceptualization is long overdue, it has been assumed to be context free. However, CSR is context specific and thus, this paper attempts to focus on CSR in challenging times. As such this leads to extension and broadening of CSR agenda to different contexts and operating environments.

In practice, CSR implementation in Kenya and elsewhere in the developing world is erratic, incoherent and unsystematic. indeed, CSR implementation in practice is multifaceted. Accordingly, CSR has been touted as a blessing, curse or necessary evil. There is need to initiate and adopt a regulatory and institutional framework to drive CSR agenda in Kenya and other contexts where CSR agenda is largely voluntary. These CSR programs and initiatives need systematic focus and consistency.

On research implications, in order to deepen and broaden CSR agenda in Kenya and elsewhere in Africa research focus should be on qualitative and quantitative aspects. More specifically, there is need for proper operationalization and conceptualization of CSR agenda, including empirical testing of inherent relationships and attendant implications for theory and practice in Africa generally and Kenya specifically. Thus, CSR concepts, constructs and instruments be developed and tested empirically in various contexts. Granted that CSR is relatively mature in the developed north than the developing south, range of tools and instruments utilized in the former be made custom specific for the latter. More research focus should extent to other relatively ignored CSR areas such as sustainability. The multidisciplinary nature of CSR should be espoused. For instance focus could extent to the economics, sociological, cultural and political aspects of CSR agenda.

References

Adamu, A. (2009). The effects of global financial crisis on Nigerian economy. Available at SSRN1397232.

Cheruiyot, T. K., & Maru, L. C. (2012). Employee social responsibility practices and outcomes in Kenya's tourist hotels. *African Journal of Economic and Management Studies, 3*(1), 23–41.

Cheruiyot, T. K., & Maru, L. C. (2014). Corporate human rights social responsibility and employee job outcomes in Kenya. *International Journal of Law and Management, 56*(2), 152–168.

Cheruiyot, T. K., & Tarus, D. K. (2016). Corporate social responsibility in Kenya: Reflections and implications. In S. Vertigans, S. O. Idowu, & R. Schmidpeter (Eds.), *Corporate social responsibility in Sub-Saharan Africa* (pp. 87–110). Cham: Springer.

Dahan, N. M., Doh, J. P., Oetzel, J., & Yaziji, M. (2010). Corporate-NGO collaboration: Co-creating new business models for developing markets. *Long Range Planning, 43*(2), 326–342.

Fernández-Feijóo Souto, D. (2009). Crisis and corporate social responsibility: Threat or opportunity? *International Journal of Economic Sciences and Applied Research, 2*(1), 36–50.

Garriga, E., & Melé, D. (2004). Corporate social responsibility theories: Mapping the territory. *Journal of Business Ethics, 53*(1–2), 51–71.

Government of Kenya. (2002). Corporate Governance Guidelines, Gazette Notice No. 3362. Government Printer, Nairobi.

Idemudia, U. (2014). Corporate social responsibility and development in Africa: Issues and possibilities. *Geography Compass, 8*(7), 421–435.

Jamali, D., & Mirshak, R. (2007). Corporate social responsibility (CSR): Theory and practice in a developing country context. *Journal of Business Ethics, 72*(3), 243–262.

Kensall, T. (2001). Donors, NGOs and the state: Governance and civil society in Tanzania. In O. Barrow & M. Jennings (Eds.), *The charitable impulse: NGO and development in East and North-East Africa* (pp. 133–148). Oxford: James Currey.

Kenya Vision 2030 (2007). A globally competitive and prosperous Kenya.

KIPPRA. (2013). Kenya Economic Report 2013, Nairobi Kenya.

Kolk, A., & Lenfant, F. (2013). Multinationals, CSR and partnerships in Central African conflict countries. *Corporate Social Responsibility and Environmental Management, 20*(1), 43–54.

Mbithi, P. M., & Rasmusson, R. (1977). *Self reliance in Kenya: The case of harambee*. Uppsala: Nordic Africa Institute.

Moon, J. (2002). Corporate social responsibility: An overview. In C. Hartley (Ed.), *The international directory of corporate philanthropy* (1st ed., pp. 3–14). London: Europa Publications.

Musau, C. N. (2014). *Understanding the impact of 'harambee' tradition on the philanthropic activities of Kenyan immigrants in the Twin Cities*. Minneapolis, MN: Saint Mary's University of Minnesota.

Muthuri, J. N., & Gilbert, V. (2011). An institutional analysis of corporate social responsibility in Kenya. *Journal of Business Ethics, 98*(3), 467–483.

Muthuri, J. N., Matten, D., & Moon, J. (2009). Employee volunteering and social capital: Contributions to corporate social responsibility. *British Journal of Management, 20*(1), 75–89.

Njoroge, J. (2009). Effects of the global financial crisis on corporate social responsibility in multinational companies in Kenya. Covalence Intern Analyst Papers, available at: www.covalence.ch/docs/Kenya-Crisis.pdf, accessed on 05 April, 2016.

Okoye, A. (2009). Theorizing corporate social responsibility as an essentially contested concept: Is a definition necessary? *Journal of Business Ethics, 89*(4), 613–627.

Opondo, M. (2006). *Emerging corporate social responsibility in Kenya's cut flower industry*. University of Nairobi, Department of Geography and Environmental Studies.

Orlitzky, M., Schmidt, F. L., & Rynes, S. L. (2003). Corporate social and financial performance: A meta-analysis. *Organization Studies, 24*(3), 403–441.

Otieno, D. (2015, December 1). World Aids Day: A snapshot of HIV in Kenya. *Daily Nation*.

Quelch, J. A., & Jocz, K. E. (2009). Can corporate social responsibility survive recession? *Leader to Leader, 2009*(53), 37–43.

Souto, B. F. F. (2009). Crisis and corporate social responsibility: Threat or opportunity? *International Journal of Economic Sciences and Applied Research, 1*, 36–50.

Tarus, D. K. (2015). Corporate social responsibility engagement in Kenya: Bottom line or rhetoric? *Journal of African Business, 16*(3), 289–304.

Utting, P. (2005). Corporate responsibility and the movement of business. *Development in Practice, 15*(3–4), 375–388.

Visser, W. (2005). Corporate citizenship in South Africa. *Journal of Corporate Citizenship, 2005* (18), 29–38.

Visser, W. (2008). CSR and the financial crisis: Taking stock. Wayne Visser Blog Briefing, *4*.

Winston, B. E., & Ryan, B. (2008). Servant leadership as a humane orientation: Using the GLOBE study construct of humane orientation to show that servant leadership is more global than western. *International Journal of Leadership Studies, 3*(2), 212–222.

World Bank. (2016). *Economic update: Kenya's economy strong in a challenging global environment*. Washington, DC: World Bank Group.

Corporate Social Responsibility in Ghana: Issues and Concerns

Sam Sarpong

1 Introduction

The concept of Corporate Social Responsibility (CSR) in Ghana has, to a large extent, been canvassed by civil society which initially focused on how multinational companies could contribute to solving pressing social and environmental problems by acting on a voluntary basis through partnerships with other stakeholders. Concerns had been expressed in industries like mining which conspicuously were impacting negatively with their externalities on rural communities. Confronted with pollution and waste, soil erosion, deforestation and the destruction in mining areas, especially, many Ghanaians began to question the prevailing beliefs that nature was to be ruthlessly subdued and controlled in the cause of industrialisation and economic growth. These concerns led to huge public disparagement and also provided the impetus for the awareness of firms to the growing expectations from stakeholders about how they should manage their social, environmental and economic impacts (Sarpong, 2010).

The considerable influence of civil society has since compelled businesses to reconsider a number of corporate issues. Clearly, it is within this line of thought that many companies are now increasingly incorporating among others social justice concerns into corporate policy commitments, often augmenting already established environmental policies. This has also happened, in part, because various tools and models have also been developed at the international level to manage social and environmental performance of multinational companies in recent years. Alongside these have been international frameworks which have been developed to guide

S. Sarpong (✉)
Centre for Communication and Entrepreneurship Skills (CENCES), University of Mines and Technology, Tarkwa, Ghana
e-mail: ssarpong@umat.edu.gh

S.O. Idowu et al. (eds.), *Corporate Social Responsibility in Times of Crisis*,
CSR, Sustainability, Ethics & Governance, DOI 10.1007/978-3-319-52839-7_10

governments and companies on how to ensure projects conform to global norms and also provide economic value to their people.

As a result, many ordinary businesses now acknowledge the need for CSR activities in their operations. They see it as a responsibility that they need to disburse to unfortunate people within communities where they operate and even to under-resourced and inefficient state institutions in the country. Such activities have largely been spearheaded by big firms as the multi-faceted problems of Ghana tend to make it almost difficult for many small-scale enterprises, many of whom are engaged in the retail and in the production of primary commodities, to undertake major social actions in the development of the country.

The chapter looks at the role of CSR in Ghana and the ways in which this has changed in the last few years. Firstly, our focus would be on what constitutes CSR and then a general view of CSR in Ghana. We would then turn our attention to the socio-economic challenges in Ghana and the CSR approach envisaged in that sphere. Eventually, we would look at the challenges corporate entities face as they operate in this particular environment. The challenge of NGOs in shaping CSR activities in Ghana would be delved into and finally an attempt would be made to look at the involvement of small and medium scale enterprises (SMEs) in CSR activities.

2 The General View of CSR

Many definitions have been offered for the concept of CSR (van Marrewijk, 2003). Besides, a variety of definitions of CSR have been proposed but a fundamental problem in the field of CSR is that there is no clear, universally accepted consensus in the ideal meaning of CSR, making theoretical development and measurement difficult (Hopkins, 2004; Ismail, 2011; McWilliams, Siegel, & Wright, 2005; Sriramesh, Ng, Ting, & Wanyin, 2007). According to Reinhardt, Stavins, and Vietor (2008), one of the challenges of examining the concept of CSR is identifying a consistent and sensible definition from among a bewildering range of concepts and definitions that have been proposed in the literature. Numerous writers have sought to give an indication that CSR amplifies business' role in society (Lea, 2002; Marsden, 2001; Pinney, 2001).

CSR, at its core, also means respecting environmental values, contributing to economic growth, promoting social progress and conducting business in a way that prevents or mitigates negative impacts and enhances or sustains positive impacts on local communities and host countries (Taylor, 2004). What cuts across a number of definitions that scholars have proposed on the concept of CSR is the general belief that beyond the quest to maximise corporate profits, corporate organisations play a crucial role in solving society's problems (Masaka, 2008). It is in this respect that Drucker's (1984) contribution comes in handy. He proposed a 'new meaning' for CSR, pointing out that profitability and responsibility are complementary notions. CSR can also be defined as a company's voluntary contribution to sustainable development which goes beyond legal requirements (Carroll, 2006; Crane & Matten, 2007). A company's commitment to CSR, thus, suggests its strategy to fulfil its social responsibilities as expected by their stakeholders (Ferrell, Fraedrich, & Ferrell, 2005). By its very nature CSR is a complex, multiform phenomenon emerging as the interface between enterprises and society.

3 Locating CSR in the Ghanaian Economy

The quest to locate the notion of CSR in the operations of all major corporate bodies these days has become almost a mantra that many recite for various reasons. Corporate bodies in Ghana profess to be pursuing CSR based on their own way of seeing it. The confusion that often arises stems, perhaps in part, from the misunderstanding with regard to what CSR is all about. When companies in Ghana make donations to the needy, hospitals and even ineffectual state institutions, they call it CSR. Some build schools and other social amenities for people in rural communities, others emblazon their brand logos and paint buildings among other things and describe them as CSR activities.

In spite of attempts made to bring the issue of CSR to the forefront, it is significant to note that the CSR concept has not been accorded its due in Ghana. Past studies including one study conducted by World Business Council for Sustainable Development (WBCSD) in 2000, entitled 'Corporate Social Responsibility—making good business sense,' all found that CSR did not seem high on the business agenda in Ghana. One of the reasons adduced by these studies was that CSR is thought to be too expensive. Lack of government control and involvement is also cited as a reason for CSR having a lower priority. The studies also found out that there is little outside pressure on companies to encourage them to take the CSR initiative. Recent studies (Ofori & Hinson, 2007; Sarpong, 2010) have also concluded that CSR is far from being given its due and that a lot more ought to be done by firms to make CSR a strategic concern.

Nowhere in particular is the impact of CSR much more felt than the mining sector in Ghana. This is in view of the huge impact that it makes in host communities. Admittedly, the notion of CSR became much more pronounced following the Structural Adjustment Programme that Ghana embarked upon, together with many other African countries in the 1980s. Following this development, Ghana's liberalised economy ushered in a number of international mining firms to exploit for minerals. The increased competition amongst foreign investors and the development and proliferation of a standard set of legal and environmental protections for both investors and host countries alike created a basis for CSR to gain root in many developing countries, including Ghana. Without effective environmental controls, developing countries like Ghana risked suffering serious, irreversible environmental harm from large-scale mining projects. The mining sector provided a good case study for the analysis of the relationship between investment and the environment in view of the environmental, economic and social importance of the sector in many foreign direct investment-host countries (Sandbroke & Mehta, 2002).

CSR in the mining industry started as a defensive and reactive measure to complaints from community members and the numerous civil society organisations operating within the sector. For years, mining companies operating in the country stood accused of paying lip-service to the concept of CSR. The new dawn of awareness has ensured they now have designated Community Relations departments and Foundations, which are involved in a range of community development works including the funding of community assistance programmes in their host communities. Outside the mining industry, many corporate bodies especially, those with some 'international connections' have accentuated the potential notion of responsibility, by placing emphasis on the moral dimension of economic globalisation; a compulsion on international corporations to take a more unambiguous stance on environmental management and human rights standards, and the promotion of new ways of thinking about the relationship between business and society.

4 The Socio-economicChallenges and the Quest for CSR

It is quite profound that globalisation is contributing to fundamental changes in the relationship between markets, states and firms (Cohn, 2003). The high degree of globalisation has placed constraints on the policyautonomy of national governments (Julius, 1994). The way in which globalisation affects nations, either by producing the obsolescence of their functions, the reconfiguration of their power, or even the continuity of their traditional sovereign status, has been the subject of a huge discrepancy among scholars in the globalisation debate (Chang, 2004). It has led to the weakening of support for social and welfare programmes as well as issues concerning the environment and matters of sustainability (Campbell, 2001).

Poverty has widened in the last few years in Ghana and job prospects remain negligible. Local communities perceiving the state to have failed in its developmental task have often become confrontational with some business firms in their localities. They see these firms as a substitute for an absentee government and very often challenge the profit-driven attitudes of businesses. The surrogate role assigned to business requires business to help in the development of society. Recognition of this role is evidenced by recent acknowledgement of business's potential to impact on development and poverty alleviation, often seen in CSR activities.

A clear case of how Ghanaian state officials see CSR can be encapsulated by the assertion of a former Deputy Minister of Trade and Industry, John Gyetuah, who indicated that CSR is a useful concept for the country because it is impossible for the government alone to provide all the diverse needs of the society. Gyetuah's assertion denotes the desire of government to work in partnership with business in meeting the needs of the populace. It also reinforces the government's position that the private sector has a key role to play in the development of the country and that, the sector remains well placed to make a significant positive contribution towards improving social, economic and environmental conditions, but the track record of big business in Ghana, at best, has been quite mixed. There are clearly examples of specific corporate complicity in political corruption, environmental destruction, labour exploitation and social disruption, stretching back many years (Sarpong, 2010). In spite of that, there is also evidence of the benefits of business generally bringing in capital investment, job creation, skills transfer, infrastructure development, knowledge sharing and social responsibility programmes, even if not adequate enough to satisfy local appreciation.

5 Challenges Faced by Corporate Entities Operating in Ghana

The private sector has now increasingly assumed an influential role as a principal engine for development in many countries (Krugman, 1995). Corporate bodies have been gradually consolidating their positions as key components for the generation of progress and development in the absence of the state's ability to do so in many places (Chang, 2004). Local communities now look up to these companies for support in view of the state's 'fragility' in many rural areas (Harrison, 2001; Lodge & Wilson, 2006). The on-going shift in power from the public to the private sector, has meant that many societies in Ghana now look up to the business sector to help solve social, environmental and economic problems that were once considered solely in the province of government.

These envisaged corporate contributions are most often discussed in terms of corporate social responsibility. Hence, CSR is enmeshed in the debate about Ghana's development in view of the economic problems besieging the country. The multi-faceted problems of the country, i.e., lack of jobs, the state's inability to pursue the developmental needs of the people, lack of resources on the part of some state institutions like the police, make it absolutely imperative for companies to come to the aid of the state in order to ensure the operations of some of these institutions. As such, CSR has acquired a new resonance among companies in the country.

What currently pertains in Ghana, however, is that majority of corporate contributions have remained dispersed and unfocused (Ofori & Hinson, 2007). The poor understanding and adhocism associated with CSR has led to a situation where 'philanthropic CSR' has emerged, though its sustainability remains questionable. Quite notably, there is also a corps of receiving communities who do not even have the wherewithal to look after some facilities that have been handed over to them. These communities have been over-reliant on corporate bodies who, it seems, are ready to accede to the demands of the community members in order to garner their CSR credentials. The support that is offered often consists of cash donations which are given out to aid local community causes or universities, hospitals and national charities in the hope of generating goodwill. Many of the projects that the companies have helped with usually have their names emblazoned on them. This conception falls in line with Klins, van Niekerk, and Smit's (2010) assertion that the creation of a positive corporate image is an important drive of CSR activities in Africa.

Whilst some companies dwell on the business case of greater brand creation and reputation as their motivation, others show that the need to help build the country's economy is more paramount. It is not, however, easy to draw a distinction between the motives. As Masaka (2008: 17) notes, 'it is often difficult to know if it is an elaborate show of CSR motivated by a genuine concern for the welfare of society and the environment they operate in and serve or [whether it is] just a gimmick to enhance their corporate standing through these seemingly good-intentioned actions.' Ghana is yet to integrate CSR into public policy. This, therefore, has led to a situation where organisations virtually isolate certain aspects of CSR and use it as a means of promoting their image, with the goals of looking good more than doing good. It has been simplified to the populist proposition of 'giving something back'.

For some corporate entities in Ghana, CSR is essentially a one-off box-ticking endeavour—cash hand-outs. It is in the light of this that CSR in Ghana has been branded as a 'haphazard indulgence in good corporate works' (Ofori, 2007: 67). But, although, CSR practices are diverse, and could range from the traditionalphilanthropic programmes to a more integrated and comprehensive set of policies and practices linked to a company's operations (Turker, 2009), it could be held that the mere attempts at organising a one-off activity for a community does not in itself provide a strong basis for CSR activities.

In technical terms, social responsibility requires a specified objective, an operational definition to guide planning and to determine success (Hubo & Mines and Geosciences Bureau, 2003). Warhurst (1998) puts this in perspective when she says, CSR implies compliance plus the active development and implementation of mainstream business strategy, supported by technological and organisational innovation, to prevent negative social impacts and optimise social benefits from the outset. The key drivers for corporate interest in, and responses to, the challenge of CSR, would, therefore, be the need to know a particular company is living out its mission and values, along with the need to protect and enhance its reputation (Gardberg & Fombrun, 2006; McCorkindale, 2008). A company's commitment to CSR, thus, suggests its strategy to fulfil its social responsibilities as expected by its stakeholders (Ferrell et al., 2005).

CSR's ambiguity, thus, raises some misunderstandings (Rana, Platts, & Gregory, 2008). To this end, corporate organisations may manipulate CSR because the companies do not freely choose to contribute towards solving society's problems, but because society expects them to do so (Masaka, 2008). Hence, 'the increased pressure on corporate organisations to engage in CSR has also induced them to become very ingenious as they rip off more' than they commit to the social cause (Masaka, 2008: 17). Manteaw (2007: 439) also denounces the impression that is often created of CSR as a mutual or symbiotic relationship serving the good interest of all parties, stressing that, 'it is the logic and backstage activities of these seemingly symbiotic relationships that have not been fully explored as a social justice issue.'

In many cases, the notion of CSR in this country has been more of a public relations tool, rather than a value-creating process. Many companies claiming to pursue CSR have in effect only used contributions to social objectives as a mechanism for carrying on profit-maximising operations. The case of infrastructural provisions in host communities is a clear example of such an act. According to Parker (2006), infrastructure is often perceived as central to operations, and not considered part of community relations. Hence, although many companies have provided considerable help for communities, the basis of this stemmed more from the benefits that accrued to the companies rather than the communities.

An example of this is the rehabilitation of Ghana's western rail line by the mining companies. For a considerable period, the Ghana Manganese Company (GMC) provided funds for a portion of the rehabilitation of the railway system in the western part of the country. It was obvious the company's involvement stemmed more from its desire to secure a reliable means of transport for the carting of its manganese to the nearby port of Takoradi, rather than a determination to develop its operational area in Nsuta or to help the under-resourced state organisation, the Ghana Railway Company. When the state-owned Ghana National Petroleum Corporation, for instance, decided to sponsor hundreds of supporters to the World Cup in Brazil in 2014, it considered it as its corporate social responsibility to the nation but the company took up the initiative because it wanted to tie-up its interest to a national passion. After all, there was not a strong desire then for people to travel to Brazil as they had the opportunity of watching all the matches at home. A lot of such cases prevail in the country.

There is also evidence to the effect that some of the assistance that have been offered in some rural communities had come about because the mining companies had polluted their environment. For instance, a number of these communities had boreholes provided for them because their sources of drinking water had been polluted by the mining companies. The companies were, thus, compelled to provide the said assistance not because they wanted to, but, rather they had to do so in view of the devastation they had caused. Incidentally, such an action is documented as an important contribution to the society under the auspices of the companies' CSR although social responsibility must be premised on fair operating practices.

Hence, if we were to accept CSR as a strategic concern then the concept as practised in Ghana today, completely lacks that delineation. It is only in the last few years that a number of companies have attempted to make their CSR activities more strategic by incorporating stakeholder concerns in their corporate strategies (Boon & Ababio, 2009). Although local companies are now familiar with the concept of CSR and do, indeed, practice some level of the concept, they subscribe less to the contemporary notion of CSR as they are less strategic, less moral and less ethical in their approach to CSR (Ofori & Hinson, 2007). Besides, CSR issues are not well understood, and are regarded as 'philanthropic add-on' (Atuguba & Dowuona-Hammond, 2006).

It is rather the internationally-connected firms in Ghana that seem to have a better grasp of the various dimensions of CSR and how these could be used to business and strategic advantage (Ofori & Hinson, 2007). It is quite pertinent to note that the prevailing approaches to CSR in Ghana are disconnected from business and strategy to the extent that they obscure many of the greatest opportunities for companies to benefit society. If indeed companies were to analyse their prospects for social responsibility by using the same frameworks that guide their core business choices, they would discover that CSR can be much more than a cost, a constraint, or a charitable deed and that it could be a source of opportunity, innovation, and a competitive advantage.

Heightened corporate attention to CSR has not been entirely voluntary in Ghana. Many companies, especially those in the mining sector, awoke to it only after being surprised by public responses to issues they had not previously thought were part of business responsibilities. Advocacy organisations have also brought public pressure to bear on many businesses especially those in the mining sector. The mining companies have long had a disputable reputation for social responsibility and it is highly perceived that communitydevelopment has been inconsequential to them. They have since come under increased pressure by environmental groups and even government officials to do more for their host communities. In a speech at the 77th Annual General Meeting of the Ghana Chamber of Mines, (an umbrella group of international mining companies in Ghana) ex-President, John Kufuor stated,

> At the moment, another major challenge facing the industry in this country is the distressed state of the communities in which your companies operate. This is an indication of the low levels of social investment which have been made within the communities since industrial mining started in this country over a century ago. I call on the industry to support government with radical and creative policies and programmes to redress this historical neglect (Kufuor, 2005: 3).

In recent year, however, the mining companies have used a variety of vehicles for their CSR initiatives. These are either direct action, through foundations established solely for that purpose, through tie-ups with government agencies or Non-Governmental Organisations (NGOs). They continue to address education, water and sanitation requirements in their host communities. However, just as it is globally, extractive industries whose operations have direct impact on the environment and local communities are always in the news for obvious reasons. They are either breaching some of the tenets of the CSR agenda or fulfilling them in earnest. Despite the elaborate CSR policies adopted by the mining companies in last few years, numerous incidents occur, which in effect reflect a discrepancy between the CSR discourse employed by them and their actual practice in relation to their host communities.

6 The Search for a National Policy Framework on CSR

As there is no national policy framework that guides the implementation of CSR in Ghana, firms are not bound by law to implement CSR activities in the country, although they do so more in response to moral convictions (Boon & Ababio, 2009). Despite that, there are a variety of policies, laws, practices and initiatives that together provide some kind of framework for CSR actions in Ghana. Constitutional provisions, local government laws and requirements for environmental impact assessments (EIA) legislation that defines minimum standards for business performance all help to provide the basis for CSR activities in the country (Atuguba & Dowuona-Hammond, 2006).

Statements by government officials, some clearly rebuking companies, especially those in the mining sector for their inadequate provision for their host communities, Non-Governmental Organisations (NGOs) and the media's frequent chastisement of some firms over their environmental degradation, have all helped to improve standards in the country. In addition to civil society pressures, there are numerous international treaties and agreements like the Universal Declaration of Human Rights, International Labour Organisation conventions, among others, that attempt to promote more responsible activity by business with respect to their social and environmental impacts.

The definition of CSR, as 'an entirely voluntary, corporate-driven initiative to promote self-regulation as a substitute for regulation at either national or international level' (Pendleton, 2004: 77), clearly suggests that in order to avoid possible state inferences in their affairs, companies resort to CSR as a means to avoid possible laws or regulations. The argument here is that one of the motivations for CSR policy and strategy development is the avoidance of a legal framework. The suggestion is that companies may not adopt such policies or strategies in the absence of legal and regulatory pressure (Davies, 2003). This debate pits adherents of self-regulation against those who advocate for an increase in the laws and control mechanisms by governments. Many Ghanaians, however, agree on the need to unite responsibility and opportunity. They feel businesses should respond to the complexity and competitiveness of a cultural environment in which, for decades, maximum profit had always been companies' only objective (Sarpong, 2010). In this regard, a company is portrayed as socially responsible in Ghana if it finds a new model for social integration by including the need to add a social and environmental capital to its traditional obligation of generating financial capital.

The lack of a national CSR policy, meanwhile, remains a concern to the government, as such, the Ministry of Trade and Industry was tasked in 2014 to put together a technical committee to draft a national CSR policy for the country. The government is of the view that such a policy would create a framework which would encourage and enable responsible behaviour of businesses and also align and streamline CSR initiatives with the national and development plans. The policy draft is expected to encourage innovative approaches and continuing development and application of best practice. It is also meant to ensure the country has at least decent levels of performance in areas such as health, safety and environment.

7 Specifics of CSR in Small and Medium-Scale Enterprises in Ghana

Some SMEs (those quite small in size) do not appreciate what CSR is all about and cannot fathom how such an initiative can assist them in their businesses. However, others (much bigger SMEs) are beginning to appreciate the importance of CSR. Though the acceptance of the concept has been gradual, the awareness is being created for businesses to take advantage of the benefits of CSR. SMEs in Ghana may not be the key players but rather followers. They are influenced by large and multinational companies because of their incorporation into global production and value chains of these corporate giants. The challenge for SMEs in Ghana is to adapt and narrow the broad CSR concept used by large and multinational companies and incorporated in the global reporting initiatives so that it has relevance for their CSR agenda.

Strengths of SMEs that should facilitate CSR implementation and reinforce their CSR can be found in the following:

- Their lean organisational structure brings lower costs on the management of the company, lower level of internal bureaucracy, and in this way more could be spent on CSR activities.
- They have the flexibility that can improve action time in individual CSR activities and can act in urgent situations—e.g. they can provide help to citizens after natural disasters.
- Their personal and informal relationship with employees and customers alike often fosters CSR in the company and provides credibility of CSR activities in the eyes of customers and employees.
- The SMEs are known to their communities and quite frequently have personal, sometimes emotional relationship to the communities they live in and operate from. These factors influence implementation of CSR in the communities and the attitude of such communities towards the companies.

In spite of the above, there are also issues that tend to hinder SMEs' efforts in implementing CSR into their operations. Primarily, they have less in terms of financial resources to invest in their CSR activities and also lack professionals with the know-how in CSR issues.

8 The Challenge of NGOs in Shaping CSR Activities

CSR has opened up for scrutiny corporate behaviour, a situation that has put civil society in good stead to judge the conscience of businesses on the environment, community development and other issues. Many companies are now under greater public scrutiny as people ferret through their mission statements and undertakings to determine their intentions and the profits they make, as against their contribution to the local area. A combination of what might be called the institutionalised environmental and human rights movement, a sympathetic media and ecological realities have all combined to generate enormous societal attention to ecological problems in Ghana, in the last three decades (Sarpong, 2010). There is also widespread recognition that NGOs are the type of organisations that enjoy the best image and credibility in society (Arenas, Lozano, & Albareda, 2009). They are the ones who have struggled in the localities where absolute poverty still erodes the security of people.

Seen as quite independent, their neutrality has earned them a good reputation with the public. Unlike the state agencies, who are deemed to be closely associated with the business community and hardly denounce inappropriate company behaviour, there is public acceptance of assessments made by the NGOs in Ghana. This, to a large extent, has ensured that many firms now pursue sustainable practices even if their commitment is suspect. As Bendell (2004) acknowledges, businesses have changed some of their policies and strategies, in part, due to social and political pressures linked to NGO networks. Campaigns by NGOs in Ghana have created substantial negative publicity for some of these companies and have provoked some of the recent shifts in corporate social and environmental policies (Sarpong, 2010). The NGOs with a leading presence in the mining communities, for instance, are made up of loose indigenous groups with international alliance in the developed world. They have been effective in applying pressure on the government, international agencies and financial institutions to impose higher standards in the mining industry. The government, media and activists have now become adept at holding companies to account for the social consequences of their activities.

9 Company-Community Relations: Two Logics in Conflict?

The CSR challenge in Ghana derives from the inability of the companies and the inaction and/or inadequate action by the state to address issues of concern to many communities. The state suffers from lack of financial resources to effectively implement a number of social programmes. Businesses have also become aware that factors such as poverty, health and other issues now have a bearing on the activities, hence their involvement in providing support for people in their operational areas (Blowfield, 2005). This, in effect, puts corporate bodies at the forefront of such considerations. But there are some complications regarding the 'surrogate role' of the companies as communities now look up to them for their unmet needs. Such companies are seen as a replacement for the state, whose role is hardly felt in many rural areas. This brings to the fore the need for the pursuance of a sustainable, strategic, and mutually beneficial set of responsible actions from both the state and corporation.

The need also arises for both the state and corporate bodies to invest in the social and infrastructural development of many areas of the country. The difficulty in this respect is where to draw the line demarcating the responsibilities of companies and that of state institutions. The firms seem to be in a disadvantaged position in that if they fail to formulate a clear policy for community improvement and involvement, they run the risk of losing social support and the 'social licence' which they need to operate effectively with. Unless a company earns that licence, and maintains it on the basis of good performance on the ground, there would undoubtedly be negative implications. Numerous cases abound in Ghana where communities have blocked projects or subjected them to legal challenges even after regulatory permits have been obtained, potentially halting projects, just because such companies failed to win the trust of the community members.

There are contradictions between the aspirations of companies and local communities. Whilst the companies are obviously and clearly motivated by business concerns and the protection of their shareholders, the local communities, in turn, are more pre-occupied with their quest for 'development,' a condition which often seems at odds with the objective of business. The nature and extent of development support required by the communities is an issue which often put the two bodies at loggerheads. Both the communities and companies have serious misgivings about each other in many localities in the country and nothing has happened over the years to clear these deep misgivings.

Notwithstanding the above issues, it is sometimes tempting to assume that companies have no compulsion to assist their host communities. Colman and Nixson's (1994) position on this is that, firms, especially, multinational companies, cannot be directly blamed for the lack of development in their host communities since their prime objective is global profit maximisation. If it is held that the primary interest and agenda of these companies have never been the development of the communities but rather to ensure the maximisation of their profit (a view heavily canvassed by the eminent economist, Milton Friedman), then the concept of development in the real sense of the term is, secondary if it ever figures at all. Under these circumstances, firms, as Kapelus (2002) notes, would be interested more in the appearance of social responsibility than actually being socially responsible.

10 Conclusion

The chapter looked at the role of CSR in Ghana and the ways in which this has changed in the last few years. It noted the encouraging signs of a more socially responsible private sector with some companies undertaking socially acceptable initiatives all in a bid to resolve problems within their host communities. As a result of corporate social interventions and regular engagements, tensions which often characterise business and society have reduced considerably as corporate bodies now continue to address issues within distressed communities or even aid state organisations which are under-resourced. But despite the elaborate CSR policies adopted by some companies, incidents occur, which in effect reflect a discrepancy between this discourse and corporate practice in relation to their host communities.

Hence, although CSR is much touted, the extent to which it is followed in practice in Ghana leaves a lot to be desired. As forward thinking companies in the industry are showing, the best approach is to become better neighbours in the local community. This helps in conflict prevention, resolution and management in Ghanaian communities.

References

Arenas, D., Lozano, J. M., & Albareda, L. (2009). The role of NGOs in CSR: Mutual perceptions among stakeholders. *Journal of Business Ethics, 88*, 175–197.

Atuguba, R., & Dowuona-Hammond, C. (2006, June). *Corporate social responsibility in Ghana report*. Prepared for the Freidrich Ebert Foundation (FES)—Ghana. Available: www.revitalization.org/csrm/Documents/CSR final report-3-8-06.doc (accessed 20 Sept 2015).

Bendell, J. (2004). *Barricades and boardrooms—A contemporary history of the corporate accountability movement*. Technology, Business and Society, Programme Paper 13, United Nations Research Institute for Social Development, Geneva.

Blowfield, M. (2005). Corporate social responsibility: Reinventing the meaning of development? *International Affairs, 81*(3), 515–524.

Boon, E. K., & Ababio, F. (2009, May 16–22). Corporate social responsibility in Ghana: Lessons from the mining sector. In *Conference proceedings, 29th annual conference of the International Association for Impact Assessment, Accra, Ghana*.

Campbell, B. (2001). The role of multilateral and bilateral actors in shaping mining activities in Africa. In *Mining development and social conflicts* (pp. 3–44). Accra: TWN.

Carroll, A. B. (2006). Corporate social responsibility: A historical perspective. In M. J. Epstein & K. O. Hanson (Eds.), *The accountable corporation: Corporate social responsibility* (3rd ed., pp. 3–30). Westport, CT: Praeger.

Chang, H.-J. (2004). *Globalisation, economic development and the role of the state*. London: Zed Books.

Cohn, T. H. (2003). *Global political economy: Theory and practice* (2nd ed.). New York: Longman.

Colman, D., & Nixson, F. (1994). *Economics of change in less developed countries* (3rd ed.). London: Harvester Wheatsheaf.

Crane, A., & Matten, D. (2007). Editor's introduction, Corporate social responsibility as a field of scholarship. In A. Crane & D. Matten (Eds.), *Corporate social responsibility, Volume 1: Theories and concepts of corporate social responsibility* (pp. xvi–xxx). London: Sage.

Davies, R. (2003). The business community: Social responsibility and corporate values. In J. Dunning (Ed.), *Making globalization good: The moral challenges of global capitalism* (pp. 301–319). Oxford: Oxford University Press.

Drucker, P. (1984). The new meaning of corporate social responsibility. *California Management Review, 26*, 53–63.

Ferrell, O. C., Fraedrich, J., & Ferrell, L. (2005). *Business ethics*. Boston: Houghton Mifflin.

Gardberg, N. A., & Fombrun, C. J. (2006). Corporate citizenship: Creating intangible assets across institutional environments. *Academy of Management Review, 31*, 329–346.

Harrison, G. (2001). Peasants, the agrarian question and the lenses of development. *Progress in Development Studies, 1*(3), 187–203.

Hopkins, M. (2004, May). *Corporate social responsibility: An issues paper* (Policy Integration Department, Working Paper No. 27). Geneva: World Commission on the Social Dimension of Globalization, International Labour Office.

Hubo, C.L., & Mines and Geosciences Bureau. (2003). *Profiles of the CSR practices of Philippine mining firms*. Report commissioned by CSR Practice of the World Bank Group. Available: http://siteresources.worldbank.org/INTPSD/Resources/Philippines/phil_CSR_Case_Studies.pdf (accessed 20 Nov 2015)

Ismail, T. N. T. (2011). Corporate social responsibility: The influence of the silver book. *International Journal of Business and Management Studies, 3*(2), 371–383.

Julius, D. (1994). International direct investment: Strengthening the policy regime. In P. B. Kenen (Ed.), *Managing the world economy* (pp. 269–286). Washington, DC: Institute for International Economics.

Kapelus, P. (2002). Mining, corporate social responsibility and "the community": The case of Rio Tinto, Richards Bay and the Mbonambi. *Journal of Business Ethics, 39*, 275–296.

Klins, U., van Niekerk, J. H., & Smit, A. (2010). Africa. In W. Visser & N. Tolhurst (Eds.), *The world guide to CSR: A country-by-country analysis of corporate, sustainability and responsibility* (pp. 2–15). Sheffield: Greenleaf.

Krugman, P. (1995). Growing world trade: Causes and consequences. *Brookings Papers on Economic Activity, 26*(1), 327–777.

Kufuor, J. (2005, May 27). Address at the 77th AGM of the Ghana Chamber of Mines, Accra, Ghana.

Lea, R. (2002). *Corporate social responsibility. Institute of Directors member opinion survey.* London: IoD. http://www.epolitix.com/data/companies/images/Companies/Institute-of-Directors/CSR_Report.pdf (23 Nov 2015).

Lodge, G., & Wilson, C. (2006). *A corporate solution to global poverty; How multinationals can help the poor and invigorate their own legitimacy.* Princeton, NJ: Princeton University Press.

Manteaw, B. (2007). From tokenism to social justice: Rethinking the bottom line for sustainable community development. *Community Development Journal, 43*(4), 428–443.

Marsden, C. (2001). The role of public authorities in corporate social responsibility. http://www.alter.be/socialresponsibility/people/marchri/en/displayPerson (accessed 23 Oct 2015).

Masaka, D. (2008). Why enforcing social responsibility (CSR) is morally questionable. *Electronic Journal of Business Ethics and Organization Studies, 13*(1), 13–21.

McCorkindale, T. (2008). Does familiarity breed contempt?: Analyses of the relationship among company familiarity, company reputation, company citizenship, and company personality on corporate equity. *Public Relations Review, 34*, 392–395.

McWilliams, A., Siegel, D. S., & Wright, P. M. (2005, May). *Corporate social responsibility: Strategic implications* (Working Papers in Economics, Number 0506). Department of Economics, Rensselaer Polytechnic Institute.

Ofori, D. (2007). Social responsibility and ethics in Ghana: Stakeholders' expectations and challenges. In J. Okpara (Ed.), *Management and economic development in Sub-Saharan Africa: Theoretical and applied perspectives* (pp. 63–97). London: Adonis and Abbey.

Ofori, D. F., & Hinson, R. E. (2007). Corporate social responsibility (CSR) perspectives of leading firms in Ghana. *Corporate Governance, 7*(2), 178–193.

Parker, A. R. (2006, November 20–22). Mining and sustainable community development: A framework for assessment of community impacts. In *2nd international academic symposium on corporate citizenship*. Accra: GIMPA.

Pendleton, A. (2004). The real face of corporate social responsibility. *Consumer Policy Review, 14*, 77–83.

Pinney, C. (2001). Imagine speaks out. How to manage corporate social responsibility and reputation in a global marketplace: The challenge for Canadian business. http://www.imagine.ca/content/media/team_canada_china_paper.asp?section=media (accessed 23 Dec 2015).

Rana, P., Platts, J., & Gregory, M. (2008, September). Exploration of corporate social responsibility (CSR) in multinational companies within the food industry. In *CRR conference*. Institute for Manufacturing, University of Cambridge. Available: http://www.crrconference.org/downloads/rana.pdf (accessed 12 Nov 2015).

Reinhardt, F. L., Stavins, R. N., & Vietor, R. H. K. (2008, April). Corporate social responsibility through an economic lens. *Resources for the Future*, RFF - DP 08-12, www.rff.org.

Sandbroke, R., & Mehta, P. (2002). *Rapporteurs' report from the OECD Global Forum on International Investment Conference 'FDI and environment—Lessons from the mining sector'*. Paris: OECD.

Sarpong, S. (2010). *Multinational and the social perspective*. PhD Thesis. Cardiff: Cardiff University.

Sriramesh, K., Ng, C. W., Ting, S. T., & Wanyin, L. (2007, July 3–5). *Corporate social responsibility and public relations: Perceptions and practices in Singapore*. Paper submitted to the 14th international public relations symposium, Bled, Slovenia.

Taylor, J. K. (2004, March 11). Mining investment: A catalyst for developing country relations. In *Conference speech, Ottawa, Canada.*

Turker, D. (2009). Measuring corporate social responsibility: A scale development study. *Journal of Business Ethics, 85*(4), 411–427.

van Marrewijk, M. (2003). Concepts and definitions of CSR and corporate sustainability: Between agency and communion. *Journal of Business Ethics, 44*, 95–105.

Warhurst, A. (1998, June 4). Corporate social responsibility and the mining industry. In *Presentation to Euromines*. Brussels, Belgium: Mining and Environment Research Network.

World Business Council for Sustainable Development (WBCSD). (2000, January). *Corporate social responsibility: Making good business sense*. Geneva: WBCSD. Available: http://www.wbcsd.org/DocRoot/IunSPdIKvmYH5HjbN4XC/csr2000.pdf (accessed 16 Nov 2015).

Corporate Social Responsibility in Challenging Times in Developing Countries

Shame Mugova, Moono Mudenda, and Paul R. Sachs

1 Introduction

There has been limited research in developing countries particularly how their unique local conditions impact CSR implementation. It has been argued that the mainstream CSR agenda is not attuned with the priorities and realities in the South, and therefore is not particularly relevant to local contexts and actors (Blowfield & Frynas, 2005; Fox, 2004; Frynas, 2005; Utting, 2003). The developing country societies are different from developed countries in many respects and managers need to address CSR issues in different cultural context (Dobers & Halme, 2009). Mainstream CSR was also criticized for universalizing a set of conditions that do not exist in developing countries (Newell, 2005). Fox (2004) argued that the treatment of the 'business case' as an undifferentiated whole, and CSR as monolithic in mainstream CSR discourse, is both problematic and misleading. CSR needs to be adapted to local issues, different cultural contexts and the weak institutional environments, such as in developing countries. There is also need to focus not only on what business has done or failed to do but on the complex relationships that inform the processes that link CSR initiatives and the desirable developmental outcomes (Idemudia, 2011).

Arli and Lasmono (2010) suggest four rationales for focusing on CSR in a developing country as distinct from CSR in the developed world: (1) developing countries represent the most rapidly expanding economies, and hence the most

S. Mugova (✉)
Durban University of Technology, Durban, South Africa
e-mail: mugovas@gmail.com

M. Mudenda
APLP Alumni, East West Centre, Honolulu, HI, USA

P.R. Sachs
NHS Human Services, Philadelphia, PA 19119, USA

S.O. Idowu et al. (eds.), *Corporate Social Responsibility in Times of Crisis*,
CSR, Sustainability, Ethics & Governance, DOI 10.1007/978-3-319-52839-7_11

lucrative growth markets for business; (2) developing countries are where social and environmental crises are usually most acutely felt in the world; (3) developing countries are where CSR principles or lack of them are likely to have the most dramatic social and environmental impacts (both positive and negative); and (4) developing countries present a distinct set of CSR agenda challenges which are collectively quite different to those faced in the developed world.

With these points in mind, this chapter examines the challenges of CSR in developing countries, with particular attention to Zimbabwe. Economically developing countries pose unique difficulties for implementation of CSR compared to economically developed countries. Moreover, when one considers challenging times, which is the focus of this book, the difficulties are compounded. Developing countries are characterized by limited institutional infrastructure, poor governance and various social, economic and political weaknesses. Arguably, challenging times are all the time in a developing country, the rule rather than the exception.

Nonetheless, businesses in developing countries are aware of and are implementing CSR strategies. That they are doing so despite great obstacles is impressive and warrants consideration in how CSR can be best incorporated into a business.

The present chapter examines CSR in developing countries with particular attention to Zimbabwe. It is felt that the examples and lessons learned from Zimbabwean businesses are representative of challenges in other developing countries and instructive for the field as a whole.

As a starting point, an overview of an ideal CSR model will be presented and then specific examples from Zimbabwean industry will be reviewed. In this chapter the term "Business" will be used to refer to many different organizations that differ in size, industry, ownership and operations.

The situation in developing countries poses unique difficulties for implementation and sustaining of Corporate Social Responsibility (CSR) principles compared to more economically developed countries. This chapter examines and discusses these challenges with the aim of describing how CSR can be maintained in developing countries despite these difficulties. This chapter examines and discusses these challenges, with particular attention to Zimbabwe.

Arguably, challenging times in developing countries are really all the time, the rule rather than the exception. After a brief overview of an ideal CSR model, this chapter examines several Zimbabwean companies and how they have or have not addressed CSR in their operations. Cases of companies are used and data is sourced from annual financial statements and news. Discussion focuses on what may be key factors that support or weaken the implementation of CSR in a developing country and what further research and practice guidelines are recommended based on this discussion. It is felt that the situation in Zimbabwe is and can be instructive for the implementation and sustainability of CSR in other developing countries. CSR practices are established despite the obstacles to doing so in developing countries and this fact supports the validity of CSR as an essential element of a business culture.

2 CSR Overview

According to Carroll (1991), corporate social responsibility (CSR) involves the conduct of a business which is economically profitable, law abiding, ethical and socially supportive. To accomplish these goals, CSR requires a balanced approach including the economic, social and legal pillars of business operations.

Challenging times include times of recession or other circumstances that have a negative effect on economic stability and economic growth. In such situations, businesses may withdraw from strategic CSR activities due to severe resource scarcity and heightened uncertainty in the macroeconomic environment. Businesses need to find ways to overcome these challenges because CSR activities not only generate favourable stakeholder attitudes and better support behaviours (e.g. purchasing, employment, investment opportunities), but also, over the long run, build corporate image, strengthen stakeholder–company relationships, and enhance stakeholders' advocacy behaviours (Du, Bhattacharya, & Sen, 2010).

The elements of CSR are illustrated in Fig. 1. These include the following:

- **Philanthropicresponsibilities**: entail making voluntary contributions to society, for example by giving time and money to social activities
- **Ethical responsibilities**: entail doing what is right, just and fair and avoiding harm

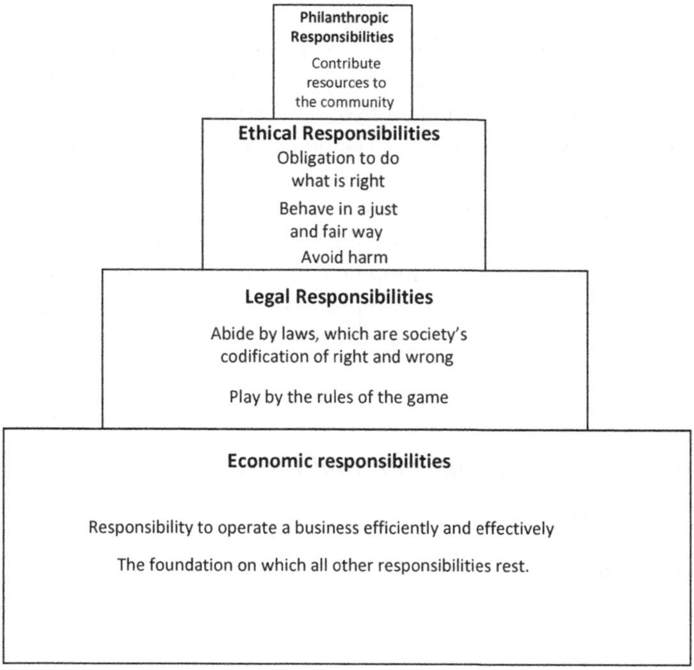

Fig. 1 The pyramid of corporate social responsibility (Carroll, 1991)

- **Legal responsibilities**: entail compliance with laws and playing by the rules of the game
- **Economic responsibilities**: entail profits for shareholders, good jobs for employees, and quality products for customers

Economic and legal responsibilities are socially required (i.e., mandatory), ethical responsibility is socially expected, while philanthropy is socially desired (Windsor, 2001).

Lantos (2001) distinguished between three types of CSR, which he referred to as ethical, altruistic, and strategic. Ethical CSR is morally mandatory and goes beyond fulfilling a firm's economic and legal obligations to its responsibility to avoid harm or social injuries, even in cases where the business does not directly benefit. Altruistic (humanitarian, philanthropic) CSR involves contributing to the common good at the possible, probable or even definite expense of the business. Strategic CSR is the fulfilment of a firm's social welfare responsibilities and it creates a win-win situation in which the corporation and one or more stakeholder benefits. Actions are taken because they are right, not merely because they are mandated by law or are profitable (e.g., money spent on product safety or pollution control) (Jamali, 2007).

2.1 General CSR Concepts

Elements of CSR can be divided into those affecting the internal and those affecting the external business environment. Internal elements include corporate structures and policies whilst external include rule of law, regulation, competition, standards, and complementary CSR institutions. The functioning of the overall CSR system is also influenced by globalization, crises and changes in the political and macro-economic environment and the government.

The following section describes the ideal implementation of CSR with attention to the internal and external factors following the model outlined by Pattberg (2009). This description serves as a basis for comparison with the actual situation of organizational CSR in the developing country of Zimbabwe.

2.2 Elements of CSR Internal and External to the Organization

2.2.1 CSR Strategy

Internal

The CSR Strategy is the initial point of the organizational CSR practice. Therefore, the CSR strategy needs to be defined clearly with the ultimate goal in view. A CSR strategy is a road map for implementing and executing CSR issues and activities. The strategy sets the firm's direction and scope over the long term allowing the firm to be successful by using its resources within its unique environment to meet market needs and fulfil stakeholder expectations (Hohnen & Potts, 2007).

Internal CSR focuses on what can be done internally to improve the well-being of human capital, the lives and productivity of employees as well as their impact on profitability at the bottom line (KPMG, 2011). A company needs to build a solid support for the strategy within the organisation before setting the goals to be achieved with this strategy (Pattberg, 2009).

CSR strategy effectiveness requires support and participation from top management in the organization and involves identifying and engaging employees that are passionate about CSR. The CSR internal support from key staff makes it easier to define the CSR Strategy for a business. When CSR strategy is aligned with the company's core business it can lead to a wide range of bottom line benefits such as increased employee motivation and retention, higher productivity, reduced absenteeism and more efficient environmental-friendly production techniques among many others (KPMG, 2011).

External

CSR Strategy is a plan of action designed to achieve the overall aim of the company which will act as the blue print for the responsible business practice (Pattberg, 2009). Stakeholders of the company and the public expect a visionary strategy that shows responsible behaviour towards the society. External CSR strategy promotes the pursuit of positive impacts on society including economic and the natural environment from corporate activities and operations. The company's CSR strategy should encourage reporting on social and environmental impacts. The system promotes accountability and transparency through the reporting process which enhances democracy, and promotes social development (KPMG, 2011). The products and services of a company are ultimately aimed at the customers. The cost and quality of products is of interest to customers, awareness and communication should be the priority of the company in order to develop customer satisfaction and loyalty. The process of producing goods and services should consider the social, environmental, supply-chain and other such impacts.

2.2.2 CSR Management System

Internal

The management system of CSR within the company will outline what needs to be done to implement the strategy successfully and achieve the desired outcome (Pattberg, 2009). At this stage engagement of more individuals within the company is required. There must be a collection of feedback on how the CSR Strategy has been received and then building on engage the most positive people. It involves setting up a team that supports CSR then pooling their knowledge and building a sound management plan on how to deliver the different elements of the strategy.

The CSR management system is customized to each organisation for it to be successful. One can go around and research about the management systems of other companies but this management system really needs to be so customized to the unique organizational context (Pattberg, 2009). Therefore, while models and past practices may provide some guidance, the unique CSR management system needs to be built from scratch.

External

The CSR management system is a vehicle to get more external recognition of your activities. When an organisation disseminates information they will in turn they receive feedback from stakeholders. An organisation also gets helpful information and tips from other organizations about how they are managing their CSR aspects. The organisation receives feedback which in turn can be used to improve the CSR management system.

2.2.3 CSR Reporting

Internal

After successful set up of the management system, the organization must report the first results of the company's CSR activities (Pattberg, 2009). CSR reporting is usually done by companies on an annual basis. CSR reporting helps to assess whether the company has created and done something interesting and worth reporting for the various stakeholders. Internally CSR reporting will help to test the company's CSR management and coordination skills. The primary goal is to produce a meaningful publication that will be of interest to stakeholders. The report includes qualitative and quantitative data which has to be gathered from people who understand the organisation's CSR strategy and initiatives.

External

CSR Reporting is a crucial tool for an organisation's CSR policy and activities. Externally the CSR Report together and the CSR management system will be the face of the organization. CSR reporting is mainly carried out through non-financial reporting which increases transparency with respect to corporate actions concerning social and environmental issues. Financial accounting now includes sustainability accounting which incorporates social and environmental performance.

2.2.4 Stakeholder Engagement and Communication

Internal

Stakeholder engagement and communication keeps all of these areas together and connected (Pattberg, 2009). Stakeholders need to be engaged on a continuous basis in order to build real long term value in a CSR Strategy. A report of what the organization has been doing is part of a practice of successful stakeholder engagement. Key to stakeholder engagement is transparency within company about what management and employees do as CSR activities. Effective stakeholder engagement starts in one's own backyard first. Only after stakeholder engagement is functional within the organisation can one turn to the external perspective of stakeholder engagement.

External

As noted earlier, only when internal stakeholder engagement and communication is running smoothly can an organisation shift focus to the external aspects (Pattberg, 2009). The engagement of stakeholders includes interest groups, the local community and other stakeholder groups that have been identified as being important to the organization. Innovations should involve use of modern communication tools and devise effective ways to identify the particular stakeholder opinion.

It is recognized that the above description represents an optimum implementation of CSR. With this background, though, the actual conceptualization and implementation in CSR can be seen higher relief. The activities of five Zimbabwean companies are presented and discussed in light of the previous narrative. This data was collected from company annual reports and other news media.

3 Presumptions That Are Challenges in Developing Countries

The business organization in a developing country faces different challenges than one in an economically developed country. For example, business owners and managers in developing countries must be willing to absorb large costs associated with inefficiencies or bear the additional cost of pushing for legal reform to remove these inefficiencies. The capital markets, product markets and supply chains are less well-established in developing countries (Gibson, 2003). The capital markets have limited resource allocation efficiency. Information is disseminated unevenly across the market.

The most important economic role played by governments is providing credible policy environment for investors and lowering the costs of bargaining, contracting, monitoring and enforcement (Henisz (2002). Businesses operating in developing economies usually incur more costs in bargaining and contracting compared to those in the developed countries because of the differences in institutional infrastructure. The process wastes resources which could be invested in the company's core business.

Institutional Infrastructure Corporate governance intersects organizations, markets, and political systems, and hence is a microcosm of the broader institutional environment (Young, Peng, Ahlstrom, Bruton, & Jiang, 2008). CSR presumes the existence of some infrastructure that may be weak or non-existent in developing countries (Examples: licensing, auditing, employee protections, environmental protections authorities). Corporate governance, emerging economies typically do not have an effective and predictable rule of law which, in turn, creates a 'weak governance' environment (Dharwadkar, George, & Brandes, 2000; Mitton, 2002; Young et al., 2008). Developing countries are affected by issues such as lack of rule of law including lack of laws relevant to key business operations and the lack of enforcement of available laws. The transition to professional management is always difficult in emerging economies because of the weak institutional environment and it is common for even the largest firms to still be under the control of the founding family (Young et al., 2008). Developing country businesses are commonly closely held by the founding family, they do a relatively inadequate job of enforcing shareholders' legal rights, and need to improve their accounting and transparency. Companies in emerging market economies are anticipated to have corporate governance problems arising from the family ownership structures and also conflicts between majority and minority shareholders. A given piece of legislation might be good in principle but its effectiveness in practice is affected by the institutional environment.

Cultural Uniformity CSR is part of, not an addition to or a layer on top of, corporate culture. Such cultural uniformity may not exist in companies within developing countries because of history of colonialism, predominance of MNCs, short history of incorporation by the company and outsize influence of government. Developing economies face additional complications because their corporate boards lack the institutional memory and familiarity that boards in developed market economies have (McGee, 2009).

Communication/Transparency Sharing of and equal access to information is not as evident in developing countries due to weak communication infrastructure, limited education of populace including but not limited to illiteracy, autonomy of press and other media. Companies need to strengthen stakeholder–company relationships, there is therefore a need for companies to communicate CSR more effectively to stakeholders (Du et al., 2010). Corporate social responsibility communication is a very delicate matter a key challenge of CSR communication is to overcome stakeholder scepticism and to generate favourable CSR attributions. Companies must also communicate CSR more effectively to various stakeholders.

Markets Institutional protection of minority shareholders sets an upper bound on the potential for expropriation by majority shareholders in developed economies with well founded market institutions, but such protection is usually lacking in emerging economies (Young et al., 2008). The general populace of a developing country poses a challenge to implementation of CSR due to still developing levels of literacy, physical health and purchasing power. Although the populace may be characterized by low individual purchasing power, its aggregate purchasing power also presents a huge market potential. Lack of education leads this group of people to be uninformed consumers with markedly fewer choices or options for advocacy than their better off, better educated countrymen and women. Most of those consumers do not know their rights as customers and are ignorant of CSR.

Environment Environmental policies, procedures, management systems and strategies as such are not enough (Halme & Huse, 1997), there is need for compliance and verification of environmental performance. Businesses operating in developing countries do not have incentives to invest in pollution control because of weak implementation of environmental regulations. The pace of industrial development may exceed the concomitant pace of environmental regulations and control systems.

3.1 The Case of Zimbabwe

Zimbabwe is a landlocked developing country in Southern Africa with a 2013 population of **14,150,000**. The first British explorers, colonists, and missionaries arrived in the 1850s, and the massive influx of foreigners led to the establishment of the territory Rhodesia, named after Cecil Rhodes of the British South Africa Company.

The current country of Zimbabwe was founded in 1980 when the country was granted independence from colonial rule by Britain. Since becoming independent, however, the country has had inconsistency in its governing and related institutions. The strong central government has centered on autocratic leader Robert Mugabe and has not allowed the concomitant development of other judicial, financial and communications institutional infrastructure. The state institutions have been privatized by the country's leader and his ruling party. There is no democracy, freedom of expression and press freedom. The governance has been characterized by human rights abuses, fraudulent elections and failure to uphold the rule of law and violation property rights which led to targeted sanctions on Mugabe and his ruling elite by Europe and the Zimbabwe Democracy and Economic Recovery Act of 2001 passed by United States Congress. Corruption and economic mis-management of the country's resources is very high and with a look east policy adopted after a falling out with the west which has not yielded the desired economic benefits.

As a result of this legacy, Zimbabwe has one of the most uncompetitive business environments and is ranked among the worst in terms of ease of doing business. The economy is shrinking with many company closures citing liquidity crises and a lot of workers are being laid off. Capacity utilisation levels in the manufacturing sector have dropped to 39.6% (Mtomba, 2014). July 2013 National Social Security Authority (Nssa) Harare Regional Employer Closures and Registrations Report for the period July 2011 to July 2013 shows 711 companies in Harare closed down, rendering 8336 individuals jobless (Zimbabwe Independent, 2013). The Registrar of Companies struck more than 176 companies off the register in 2013 (Mbiba, 2014). Zimbabwe's social economic problems are largely blamed on political mismanagement and poor governance (Zimbabwe Independent, 2013).

Currently, Zimbabwe has very high unemployment levels estimated above 80% (Mtomba, 2014). Companies are either downsizing or closing down, sending thousands of workers out of employment, making it virtually impossible for graduates and other qualified people to get jobs, this has led to a massive exodus of qualified people to other countries.

The political and economic crises that battered the Zimbabwean economy from 1998 contributed to the sharpest drop in gross domestic product (GDP), hyper-inflation and poverty rates rising to over 72%, with over a fifth of the population in extreme poverty. Zimbabwe abandoned the now infamous Zimbabwe dollar in 2009 after years of record high levels of inflation. The country is now using the US dollar as its currency and also the South African rand.

Life expectancy is 56 years. In 2014 the GDP (PPP) of Zimbabwe was estimated at US$13.66 billion. The GDP per capita was $931.2 World Bank (2013). Agriculture, foreign trade, and tourism are the main currency earners of Zimbabwe. Mining is a significant sector; the country has some of the largest platinum and diamond reserves in the world. Zimbabwe exports tobacco, gold, nickel, cotton, ferro-alloys, textiles, and agricultural products.

Zimbabwe's economy remains in a delicate state, with an unsustainably huge external debt and extensive deindustrialisation and informalisation according to the World Bank. In 2009, Zimbabwe commenced on a period of stabilization and growth ushered in by the Global Political Agreement and the adoption of a multi-currency regime, with the US dollar becoming the major trading currency. During 2009–12, the economy recuperated with growth rates averaging around 8.7% according to World Bank statistics. Since 2012, however, the economy has experienced a sudden slowdown owing to deteriorating terms of trade, adverse weather and increasing policy instability. The World Bank states that growth slowed to 4.5% in 2013, 3.2% in 2014, and was projected at 1.5% in 2015 by the Ministry of Finance. The banking sector saw an acute increase in non-performing loans as business adjusted to new external conditions resulting in contraction of banking activities and reduced liquidity in the market. The liquidity crises have been compounded by unfavorable trade with Zimbabwe importing more than it is exporting and particularly the use of the US dollar with major currency inflows coming from international trade. Liquidity challenges have led to fall in demand for goods and services, downsizing, closure of companies and more job losses. In 2013 Zimbabwe adopted a new constitution and is in the process of aligning its laws and regulations to this new framework. The Zimbabwean industries are struggling in the face of foreign competition, mainly from South Africa and China which are cheaper than locally produced goods.

Zimbabwe recorded a trade deficit of US$290.99 Million in December of 2014. Balance of Trade in Zimbabwe averaged –US$422.44 Million from 1991 until 2014 (Fig. 2).

In recent years, Zimbabwe has run systemic trade deficits due to decline in exports. Statistics from Zimstat show that in 2014 Zimbabwe imported $2.5 billion worth of goods from South Africa, and exported goods worth about $170 million to Africa's second largest economy (Kachembere, 2015).

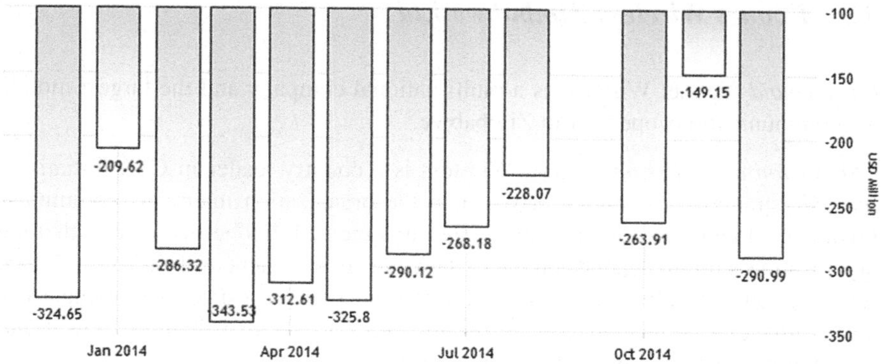

Fig. 2 Zimbabwe balance of trade. Source: www.zimstat.co.zwl ZIMSTAT, ZIMBABWE

Table 1 Labor force by occupation

Country	Agriculture	Manufacturing/construction	Service	Year
Brazil	15.3	13.7	71	2013
Canada	2	19	76	2006
Germany	1.6	23.6	73.8	2011
India	49	20	31	2012
Nigeria	70	10	20	1999
Pakistan	44	24	33	2013
Zimbabwe	66	10	24	1996

Source: Central Intelligence Agency (2015), USA: The World Fact Book https://www.cia.gov/library/publications/resources/the-world-factbook/fields/2012.html#84

Table 2 GDP distribution by sector

Country	Agriculture	Industry	Services	Year
Brazil	5.9	22.2	71.9	2015 est
Canada	1.6	28.9	70.5	2015 est
Germany	0.7	30.2	69.1	2015 est
India	16.1	29.6	54.4	2015 est
Nigeria	20.3	23.6	56.1	2015 est
Pakistan	25.5	19	55.5	2015 est
Zimbabwe	20	26	53.3	2015 est

Source: Central Intelligence Agency (2015), USA: The World Fact Book https://www.cia.gov/library/publications/resources/the-world-factbook/fields/2012.html#84

Typical of a developing country, Zimbabwe has the largest portion of its workforce involved in agriculture rather than manufacturing or service. The actual GDP contribution of agriculture is lower than the workforce participation, but still higher than that of developed countries. This distribution is important for consideration of CSR because it speaks to the role that the general population, the country's marketplace, plays in a business' employee and community relations (Tables 1 and 2).

3.2 Econet Wireless Zimbabwe Ltd

Background Econet Wireless is a multi-national company and the largest mobile telecommunications operator in Zimbabwe.

CSR Vision and Mission Econet Wireless is a country leader in CSR initiatives. Its CSR strategy and policy drives how the organisation maintains community involvement and how such initiatives are implemented. Econet recognizes that the organisation's future depends on the sustainable development of the communities it serves. Econet Wireless believes that the true measure of a successful company is not its financial performance alone but its ability to use that financial power and community prestige to positively transform its communities.

CSR Activities Econet Wireless is now moving beyond corporate social responsibility to social innovation. Econet believes that technology that does not transform lives is irrelevant. Econet Wireless constantly reviews its social role is in order to remain relevant to the communities it serves. The company offers education scholarships for students up to university level, supports football in many African countries and participates in HIV/AIDS education programmes. "I believe the issue of giving back to community is not about how much in terms of resources a company has, but it should be a gesture of goodwill and willingness that is sincere and aimed at impacting positively on the lives of the recipients," said Mr Mboweni the Econet Wireless Zimbabwe chief executive officer.
The company reported on CSR in its 2013 annual report that as part of its community involvement and bringing community transformation, it has established the Higher Life Foundation which actively manages the three corporate responsibility vehicles: Capernaum Trust, caring for orphaned and marginalised children, Joshua Nkomo scholarship, sponsoring socially responsible academically gifted students, and National Health Care Trust, promoting the health of the nation.

The company also addressed responsibility internally. The 2014 annual report noted environmental and social accidents in Zimbabwe that led it to develop processes for accident and incident reporting. The Econet Group in 2013 reviewed its safety, health and environmental policy to formulate broader commitments to Environmental and Social issues in line with International Finance Corporation (IFC) requirements.

3.3 Zimbabwe Platinum Mines (Zimplats)

Background Platinum giant, Zimbabwe Platinum Mines (Zimplats), is a subsidiary of South Africa's Impala Platinum Mines (Implats). The company runs four mines in Mhondoro-Ngezi, Mashonaland West Province.

CSR Vision and Mission Zimplats estimates that they spent US$200 million by 2010 in their CSR programmes for the 12 years they had been operating in the country. Zimplats as a multi-national company is one of the few companies in Zimbabwe with CSR policies and strategies.

CSR Activities Most of the CSR investment has been towards infrastructure development. Zimplats is committed to exploring, mining and processing operations in a responsible manner and to upholding the human rights of employees and stakeholders in line with legislation and the United Nations Guiding Principles on Business and Human Rights.

Although overall Zimbabwe has been characterised by human rights violations, Zimplats seeks to influence human rights practice within the company's sphere of influence in a sustainable manner. Zimplats is committed to treating all stakeholders fairly and with dignity and they also recognise the imperatives for corporate social responsibility in the communities of their operations. The company seeks to have open dialogue with stakeholders and participate in community engagement activities.

Zimplats' corporate social initiatives fall within education, health, water and sanitation and income generating projects. The Zimbabwean government is on record as accusing foreign owned mining companies of exploiting natural resources without ploughing back to the local community. The indigenization of especially mining companies across Zimbabwe and the subsequent benefits of such a process to rural communities living near mine sites have been the major issues affecting multi-national mining companies operating in Zimbabwe.

As a result, Zimplats has responded to activism among communities living near mines by implementing a wide variety of schemes which have brought socio-economic benefits to the local area. A large number of local youths from the wider Mhondoro Ngezi area are now employed as wage labourers at the mine as a result of the company employing local people. The mine's wider corporate social responsibility programs such as the drilling of boreholes, construction of roads, schools and clinics have made a significant contribution to the local economy at a time when many rural communities are facing socio-economic challenges.

Zimplats is of particular interest because it has managed to implement its CSR policies and adherence to business ethics in a country which is characterized by wide economic swings and overall hardships, and an ingrained political corruption culture.

3.4 Innscor Africa Limited

Background Innscor is a diversified holding company which manufactures, distributes and retails food. It is a fast moving consumer and durable goods supplier to the mass market.

CSR Vision and Mission Innscor's CSR incorporates sustainability strategy and governance, social cohesion and community investments, work place wellness, environmental and product safety. The Group's strategic CSR approach includes striving to operate its business in a sustainable manner that recognises environmental and social impacts. The Group believes that identifying, measuring and being accountable to its stakeholders through sustainability reporting (Economic, Environmental, Social and Governance) has potential for long term business success. In 2014, the Group adopted the Global Reporting Initiatives (GRI)'s G4—Sustainability Reporting Guidelines to guide how it manages material sustainability issues. Sustainability strategy and governance incorporates the strategic approach, governance, inclusivity and responsiveness, supply chain and sustainable capital management. The Group strives to continuously improve and maintain human capital and community investments at appropriate standing for the purpose of ensuring the Group' s long term business success and sustainability.

Innscor believes that stakeholders are an integral capital contributing to the long term value creation and sustainability of the business. The Group's Stakeholder Engagement strategy is integrated with its risk and business development management. Material issues discussed during stakeholder engagement are assessed and weighted for appropriate action or responses. The issues are further collated at company level for Group management attention in defining material issues for the report content and boundaries. Stakeholder engagement provides a crucial business strategy for identifying key risks and opportunities for responsiveness and value creation respectively.

CSR Activities SPAR, one of Innscor subsidiaries, has been involved with sponsorship of Childline Zimbabwe and of various charities and orphanages. In support of the company's first brand pillar—the promotion of a healthy and active lifestyle, SPAR sponsored the national 7s rugby side who qualified for the 2013 Rugby 7s World Cup. Also the company sponsors TAG Rugby which focuses on the introduction and development of rugby as a sport at the grassroots level.

The Group believes that partnerships are an effective means for addressing economic, environmental and social challenges. Through subsidiaries such as SPAR, Innscor, offers support to communities through giving out grocery hampers to vulnerable members of the community such as widows. Other initiatives include the sponsorship of Childline Zimbabwe and various other charities as well as orphanages. Women are supported through the "mother of the year" annual award. Furthermore, clean up drives are often organized. Irvines, another Innscor subsidiary, provides chicken farming trainings. Eggs and poultry are donated to local communities .

3.5 NMB Bank

Background NMB Bank is a commercial bank established in 1993 in Zimbabwe. NMB was the first locally-owned bank to be established in Zimbabwe, and has a strong brand with a reputation for excellent customer service. The bank was established by a group of Zimbabwean entrepreneurs who had previously held senior positions in renowned international financial institutions such as the World Bank and the International Finance Corporation. The impetus for the creation of

NMB Bank came from the liberalisation of the Zimbabwean economy and financial sector in the early 1990s. NMBZ has a dual listing on Zimbabwean stock exchange and London Stock Exchange.

CSR Vision and Mission Although serving a large number of citizens and businesses, the bank does not evidence strong CSR activities and those that are shown are limited to philanthropic activities. Moreover, internally there is only one woman listed in the 14 member company leadership team. There is no information about how they treat their employees, the ratio of men and women, number of employees who have disabilities or benefits for employees.
Nonetheless, the bank's philanthropic contribution to communities should be noted. NMB sponsors the arts and sports and also donates to deserving charities from time to time. The activities and charities supported during the year ended 31 December 2014 included special education needs, health and social services, the environment and the arts.

NMB bank does not support or provide micro-finance. But the bank does have a loan facility open to schools that facilitates payment of school fees for those school children who would otherwise be unable to afford a decent education. Through this initiative, NMB therefore supports women, families and consequently local communities .

NMB has a commitment to its clients and the community and is conscious of their corporate responsibility to *sustainability* and social justice. However, the NMB website does not have a comprehensive CSR policy and strategy documented and newspaper sites do not have a news articles about the banks CSR initiatives. NMB bank CSR is not well developed despite their impressive success in their sector.

3.6 Sino Zimbabwe

Background Sino-Zimbabwe Cement Company (SZCC) is a joint business venture between a Chinese Foreign Direct Investment partner, China Building Material Corporation for Foreign Econo-Technical Co-operation (CBMC), and the Industrial Development Corporation of Zimbabwe Limited (IDC). CBMC contributed 65% of the original funding. Sino Zimbabwe's vision is to be the preferred manufacturer of high quality cement and related construction products.

CSR Vision and Mission Sino Zimbabwe Cement Company policies include safety, health and the environment, within those policies, Sino Zimbabwe aims to foster ethical ways that increasingly benefit society, the economy as well as the environment. At Sino Zimbabwe, all employees are obligated to prevent or to refuse to perform any task which is considered unsafe or which cannot be performed safely.

CSR Activities In addition to the abovementioned safety culture and policies, the company offers assistance to the community in the form of cement donations for local construction projects, employment of casual and contract labour from the local community, supporting schools by sponsoring transportation to and from sporting activities and transport if also offered to teachers from local schools. Finally the Sino Zimbabwe company clinic offers voluntary counselling and testing for HIV, provides essential drugs and surgical sunrise to the community, etc.

4 CSR Problems and Issues in Marange Diamond Mining Fields

Apart from review of particular companies in Zimbabwe, it is worthwhile to examine a particular economic and political case within the country because it illustrates some of the challenges and pitfalls of CSR implementation. This case involves the Marange diamond mines.

Background Zimbabwe's wealth of natural resources is a source of national pride and also an economic asset. Therefore, natural resources are frequent lightning rod for conflict between business and community. The Marange diamond fields are an area of extensive small and large scale diamond mining in Manicaland Province on the eastern side of Zimbabwe. Alluvial diamond deposits in the Marange were discovered here in June 2006 by African Consolidated Resources. Currently the Marange mines are among the largest producing diamond mines in the world. Since 2006 control of the mining has been a source of contention between various stakeholders including the developer, other business interests, the national government and the local villagers.

A brief chronology highlights some of the conflicts that have been noted in the 10 year history of the mines:

- 2006: African Consolidated Resources (ACR) discovered the deposits and developed them.
- 2006: ACR had claims taken from them by government.
- 2006–2008: Zimbabwe President Robert Mugabe and a small group of insiders and generals, known as the Joint Operations Command (JOC), took over mine and control operations.
- July 2009, a Kimberley Process Review Mission visited Marange and concluded there were 'credible indications of significant non-compliance' with the minimum requirements of the Kimberley Process Certification Scheme (KPCS). The major concerns were the evidence of government involvement in human rights abuses, smuggling and lax controls that compromised the entire chain of production.
- 2010: Ministry of Mines accepted expressions of interest from companies willing to enter into joint ventures agreements to mine in Marange under the auspices of the Zimbabwe Mining Development Corporation (ZMDC).
- 2010: Diamond mining firms were formed as joint ventures between the state-owned ZMDC and Marange Resources with Chinese firms. The Chinese owned

stakes in mining companies which were exempt from the indigenisation policy. It was noted that China and Zimbabwe have a relationship whereby Beijing supplies weapons and Chinese firms are allowed to mine diamonds (Chimonyo, Mungure and Scott, 2013).

- 2012: Former Finance minister Tendai Biti confirmed that he had trouble with some government officials who were helping the Chinese to siphon money out of the country. Biti said Chinese diamond firm, Anjin failed to remit revenue from its operations in Zimbabwe's Marange fields in the first quarter of 2012 amid fears that the money was used clandestinely. "During that time, we did not receive a single cent from Anjin, yet Anjin is seven times bigger than some of the other [diamond] businesses," he said. Biti said during his tenure there were suspicions that the funds were funnelled elsewhere, outside the government central coffers. "Clearly, there might be a parallel government somewhere in respect of where these revenues are going, and are not coming to us," he said
- 2014: 1000 cattle died and it was discovered that the river was being contaminated by substances discharged by mines operated by Anjin Investments and Marange Resources.
- 2015: Marange villagers accused the government's Environmental Management Agency (EMA) of folding its hands whilst diamond mining businesses were destroying the environment. The businesses operating in the diamond mining fields were not forthcoming with regards to honouring their obligations to the Marange-Zimunya Community Share Ownership Trust. The villagers further accused the state watchdog of making profits from pollution fines paid by the businesses.
- 2016 the government of Zimbabwe has not renewed diamond mining licences for the firms and they want to consolidate all mining operations under one mining firm owned and controlled by the state. President Robert Mugabe revealed that diamond worth US$15 billion dollars could have spirited away by cartels which were working cahoots with the seven mining firms for the past 6 years. Yet the firms which were all mining in Marange had a government shareholding of 50%.
- 2016: Manicaland has remained one of the poorest provinces in Zimbabwe despite vast diamond fields.

5 Lessons Learned: The Specific Challenges of CSR in a Developing Country

In the latter twentieth century the primary social responsibility of a business was claimed to be maximizing profits and shareholder value (Friedman, 1970). Since then, social responsibility has evolved to include a greater recognition of the business's duty to its employees, the markets it serves and communities in which it operates. As noted, this evolution has led to CSR best practices for both internal and external business operations.

Yet, this evolution and practice has primarily been a First World or, more specifically, Euro-centric process. The current model of CSR has been refined in countries that have well-established institutions, a history of political stability and a well-educated populace with a long life expectancy.

In the present day international business environment, with interconnection that will only intensify in the future, ideal CSR principles are being pressed on businesses in other parts of the world, where these preconditions noted above do not apply. Is this a benevolent partnership seeking to establish international standards of business operations and social concern? Or is it another example of paternalism and latent colonialism, where richer nations claim they know what is best for poorer nations?

The success of CSR principles in the developing world depends on partnership not paternalism.

The challenges of developing countries must be recognized as givens, not idiosyncrasies. The burden on a business in a developing country is huge. The business must often create infrastructure within itself because this infrastructure does not exist externally in the developing nation.

Yet, over time, if this occurs, there can be a natural disincentive for a business to cede power to a growing national infrastructure. Why should the business abide by the central government's auditing standards when it has managed perfectly well without them for years? In fact, where was the central government when the business was trying to manage changes in supply chains, labor force development and recession on its own? How can a business trust the governmental infrastructure when the government's own integrity and accountability is in question?

In the small sample of Zimbabwean businesses reviewed here, it can be seen that CSR practices are established despite the obstacles to doing so. This fact supports the validity of CSR as an essential element of a business culture, rather than a frill added as existing culture. If CSR were a luxury, then businesses in a developing country would have every reason not add it.

The existence of CSR practices in businesses within the extremely challenging business environment of Zimbabwe raises the question about what factors led to CSR implementation. Consideration of the special case of the Marange diamond mines highlights what can happen when these factors are ignored. Several factors are proposed here for consideration and as suggested for future study.

Leadership

Commitment and charisma of leader helps to drive the development and sustainability of CSR.

Such a leader could be more critical to the establishment of CSR in a developing country than a developed country. In a developing country resources and opportunities are not distributed evenly among the general population. There may also be less precedent for autonomy and advocacy. Therefore it requires an exceptional leader to demonstrate these qualities and lead a business-wide CSR implementation.

Strive Masiyiwa is such a charismatic and visionary leader. The electrical engineer who founded the Econet Wireless Group is probably best known as a kind of African David, a CEO who fought off the Goliath of government monopoly repression to take the mobile phone revolution to Zimbabwe in 1998. Since then he has expanded into Botswana, Nigeria, Kenya, Cameroon, Malawi, Burundi and Lesotho as well as taking on ventures in the UK and New Zealand. In 2002, he was crowned one of Time's Global Influentials—1 of 15 selected in a global poll from a list of more than 100—for his potential as well as achievements as an entrepreneur.

Strive Masiyiwa took the Zimbabwean government to court in a grueling 5-year legal battle (Nsehe, 2013). Masiyiwa had been planning to launch his own mobile telecoms company, but the government, which claimed it held an exclusive monopoly to manage and operate telecommunications in Zimbabwe, was desperately trying to thwart his ambitions. After the High Court ruled in favor of Masiyiwa, the Zimbabwean tycoon went on to build Econet Wireless. Joshua Nkomo scholarship one of Econet Wireless CSR initiatives was inspired by the help he got from Joshua Nkomo who was the country's then vice president who ensured that the 5-year persecution came to an end.

Yet, in the absence of a strong business leader, there is no countervailing force to governmental power. The Marange case illustrates how the Zimbabwean government, in collusion with foreign based companies, enriched itself and ignored the needs of other stakeholders.

Predominance of Small and Medium Businesses

Business size may limit the implementation of CSR in a developing country. The examples provided in this chapter are businesses with a large capitalization. Such larger businesses have the advantage of scale in order to implement CSR. Moreover because of their size and the complexity of their operations, CSR may provide the best means of managing business operations. It may, for example, be sufficient to have rudimentary tracking of production and accounts receivables when a business is small. As the business grows, however, it outstrips the simple system and requires more people and more sophisticated tracking in order to assure that quality and accountability are maintained. CSR standards for accounting principles and for quality control provide a model for a larger business.

In any case, within a developing country, where there may be few larger businesses, a small or medium enterprise does not have relevant models of CSR upon which to base its own CSR plan. CSR policy and strategies are modern management techniques that any company striving to reach world-class status has to incorporate into its operations. A business with a global vision and growth strategy will be strongly influenced to establish comprehensive CSR strategies. Yet businesses in a developing country operate, regardless of their vision and growth strategy, operate in challenging market environment. The asymmetrical distribution of information in the market place and the limited choices of large numbers of people within the country's market reduce the ability of consumers to drive CSR implementation in local businesses. Therefore, possibly the best way to strengthen CSR in developing countries is to start with models that are suitable for SMEs, models that can be implemented on a small scale before the business grows, rather than trying to foist a CSR plan and philosophy onto a large company.

Lack of Indigenous Business Ownership

As is evident in the narrative about the Marange diamond mines and the travails of the local communities near the mines, when a business does not have a close collaborative relationship with the local community it is an invitation to exploitation. As a corollary to the abovementioned point about business size, lack of extensive indigenous business ownership can reduce the implementation of CSR in a developing country. As noted previously in this chapter, the resources of Zimbabwe have been exploited rather than protected by the person and the government of Mr. Mugabe. This exploitation has occurred in collusion, overtly and indirectly, with multinational companies, particularly those from China.

The multinational business's lack of connection to the local community and the inability or unwillingness of a central government to assure such a connection reduces accountability and responsibility—core elements of a CSR program.

Variations by Industry

Businesses from several different industries were examined in the current review. Because of the small sample size it is difficult to make any hypotheses about the connection between industries and CSR in developing countries. The economy of the developing country is less diversified than that of a developed country. Also, the relative percentage of agriculture is higher and the relative percentage of services compared to manufacturing is lower when undeveloped countries are compared with developed countries.

Is there a connection between CSR and diversification in the economy? Is it easier to implement CSR within a service, agricultural or manufacturing industry? Lack of diversification certainly minimizes the opportunities for cross-industry learning regarding CSR concepts. Again, there is a need to consider the special circumstances of a business within a developing country and the resources and supports that the business has or does not have in order to implement and sustain CSR.

Competitive Government Operations

The central government in a developing country has significant stake in businesses through state ownership. In Zimbabwe a parastal Zimbabwe Mining Development Corporation (ZMDC) is a shareholder in eight companies that operate in Marange diamond fields, they own 100% of Marange Resources, 50% of Mbada Diamonds, 50% of Diamond Mining Corporation, 10% of Anjini Investments, 50% of Jinan Private Limited, 60% of Rera Diamonds Pvt Ltd, 50% of Gye Nyame Resources (Pvt) Ltd and 100% of Kusena Diamonds. Management of state owned enterprises are usually political appointments or done on a partisan basis mainly to maintain a patronage of corruption.

Marange diamond mining businesses have significant state ownership through state enterprises and foreign ownership from Chinese businesses as a result of the country's "look East" policy. The government in a developing country may be driven to focus on short-term gains for the country and its leaders rather than social responsibility which is in the country's long term interests. Another reason for the short-term focus is the imbalance in capital across the globe and the current model of reporting in publicly traded businesses. A short-term increase in assets may offer the chance for a reduction in loan rates for a developing country, even if that short-term increase comes at the expense of selling out natural resources or exploiting workers.

6 Conclusion

In sum, CSR goals and strategies can be maintained in acute challenging times and also in chronic challenging times such as those which characterize developing countries. CSR provides ultimate benefits to all stakeholders especially to the communities in which the businesses operate. CSR challenges facing developing countries such as Zimbabwe, given its weak institutional environment and poor governance structure require innovation, integrity and ethical behaviour on the side of business to champion their CSR initiatives. Research which focuses on the unique situations of developing countries can help to broaden and deepen CSR models, and strengthen the sustainability of CSR worldwide.

References

Arli, D. I., & Lasmono, H. K. (2010). Consumers' perception of corporate social responsibility in a developing country. *International Journal of Consumer Studies, 34*(1), 46–51.

Blowfield, M., & Frynas, J. G. (2005). Editorial setting new agendas: Critical perspectives on corporate social responsibility in the developing world. *International Affairs, 81*(3), 499–513.

Carroll, A. B. (1991). The pyramid of corporate social responsibility: Toward the moral management of organizational stakeholders. *Business Horizons, 34*(4), 39–48.

Central Intelligence Agency. (2015). https://www.cia.gov/library/publications/resources/the-world-factbook/fields/2012.html#84 (accessed 06 June 2016).

Chimonyo, G. R., Mungure, S., & Scott, P. D. (2013). *The social, economic and environmental implications of diamond mining in Chiadzwa*. Accessed February 10, 2017, from http://archive.kubatana.net/docs/demgg/crd_implicatns_diamond_ming_chiadzwa_130416.pdf

Dharwadkar, B., George, G., & Brandes, P. (2000). Privatization in emerging economies: An agency theory perspective. *Academy of Management Review, 25*(3), 650–669.

Dobers, P., & Halme, M. (2009). Corporate social responsibility and developing countries. *Corporate Social Responsibility and Environmental Management, 16*(5), 237–249.

Du, S., Bhattacharya, C., & Sen, S. (2010). Maximizing business returns to corporate social responsibility (CSR): The role of CSR communication. *International Journal of Management Reviews, 12*(1), 8–19.

Fox, T. (2004). Corporate social responsibility and development: In quest of an agenda. *Development, 47*(3), 29–36.

Friedman, M. (1970, September 13). The social responsibility of business is to increase its profits. *New York Times Magazine*.

Frynas, J. G. (2005). The false developmental promise of corporate social responsibility: Evidence from multinational oil companies. *International Affairs, 81*(3), 581–598.
Gibson, M. S. (2003). Is corporate governance ineffective in emerging markets? *Journal of Financial and Quantitative Analysis, 38*(01), 231–250.
Halme, M., & Huse, M. (1997). The influence of corporate governance, industry and country factors on environmental reporting. *Scandinavian Journal of Management, 13*(2), 137–157.
Henisz, W. J. (2002). The institutional environment for infrastructure investment. *Industrial and Corporate Change, 11*(2), 355–389.
Hohnen, P., & Potts, J. (2007). *Corporate social responsibility: An implementation guide for business.* Manitoba: International Institute for Sustainable Development = Institut international du dévеloppement durable. http://www.zimstat.co.zw/
Idemudia, U. (2011). Corporate social responsibility and developing countries moving the critical CSR research agenda in Africa forward. *Progress in Development Studies, 11*(1), 1–18.
Jamali, D. (2007). The case for strategic corporate social responsibility in developing countries. *Business and Society Review, 112*(1), 1–27.
Kachembere, J. (2015, April 11). Zim can reduce trade deficit with SA. *Daily News*. Retrieved from https://www.dailynews.co.zw/articles/2015/04/11/zim-can-reduce-trade-deficit-with-sa
KPMG. (2011). KPMG International Survey of corporate responsibility reporting. Retrieved from http://www.csr.ee/wp-content/uploads/2016/03/KPMHG-International-Survey-of-CSR-2011.pdf
Lantos, G. P. (2001). The boundaries of strategic corporate social responsibility. *Journal of Consumer Marketing, 18*(7), 595–632.
Mbiba, L. (2014). Hundreds of companies shut down. *Daily News*. Retrieved from https://www.dailynews.co.zw/articles/2014/04/14/hundreds-of-companies-shut-down
McGee, R. W. (2009). *Corporate governance in developing economies*. New York: Springer.
Mitton, T. (2002). A cross-firm analysis of the impact of corporate governance on the East Asian financial crisis. *Journal of Financial Economics, 64*(2), 215–241.
Mtomba, V. (2014). Zimbabwe records 75 company closures, over 9 000 job losses. *News Day*. Retrieved from https://www.newsday.co.zw/2014/02/21/zimbabwe-records-75-company-closures-9-000-job-losses/
Newell, P. (2005). Citizenship, accountability and community: The limits of the CSR agenda. *International Affairs, 81*(3), 541–557.
Nsehe, M. (2013). Five lessons from Zimbabwe's richest man, Strive Masiyiwa. Retrieved 4/01/2016, from http://www.forbes.com/sites/mfonobongnsehe/2013/02/24/five-lessons-from-zimbabwes-richest-man-strive-masiyiwa/
Pattberg, F. (2009). From http://3blmedia.com/CSR-Blogs/Key-Elements-Corporate-Social-Responsibility-Organization-0, http://www.managers.org.uk/insights/opinion/2009/july/key-elements-of-corporate-social-responsibility-csr-in-an-organisation
Utting, P. (2003). Promoting development through corporate social responsibility-does it work. *Global Future, Third Quarter*, 11–13.
Windsor, D. (2001). The future of corporate social responsibility. *The International Journal of Organizational Analysis, 9*(3), 225–256.
World Bank. (2013). From http://data.worldbank.org/indicator/NY.GDP.PCAP.CD
Young, M. N., Peng, M. W., Ahlstrom, D., Bruton, G. D., & Jiang, Y. (2008). Corporate governance in emerging economies: A review of the principal–principal perspective. *Journal of Management Studies, 45*(1), 196–220.
Zimbabwe Independent. (2013). 700 Harare firms close. *Zimbabwe Independent*. Retrieved from http://www.theindependent.co.zw/2013/10/18/700-harare-firms-close/

Effects of Dwindling Gold Prices on Corporate Social Responsibility (CSR) Performance in Ghana's Mining Sector

Kwesi Amponsah-Tawiah and Francis Xavier Dery Tuokuu

> *Gold has motivated entire societies, torn economies to shreds, determined the fate of kings and emperors, inspired the most beautiful works of art, provoked horrible acts by one people against another, and driven men to endure intense hardship in the hope of finding instant wealth and annihilating uncertainty* (Bernstein, 2000, p. 1).

1 Introduction

The above quotation succinctly captures the important role gold has played and continues to play in the lives of people, kingdoms, empires, and states across the world. It has made and unmade people and kingdoms. The Gold Coast, known today as Ghana, played an important role in the history of the gold trade dating back to several centuries. This is because of the central role this precious yellow mineral had played in the emergence of kingdoms and powerful kings long before the arrival of the Europeans in the fifteenth century.

Ghana's gold attracted the Arabs and Western European merchants long before colonization. The trade in gold led to the emergence of major states and kingdoms in the forest and coastal belts. The Arabs, including the Egyptians and Carthaginians before the seventh and eighth centuries, could not engage in trade with those in the interior because of the threat the Sahara desert posed. However, they later employed the use of camel because of its resistance to harsh conditions and being able to transport heavy loads (Hilson, 2013).

Recent studies have confirmed that before the Portuguese arrived in 1471, the forest region of Ghana was a major center of gold supply for the Trans-Saharan

K. Amponsah-Tawiah (✉)
Department of Organisation and Human Resource, University of Ghana Business School, Accra, Ghana
e-mail: KAmponsah-Tawiah@ug.edu.gh

F.X D. Tuokuu
Antioch New England Graduate School, Keene, NH, USA

S.O. Idowu et al. (eds.), *Corporate Social Responsibility in Times of Crisis*,
CSR, Sustainability, Ethics & Governance, DOI 10.1007/978-3-319-52839-7_12

Caravan Trade. During the Tran-Saharan Trade, lands for mining were acquired by miners through the local chiefs and they (miners) were expected to pay one-third of their gold proceeds as royalties to the chiefs (Dumett, 1998).

Ghana before independence was known as the Gold Coast because of her abundant gold reserves and mineral resources the first Europeans discovered on their arrival on the West African Coast. The Gold Coast was endowed with rich geological gold-mining zones including the "blanket" deposits of the Wassa West District and the conventional reef deposits of Southern Asante (Dumett, 1998). In all, the country is covered by two main geological zones, the Paleoproterozoic rocks of the Birimian and the Tarkwain formations (Hilson, 2013). These formations together have five gold belts namely, "Kibi-Winneba belt, Ashanti belt, Sewfi belt, Bui belt and Bole-Nangodi belt" (Mireku-Gyimah, 2011, p. 4). In addition, the Birimian and Tarkwain formations intersect and move northwards into Burkina Faso and westwards into Senegal and Mauritania ranging in width from 15 and 40 km featuring abundant gold, with Ghana accounting for 70 % of total gold output in West Africa (Hilson, 2013).

Gold mining in Ghana can be categorized into large scale mining and small scale mining. The large scale mining is usually done by big corporations such as Newmont, Anglo-Gold Ashanti, Goldfields, and Golden Star Resource using sophisticated and modern machines whereas the small scale mining is done mostly by indigenous people using simple basic tools and machinery.

Until recently, gold mining contributed about 45% of total foreign exchange to Ghana's economy injecting over $300 million to the country between 1986 and 1988 (Aryee, 2001; Garvin, McGee, Smoyer-Tomic, & Aubynn, 2009). According to Ghana's Minerals Commission, 3% of gross sales by mining companies are paid to government as royalties. From this, 80% goes to consolidated fund, 10% to administrative departments related to mining and the other 10% is distributed to local communities through the Office of the Administration of Stool Lands (ibid, 2001). Apart from the contributions of mining to Ghana's economy through the payment of royalties and taxes by mining companies, mining companies have been engaged in Corporate Social Responsibility (CSR) projects over the years in their catchment areas.

CSR has become increasingly important in mining operations (Garvin, McGee, Smoyer-Tomic, & Aubynn, 2009) particularly in developing countries where governments alone cannot provide the needs of the people. Corporations have therefore found it necessary not only to compliment governments' effort in providing social services for the people, but, as an obligation to give back to society the resources they are exploiting from their lands.

Some argue that corporations adopt CSR to enhance their corporate image and to establish good relationship with the communities in which they operate. Others say CSR is undertaken to increase the profit margins of corporations in the long run (Levis, 2006). Whatever the reasons are, every CSR program should bring a win-win outcome between business and society.

Interestingly, the prices of gold across the globe in recent times have witnessed downward trends mainly as a result of the global economic downturns and the financial debt crisis in the Eurozone (Agei & Asare, 2013). In Ghana, over 3000 employees of various mining companies were relieved off their jobs due to the fall in gold price (www.myjoyonline.com, 2013, accessed: December 10, 2015). The price of gold peaked in 2011 at $1925.15 an ounce and fell to $1180.50 an ounce in 2013 (ibid). The price of gold dipped further in December 2015 to $1071.99 an ounce with analysts estimating a further fall in the coming year (www.gold-eagle.com, 2015, accessed: December 10, 2015). This study sought to examine how the fall in the price of gold is affecting the performance of mining companies in carrying out their CSR obligations in Ghana.

2 Background of CSR in Ghana

Multinational corporations (MNCs) over the last few years have been trumpeting their corporate social responsibility (CSR) 'success stories' particularly in low income countries such as those of Africa. The reasons for communicating such 'success stories' are not far-fetched and they include, the global economic conditions and the negative perceptions consumers have about the activities of MNCs (Wagner, Lutz, & Weitz, 2009).

The origin of CSR has been a subject of debate. Some argue that the concept originated from the west, and as such, it is a capitalist ideology. Yet, others say it came from the east and for this reason, it is a socialist conception (see Idowu, 2012). What is common to both schools of thought is that CSR is meant to help a social course and to establish a good relationship between a host community where a business activity is been run and the business or a corporate entity. The outcome of such relationship should lead to a win-win outcome between the community and the business or corporate entity.

Idowu (2012) argues that it is a misconception to assume that the word 'social' in CSR makes it a socialist conception. According to him, CSR may just be a capitalist ideology with the word social attached to it. This may be true considering that many corporations are beginning to drop the 'social' component of CSR. Lately, it is not uncommon to see on corporate websites or CSR reports with the following; Corporate Responsibility Reports, Sustainability Reports, and Corporate Accountability Reports, all supposedly referring to CSR reports.

The practice of CSR in Africa is based on the culture of giving or sharing, morality and communalism (Dartey-Baah & Amponsah-Tawiah, 2011). Simply, this is known as 'ubuntu' or African humanism (humanity towards others) in South Africa (Visser, 2005). 'Ubuntu' therefore, means, making oneself available to the service and needs of others with the mindset that one belongs to a larger society or community (Dartey-Baah & Amponsah-Tawiah, 2011).

In Ghana, the socialist orientation of the first president, Dr. Kwame Nkrumah, saw the government as the sole provider of all social services and this made the concept alien to the Ghanaian until the early 1990s when some large scale multinational corporations (MNCs) extended some social responsible practices to their areas of operation. Therefore, CSR in Ghana is spearheaded by large scale MNCs particularly in the mining industry. This confirms with a study conducted on Africa by Visser (2005) which reveals that CSR is often practiced by large MNCs with recognizable global brand.

The world Business Council for Sustainable Development (WBCSD) defines CSR as, “the ethical behavior of a company towards society; management acting responsibly in its relationship with other stakeholders who have a legitimate interest in the business, and it is the commitment by business to behave ethically and contribute to economic development while improving the quality of life of the workforce and their families as well as the local community and society at large” (Idowu, 2012, p. 243). For the European Union, CSR is the “concept whereby companies integrate social and environmental concerns in their business operations and in their interaction with their stakeholders on a voluntary basis” (European Commission, 2001, p. 6). Some may call it corporate citizenship, sustainable development or what is referred to as ‘ubuntu’ in the African parlance, but, the goal is to ensure a win-win outcome between business and society while protecting the environment.

3 Changes in CSR Activities over the Years

In a landmark article published in the *New York Time Magazine* on September 30, 1970, titled, “The Social Responsibility of Business Is to Increase Its Profits”, Milton Freidman argued that the duty of every business entity is to make profits for shareholders, and that, CSR initiatives “reveal a suicidal impulse.” He also referred to CSR as “hypocritical window-dressing” arguing that “the business of business is business” and for this reason, businesses should concentrate their efforts in making profits for their shareholders (Friedman, 1970).

There have been massive changes to CSR policy by businesses over the past two decades (Runde, 2011) throughout the world. Previously, companies were only interested in making huge profits for their shareholders like Friedman advocated for in the 70s and not worried about other stakeholders. Even where attempts were made to implement CSR projects to support the needs of communities, those attempts were reactionary to help promote their corporate image rather than initiatives meant to promote the development of their host communities (Ibid, 2011).

Most Fortune 500 companies now have CSR or Sustainability Reports which they produce annually. Over 8000 businesses have signed up onto the UN Global Compact Network program across the globe with the aim of promoting human rights, labor standards and environmental protection (knowledge@wharton, 2012). Ghana is a signatory to this network and a number of Ghanaian companies have signed up to the billing of living as good corporate citizens.

Ghana’s Civil Society and Non-Governmental Organizations have been very active and usually play watchdog roles to ensure that MNCs live up to their corporate responsibilities particularly on the environment. They ensure the government commitment to development is achieved (Okpanachi & Andrews, 2012).

They also play roles such as advocacy, social justice, equity and sustainable development. In the mining sector, civil society organization's such as ISODEC, Friends of the Earth-Ghana, Centre for Responsible Business-Ghana among others ensure that mining revenues accruing to government are utilized efficiently for the benefit of the people. They also ensure that MNCs pay the right royalties to the government and practice sustainability. Ghana's civil society is regarded as an active one because they are able to organize important activities such as public debates and engagements with major stakeholders, seminars and conferences.

The media is seen as the fourth arm of government in Ghana after the executive, legislature and judiciary. It plays a critical role in information dissemination and education as well as holding leaders and corporations accountable. The media also serve as a check on both government and MNCs in ensuring transparency and accountability of contracts and revenue generated from the mining sector and has been very useful in that regard especially over the past two decades.

There is however no law or policy to compel mining companies to perform their CSR obligations. As a result, most of the CSR initiatives undertaken by mining companies are voluntary. Mireku-Gyimah (2011) recommends that mining companies engage with their host communities to identify their development priorities and formulate programs and policies to ensure that they are implemented to the letter. He cites example of a similar program implemented by the Ahafo Mine of Newmont Ghana Gold Limited (NGGL) as a successful CSR program, and recommends other mining companies to learn from.

4 CSR in Ghana's Mining Industry

The mining sector in Ghana was in crisis from the 1950s to the 1980s until the 'Twin Doctors' of Bretton Woods, that is, the International Monetary Fund (IMF) and the International Bank for Reconstruction and Development (IBRD), otherwise known as the World Bank intervened. An Economic Recovery Plan (ERP) was launched in 1983, which supported the private sector in two ways: it enabled the government to remove barriers hindering the private sector; and it established an efficient export growth economy (Akabzaa & Darimani, 2001).

The IMF and World Bank's reforms yielded the desired results and Ghana bounced back and became a vibrant mining country again. Massive support was received from the International Development Association (IDA) who provided US $2.86 billion by the close of 1994 for the rehabilitation of gold mines in Tarkwa, Dunkwa and Prestea. Output increased to about 250% from 1990 to 1994 (Hilson, 2013). Despite these aforementioned benefits at the macro-level, the situation has been different over the years at the micro-level as mining communities in Ghana are some of the most deprived areas in terms of infrastructural and social development.

The environmental impacts of gold mining are enormous and include; water pollution, land degradation, air pollution, and dust pollution. These have devastating effects on the health of communities in which the resource is exploited (Garvin et al., 2009). It is against this backdrop that MNCs sometimes use CSR as a tool or vehicle to compensate for some of the destructions they cause in mining communities (Ibid, 2009).

Most of the CSR initiatives are performed by transnationals and they focus mostly on health, education, environment and social enterprise projects (Oppong, 2016). Anglo-Gold Ashanti for example uses its Malaria Control Program as its CSR contributions to the people across the country (Ibid, 2016).

Similarly, Anaman (2008 cited in Oppong, 2016) reports that Goldfields Ghana makes an annual contribution of $1 million through its foundation it established in 2002 as a way of giving back to society. Mining companies in Ghana spent $26 million in 2002 in carrying out their CSR obligations in their catchment areas (GEITI, 2015 cited in Oppong, 2016). All these initiatives are meant to complement government's effort in providing social amenities for the people and also to enhance the corporate image of the companies.

Generally, there are two types of mining in Ghana, large scale mining and small scale mining.

Large scale mining is usually carried out by transnational corporations such as AngloGold Ashanti, Chirano Gold Mines, Golden Star Resource Limited and Newmont Ghana, who use sophisticated and modern equipment to extract minerals whereas the small scale miners use simple equipment and tools such as pick axes, pans, hoes, shovels, among others. Apart from the fact that large scale mining companies use sophisticated equipment or technology to extract gold, they have stronger financial muscle and would naturally undertake CSR activities than those engaged in small scale artisanal mining.

The small scale mining law of 1989 permits only Ghanaian citizens who want to engage in mining to register with the appropriate institution to do so but most of them do not register due to the bureaucratic processes involved in the registration process (Andrews, 2015). This means small scale mining is divided into legal and illegal activity or what is known in Ghanaian parlance as *galamsey*. Since the activities of these illegal or *galamsey* miners are not regulated, they destroy with impunity large tracts of forest lands for minerals. That explains why (Hilson & Hilson, 2015) describe small-scale artisanal mining (ASM) as informal, hazardous and environmentally degrading. Despite the fact that the laws of Ghana permit only citizens to engage in small scale mining activity (legal), there is an upsurge in illegal Chinese gold miners in recent times. With an estimated number of 50,000 since 2005, some of them have no visas whereas those who have visas are either holding expired documents or do not have the requisite permit to work in Ghana, and yet these illegal workers have destroyed large tracks of lands and rivers in their bid to extract gold (Hilson, Hilson, & Adu-Darko, 2014).

5 Methodology

The qualitative research method was used to understand the CSR concept amongst the various gold producing large scale mining companies operating in Ghana and to further examine the effects of the dwindling fortunes of gold prices in the world market on the CSR activities of these mining companies. Data collection took place in Ghana between September and December 2015.

Qualitative research methodology provides an appropriate framework for this kind of in-depth exploration as it has a wider range of possible empirical materials (Punch, 1998, p. 60) and also offers the opportunity for the understanding of phenomena in all its complexity and within particular situations and environments (Maykut & Morehouse, 1994, p. 12).

5.1 *Study Population*

The study population comprised all large scale mining companies operating in Ghana. Specifically, the researchers targeted large scale mining companies engaged in the mining of gold. Thus five mining companies (Anglogold Ashanti, Goldfields Ghana Limited, Golden Star Resource, Chirano Gold Mine and Newmont Ghana Limited) formed the cohort from which participants were selected to respond to questions relevant to the study objectives.

5.2 *Sample Technique*

As researchers seek greater depth of understanding there is an inevitable compromise in the number of people able to be included in the study (Bell, 2005; Bryman, 2004; Hammersley & Atkinson, 1983). Consequently, this may affect the sample representation and in effect place some limitation on the level of generalization. However, some scholars suggest that participants "in qualitative research are not meant to be representative of a population" and insist that the aim of a qualitative research is to present theoretical inferences (Bryman, 2004, p. 285).

In the current study participants were key personnel at the interface of five large scale mining companies operating in Ghana for the last 5 years. The rationale for targeting key personnel at the interface of identified organisations stems from leadership theories, which describe effective leaders as managing from the boundary (Druskat & Wheeler, 2003). Thus persons holding positions at the interface of employees/community and management/government are satisfactory sources of information regarding the policies and operations of their organisations, which happened to be the essence of this exploratory study. Indeed, these individuals lobby management/government about the health and welfare needs of employees'/

Table 1 Response rate for interview schedules

Name of organization	Portfolio interviewed	Interview duration (min)
AngloGold Ashanti	Sustainability Manager	45
Goldfield Ghana Ltd.	Sustainability Manager	30
Newmont Ghana Ltd.	Sustainability/External Relations Manager	35
Chirano Gold Ltd.	Community Relations superintendent	30
Golden Star Resource	Corporate Affairs Manager	20

community members and at the same time execute upper management/government policies by introducing and implementing them in the daily practice of their organisations.

The purposive sampling technique was used in the selection of the organisations that participated in the study. The researchers' focus was on large scale mining companies which have operated in Ghana for more than 5 years. This was because small scale mining companies operating in Ghana had little commitment to CSR. Indeed, they tend to be the cause of the woes of host communities through the degradation and pollution of the environment as a result of their unsafe practices. The 5 year operational cut off point was to enable the researchers examine the operations of companies which have operated in the country for some time and have had the opportunity of re-examining their initial policies.

5.3 Sample

A total of five participants were identified in terms of job positions; one each from the five mining companies Table 1 provides a graphic picture of the number of participants interviewed their designation and the average duration for each interview session.

5.4 Sample Characteristics

The backgrounds of organizations interviewed ranged from Canada, South Africa to America. Thus all the mining companies interviewed had an international background with the majority of them tracing their roots to Canada and South Africa. Participants had varied years of experience and association with the mining industry with the least experienced having 5 years of experience in the industry and the most experienced having been associated with the industry for 16 years. All the participants were males.

6 Procedure

Key personnel holding positions at the interface of employees/community and management within the identified organizations were targeted in an electronic mail communication detailing the purpose of the study and requesting their participation. All the targeted participants agreed to participate in the study with the condition that they received copies of the final document, which the researchers consented.

Appointments were scheduled with the various participants at their convenience and duly honored.

6.1 Form of Enquiry

6.2 Study Tour/Observations

Although not an observational study, the researchers had the opportunity to observe at first hand mining operations on each of the mine sites visited. This was made possible through the various tours organized for the researchers at the end of the interview sessions at the various mining companies. Although it was not a pre-determined form of enquiry for the study, observations from the various tours provided additional information, which helped to triangulate the responses provided by the participants during the interview sessions. In the company of community relations managers of the participating companies, the researchers were shown round various sections of the mining operations as well as the various CSR projects being undertaken by the companies. The average duration of the tours was 45 min; the longest tour lasted for almost an hour and the shortest 30 min.

The following observations were made:

- Most of the mining companies had trucks with sprinklers attached, which sprinkled water on the dusty working environments.
- As part of their CSR projects, most of the mining companies engaged community members to cultivate large tracts of lands mostly with oil palm and the Jathropha plant.
- Infrastructure such as school buildings, police stations and the construction of roads were also observed as some of the CSR projects the mining companies undertook.

It must be stated that all these activities were undertaken before the fall in the price of gold at the world market.

6.3 Interviews

Naturalistic enquiry employs qualitative research methodology in the environment of the people being studied in order to interact with them and learn from them (Lincoln & Guba, 1985). Interviews with participants were conducted on-site. This situation offered the researchers the opportunity to tour some of the mining sites to observe at first-hand what was being done in terms of the broader concept of CSR. All interviews were tape recorded on the understanding that they would be used for research purposes.

6.4 Data Analysis

Data analysis is the process of bringing order, structure and meaning to the mass of collected data. It is a messy, ambiguous, time-consuming, creative, and fascinating process. It does not proceed in a linear fashion; it is not neat. Qualitative data analysis is a search for general statements about relationships among categories of data (Marshall & Rossman, 2006). Hitchcock and Hughes (1995) on the other hand describe the qualitative data analysis as "the ways in which the researcher moves from a description of what is the case to an explanation of why what is the case." (1995, p. 295). Thus it involves some immersion of the researcher into the data collected to better explain a phenomenon or various phenomena (Patton, 2002). This however, depends on the level of involvement in the data collection and transcription process.

Interview data were analysed using thematic-content analysis (LeCompte & Preissle, 1993; Miles & Huberman, 1994), in which the explanations and thoughts shared by participants during the interview were related to different themes emanating from the study. This involved a verbatim transcription of the audio recorded interview data together with field notes and observations during the study, reading and re-reading the transcripts looking for similarities and differences in order to find themes and to develop categories. This was helpful in reducing the very large amount of data that emerged from the rapport interviews to manageable proportions.

Guided by Spradley's (1979) guidelines to ethnographic studies, three steps were taken in the data analysis process. First, participants' responses to each question were listed in order of interviewee. Second, the diverse responses to each of the questions were assessed using the following questions. Were there dominant themes in their answers? Was there a recurrent opinion expressed by most of the participants? Is there a single statement that vividly illustrates the point of views expressed by participants? Though a tedious process it was achieved by reading through all the data several times to gain familiarity. This was then followed by the clustering and categorisation of the various responses under emerging themes. Themes which are defined by Taylor and Bogdan (1984) as "units derived from patterns such as conversation topics, vocabulary recurring activities, meanings, feelings or folk sayings and proverbs" (1989, p. 131) were thus induced from the interview transcripts.

7 Findings and Discussions

To ascertain the impact of the fall in gold price on the CSR activities of the five large scale mining companies operating in Ghana (Anglogold Ashanti, Chirano Gold Mines, Goldfields, Golden Star Resource Limited and Newmont Ghana), a qualitative approach was used to gather data along three thematic areas; how they define CSR, what constitute their CSR focus and how the fall in gold prices have affected their CSR activities.

7.1 *Definition of CSR*

The definition of CSR by the various companies depicted what they considered as their contribution to society. Most companies lately are moving away from the philanthropic philosophy which champions a one-way stream of giving back to society without any explicit benefit to the company, to a more shared valued way of doing CSR. For instance, the CSR manager for one of the mining companies defined CSR as follows;

> ...for us, we look at CSR as a means of making sure that the environment in which we work is safeguarded and also, that all stakeholders benefit from our business. Thus, CSR activities go beyond philanthropy work. We aim at achieving a sustainable business.

This clearly shows how the company sees CSR as more of a shared value activity, though there are still some lingering bits of philanthropy when it points to safeguarding of the environment and benefit of all stakeholders. Regardless, the argument still holds that even within the shared value principle to doing CSR, there is still an aspect of giving which invariably is philanthropy in itself with benefits to all. A respondent from another mining company however drove the shared value argument more strongly in his definition of CSR;

> ...building relationships with key stakeholders based on trust and creating shared value by contributing to economic development while improving the quality of life of the workforce and the families as well as of the local community and society at large.

This definition is more emphatic on the shared value approach to CSR. A critical look however also unearths a more strategic approach to CSR where there is a mutual benefit from both ends of the company and all its interested parties from the workers through to the nation at large. This encompassing approach to doing CSR will create an enabling and sustainable environment in the long run.

Given the nature of mining activities and their impact on the areas they operate, most companies are still more of the view of paying back for the loss that their activities bring to the environment and the people in their area of operation. For instance, in defining CSR one respondent indicated the following:

> …CSR is defined as giving back to the community, activities that are aimed at improving the lots of the people in the area where our mining takes place.

Other mining companies' definitions however look at maintaining the wholesomeness of the environment which they operate by abiding by mandatory legal requirement.

> …strict compliance to all mandatory legal requirements with regards to the disposal of toxic and hazardous wastes, the company applies good management practices to minimise risk to the people and the environment by ensuring that all the environmental quality standards for toxics are complied with.

Although it might be argued that abiding by legal requirements are more obligatory than a voluntary act that can amount to some form of CSR, the aspect of abiding by the right way of treating the environment still sings a song of benefits to the people. This agrees with Wood's (1991) argument that business should not be separated from society since the two are not mutually exclusive.

7.2 CSR Focus of Companies

The focus of a company's CSR activities dovetails into its CSR philosophy. Most mining companies are more interested in keeping a clean sheet when it comes to environmental degradation issues, thus most of them will focus their attention on minimising the impact of their mining activities on the environment they operate in. Another focus area of CSR with regards to mining companies is towards the people who occupy their area of operations. Mining activities have some consequence for the people within their place of operation, some of which include pollution of water bodies, displacement of settlers and destruction of farm lands. These have given many mining companies some bad reputations especially in our part of the world where there is really no strict monitoring and law enforcement on mining activities. Because of these reasons, some CSR activities of mining companies are tailored towards redeeming the image of the mining companies by putting in place measures to augment the damage caused as a result of their activities. This is evident in the focus of some of the companies interviewed;

The first respondent from our study indicated that

> …we aim at ensuring that the impact of our mining activities on the environment in which we operate will be minimised and that all stakeholders benefit from our operations.

Another respondent said

> the company is not only interested in production but the welfare of the employees and the community. The CSR agenda sets the tone for all in the organisation to follow, thus the environment is important as the people engage in the company and the bottom line.

A third respondent had this to say

> the company seeks to achieve dynamism, innovation, and passionate about who they are and how they work, and do not compromise when it comes to doing the right thing in the communities where they operate. The company prides itself with skilled, responsible operators who strive not only to meet, but to exceed, regulatory requirements for both environmental and health and safety performance, where they operate.

The fourth respondent said

> our CSR agenda is to:
>
> (i) Reduce social risk
> (ii) Avoid organisational cost
> (iii) Demonstrate policy commitment

The fifth respondent said

> our CSR focus is encapsulated in our vision and mission statement.
>
> Vision: We will be recognised and respected for exceptional economic, environmental and social performance.
>
> Mission: We transform mineral resources into shared value for our stakeholders and lead the industry in shareholder returns, safety, social responsibility and environmental stewardship.

From the interaction with the various companies, it is observed that for most of the mining companies the focus of their CSR activities go beyond the immediate environment of operation to include issues such as employees' safety and their economic impact on other stakeholders such as the larger society.

Indeed, juxtaposing the observed CSR projects of the participating mining companies on their CSR focus/agenda, it is clear that all of them are "entrapped" in more sustainable projects, partnering with the local communities and the government, thus giving true meaning to their various portfolios-"Sustainability Manager", "Community Relations Superintendent" among others.

8 The Impact of Fallen Gold Prices on the CSR Activities of Mining Companies

Most mining companies' CSR activities are budgeted on the revenue they make on profits. Virtually all the mining companies that were interviewed asserted to the fact that the fallen price of gold at the world market has affected their CSR activities;

The first respondent from our study indicated that

> ...falling gold prices have affected our CSR activities. Golden Star Development Fund (GSDF) depends on the productivity of the mine which also depends on the price of gold on the world market. The lower the price, the smaller the mining pits become and therefore the lower the gold produce since the mine is a low grade mine.

Another respondent said

> . . .a fall in gold price affect our CSR activities. Our CSR depends on the production of the gold and our profitability. The falling prices in gold has therefore impacted on our CSR activities.

A third respondent had this to say

> . . .the falling prices in gold impacted on our CSR activities. The mine, due to decades' loss of profit and declining gold prices, has been snoozed to a care and maintenance mode where much focus is on budget re-alignments and labour restructuring.

The fourth respondent said

> . . .fallen prices have impacted our ability to execute our plans as desired but it has not affected our CSR philosophy. The contributions that we are able to make largely depends on our profitability and with the fallen gold price this implies that a squeeze in margins and profitability. Also, we make a commitment to employ people from our host communities but a fallen price regime has forced us to restructure the business which involved making people redundant and people in the host communities were affected thereby unfavorably impacting on our ability to honor our commitment in some respect.
>
> But for the fifth respondent from one of the mines who seem not to have their CSR activities so much affected by fall in gold prices.

The fifth respondent said

> . . .fall in gold prices have not really impacted on our CSR activities. The company maintains their CSR activities as stated in their 2014 annual financial report.

CSR is gaining a lot of momentum in the face of social and economic challenges. These factors have played a significant role in the metamorphosis of CSR philosophies across the globe. As Henderson (2001), puts it, the ideas on CSR are still evolving. Although there has been a tremendous urge for strategic CSR in helping a social course and to establish a good relationship between a host community where a business activity is being run and the business or a corporate entity. The outcome of such relationship should lead to a win-win outcome between the community and the business or corporate entity; most dominating mindsets of CSR are however bent towards philanthropy. For instance, the practice of CSR in Africa is based on the culture of giving or sharing, morality and communalism (Dartey-Baah & Amponsah-Tawiah, 2011). The philanthropic approach however would benefit society more than the business or corporate entity. WBCSD however defines CSR as, "the ethical behavior of a company towards society; management acting responsibly in its relationship with other stakeholders who have a legitimate interest in the business, and it is the commitment by business to behave ethically and contribute to economic development while improving the quality of life of the workforce and their families as well as the local community and society at large" (Idowu, 2012,

p. 243). It would have thus been appropriate for businesses to adapt a more sustainable approach in doing CSR in this era of economic hardship which is threatening business survival. It is against this backdrop that this study sought to find out how CSR activities of MNCs in the mining sector are affected by fall in gold price. The plummeting gold prices have had adverse impact on the mining industry. This has seen a number of workers being laid off and cuts in production as well. For instance, in Ghana, over 3000 employees of various mining companies were relieved off their jobs due to the fall in gold price (www.myjoyonline.com, 2013). The impact has not only affected the corporate entities core operations and profit margins but has invariably affected their peripheral activities which include CSR.

The findings, suggest that most mining companies especially the MNCs' CSR activities are supported by a percentage of their profit. This obviously makes CSR activities more conditional than being incorporated into the daily business of the MNC. Although most MNCs try to paint a picture of doing strategic CSR, the manner in which they tackle CSR issues within their various organizations make it more obvious that they are treating CSR activities with the least of attention and also detaching CSR from their core activities. With this approach of handling CSR, it will be inevitable to maintain a vibrant hand in doing CSR in the current economic crisis which has hit major parts of economies not precluding the mining sector.

The philosophy of strategic CSR needs to be looked at from a more incorporative approach if the MNC want to maintain a constant hand in CSR and also to enhance their corporate image and to establish good relationship with the communities in which they do business, this might increase the profit margins of corporations in the long run (Levis, 2006).

There is hence the need to move away from seeing CSR activities as basically making budgetary allocations for providing social amenities to extend it to activities that are part of the day-to-day running of the company. This kind of activities are much more considered as strategic CSR which will save the MNC cost and also can be sustained all year round regardless of the financial standing of the company in question. Hence it will be recommended that MNC reconsider their modus operandi to incorporate a more socially friendly way of doing businesses, which will go a long way to save their faces in the light of the constrains that they face when they employ a philanthropic approach to doing CSR in the current economic turmoil.

It is however worth noting that from the observations made at the research sites, there were certain positive actions that if encouraged and maintained will augment for the impact of the fall in gold prices on CSR activities. For instance, most of the mining companies had trucks with sprinklers attached, which sprinkled water on the dusty working environments. This would reduce air pollution and hence maintain a safer work environment which to a large extent will save the company cost associated with the health of employees. Some MNC also engaged community members to cultivate large tracts of lands mostly with oil palm and the jethropha plant. This also is a step in the right direction to create more economic independency for members of the community.

9 Conclusion

Findings from the current study point to the fact that CSR activities of mining companies have been highly affected by the fall in gold price. Thus, the volume and nature of CSR activities engaged in by mining companies is defined by the level of profitability of their operations. The notion of CSR being a strategic tool to propel corporate value is seen in the current study as a façade than a reality as CSR activities are not necessarily driven by strategy but by the level of profitability. Indeed, this brings to sharp focus the conceptualization of the construct as a philanthropic gesture, which belongs to history other than the current conceptualization as a strategic tool. For if CSR is a strategic tool, strategy precedes operations. However, the indications from the current study points profitability determining the nature and volume of CSR activities. Indeed, the dwindling prices of gold is re-defining the CSR agenda of mining companies, moving the construct from the main stream where it forms an integral part of corporate strategy to the margins, where profit is the main determinant of CSR activities. This approach to the operationalization of CSR activities is not sustainable. Thus to ensure a more sustainable CSR agenda, mining companies should integrate CSR projects, which emanate from the broader consultation of stakeholders, particularly host communities into the broader strategic plan of the organization to generate commitment from the board and not to be at the mercy of the levels of profit realized.

References

Agei, E., & Asare, P. K. (2013). *Fall in gold price to affect Ghana's export earnings*. Accessed December 10, 2015, from www.myjoyonline.com

Akabzaa, T. M., & Darimani, A. (2001). *Impact of mining sector investment in Ghana: A study of the Tarkwa mining region*. Draft Report for SAPRIN. Washington, DC.

Andrews, N. (2015). Digging for survival and/or justice? The drivers of illegal mining activities in Western Ghana. *Africa Today, 62*(2), 2–24.

Aryee, B. N. A. (2001). Ghana's mining sector: Its contribution to the national economy. *Resources Policy, 27*(2), 61–75.

Bell, J. (2005). *Doing your research project: A guide for the first-time researchers in education and social science*. Buckingham: Open University Press.

Bernstein, P. L. (2000). *The power of gold: The history of an obsession*. London: Wiley.

Bryman, A. (2004). *Social research methods*. Oxford: Oxford University Press.

Dartey-Baah, K., & Amponsah-Tawiah, K. (2011). Exploring the limits of Western Corporate Social Responsibility theories in Africa. *International Journal of Business and Social Science, 2*(18), 126–136.

Druskat, V. U., & Wheeler, J. V. (2003). Managing from the boundary: The effective leadership of self-managing work teams. *The Academy of Management Journal, 46*, 435–457.

Dumett, R. E. (1998). *El Dorado in West Africa: The gold-mining frontier, African labor, and colonial capitalism in the gold coast, 1875–1900*. Athens, Oxford: Ohio University Press, J. Currey.

European Commission. (2001). *Promoting a European framework for corporate social responsibility* (p. 6). Accessed December 4, 2015, from bit.ly/p2rUw

Friedman, M. (1970, September 13). The social responsibility of business is to increase its profits. *The New York Times Magazine*, 32.

Garvin, T., McGee, T. K., Smoyer-Tomic, K. E., & Aubynn, E. A. (2009). Community–company relations in gold mining in Ghana. *Journal of Environmental Management, 90*(1), 571–586.

Hammersley, M., & Atkinson, P. (1983). *Ethnography: Principles in practice*. London: Routledge.

Henderson, D. (2001). *Misguided virtue: False notions of corporate social responsibility*. London: Institute of Economic Affairs.

Hilson, G. (2013). Harvesting mineral riches: 1000 years of gold mining in Ghana. *Resources Policy, 28*(1), 13–26.

Hilson, G., & Hilson, A. (2015, April). Entrepreneurship, poverty and sustainability.

Hilson, G., Hilson, A., & Adu-Darko, E. (2014). Chinese participation in Ghana's informal gold mining economy: Drivers, implications and clarifications. *Journal of Rural Studies, 34*, 292–303. doi:10.1016/j.jrurstud.2014.03.001.

Hitchcock, G., & Hughes, D. (1995). *Research and the teacher: A qualitative introduction to school-based research* (2nd ed.). New York: Routledge.

Idowu, S. O. (2012). Corporate social responsibility: A capitalist ideology? *International Journal of Social Entrepreneurship and Innovation, 1*(3), 239–254.

Knowledge@Wharton. (2012). *Why companies can no longer afford to ignore their social responsibilities*. Accessed December 27, 2015, from www.business.time.com

LeCompte, M. D., & Preissle, J. (1993). *Ethnography and qualitative design in educational research* (2nd ed.). San Diego, CA: Academic Press.

Lewis, J. (2006). Adoption of corporate social responsibility codes by multinational companies. *Journal of Asian Economics, 17*, 50–55.

Lincoln, Y. S., & Guba, E. G. (1985). *Naturalistic inquiry*. Beverly Hills, CA: Sage.

Marshall, C., & Rossman, G. B. (2006). *Designing qualitative research* (4th ed.). Thousand Oak, CA: Sage.

Maykut, P., & Morehouse, R. (1994). *Beginning qualitative research-A philosophic and practical guide*. London: Falmer Press.

Miles, M. B., & Huberman, A. M. (1994). *Qualitative data analysis: An expanded sourcebook* (2nd ed.). Thousand Oaks, CA: Sage.

Mireku-Gyimah, D. (2011). *To mine or not to mine? The economic controversy and its resolution*. Inaugural Lecture. Ghana Academy of Arts and Sciences. ISSN: 9964-950-12-3.

Okpanachi, E., & Andrews, N. (2012). Preventing the oil "resource curse" in Ghana: Lessons from Nigeria world futures. *The Journal of Global Education, 68*(6), 430–450.

Oppong, S. (2016). Corporate social responsibility in the Ghanaian context. In S. O. Idowu (ed.), *Key initiatives in corporate social responsibility*. CSR, Sustainability, Ethics & Governance. Cham: Springer International Publishing Switzerland. Accessed December 28, 2015.

Patton, M. Q. (2002). *Qualitative evaluation and research methods* (3rd ed.). Thousand Oaks, CA: Sage.

Punch, K. (1998). *An introduction to social research*. London: Sage.

Runde, D. F. (2011). *The evolution of corporate social responsibility between the Private Sector and the U.S Government*. Accessed December 27, 2015, from www.csis.org

Spradley, J. (1979). *The ethnographic interview*. New York: Holt, Rinehart and Winston.

Taylor, S. J., & Bogdan, R. (1984). *Introduction to qualitative research methods: The search for meanings*. New York: Wiley.

Visser, W. (2005). Revisiting Carroll's CSR Pyramid: An African perspective. In E. R. Pedersen & M. Huniche (Eds.), *Corporate citizenship in a development perspective*. Copenhagen: Copenhagen Business School Press. (Journal of Business Ethics, 56(2), 163–183).

Wagner, T., Lutz, R. J., & Weitz, A. B. (2009). Corporate hypocrisy: Overcoming the threat of inconsistent corporate social responsibility perceptions. *Journal of Marketing, 73*, 77–91.

Wood, D. J. (1991). Corporate social performance revisited. *Academy of Management Review, 16*(4), 691–718.

www.gold-eagle.com. (2015). Accessed December 10, 2015.

www.myjoyonline.com. (2013). Accessed December 10, 2015.

Corporate Social Responsibility Strategies Adopted by Micro Finance Institutions: A Case Study

Daniel Agbeko, Vincent Blok, S.W.F. Omta, and G. Van der Velde

1 Introduction

Most microfinance Institutions (MFI's) differ from traditional financial institutions in that these provide loans to the poor without inhibitive traditional qualification criteria, i.e., collateral assets, cash flow or statement of accounts, client's capacity to repay and loan repayment track records or credit history (Karlan & Valdivia, 2011). In addition, MFI's provide investment programmes to socially empower their loan clients. The combination of these Corporate Social Responsibility (CSR) investments and loans is known as the Grameen model. The Grameen model was adopted from 1976 onwards by most MFI's all over the world and successfully alleviated poverty among the poor.

The social empowerment programme investments caused MFI's to yield only moderate returns on investment. From 2008 onwards MFI's therefore downsized CSR investments and focused on providing loans. MFI's abandoned the Grameen model and shifted their focus from social responsibility to economic sustainability (Wichterich, 2012). The model collapsed because the credit crunch led to a decrease in profits for the MFI's. In order to remain economically sustainable MFI's had to shift their focus to profitable business activities. This quest for survival led to mergers and acquisitions among many MFI's.

One example is the merger between Procredit, a major German micro finance institution in Ghana, and Fidelity Bank (Joy-online, 2014). The purpose of this merger was to include corporate lending in the new company which was believed to yield more profits. The corporate social responsibility of providing small loans to small businesses was abandoned and replaced by strategies aimed at providing big loans to large and profitable businesses.

D. Agbeko • V. Blok (✉) • S.W.F. Omta • G. Van der Velde
University of Wageningen, Wageningen, The Netherlands
e-mail: vincent.blok@wur.nl

S.O. Idowu et al. (eds.), *Corporate Social Responsibility in Times of Crisis*,
CSR, Sustainability, Ethics & Governance, DOI 10.1007/978-3-319-52839-7_13

This shift of focus from small to large firms therefore raises the question how MFI's can successfully invest in social empowerment programmes without jeopardizing MFI financial performance. We explore the hypothesis that CSR investments are feasible for any MFI and should be considered a business opportunity rather than an expense only. We analyse the case of uniCredit, an MFI in Ghana and assess best practices to realise both social responsibility and economic sustainability objectives.

In this paper we discuss the literature on MFI and CSR investments (Sect. 2) and review challenges faced by MFI stakeholders (Sect. 3). We rely on case analysis (Sect. 4) to examine CSR best practices (Sect. 5). In Sect. 6, we draw our conclusions.

2 Literature

This section defines CSR (Sect. 2.1), discusses the literature on alleged advantages of the Grameen model over the traditional MFI concept (Sect. 2.2) and elaborates on the relation between CSR investments and MFI financial performance (Sect. 2.3). The relationship between CSR and information disclosure is discussed in (Sect. 2.4). The entrepreneurial climate in Ghana will be reviewed in Sect. 2.5.

2.1 CSR Definition

The provision of financial services in combination with social responsibility activities are collectively referred to as Corporate Social Responsibility (Tang & Tang, 2012). The financial services include the provision of credit and loan monitoring. Social responsibility activities involve societal cohesion, the provision of health and stock insurance, client and customer training and advisory services.

2.2 The Grameen Model

A key element of the Grameen model is its concept of group lending. Loans are applied for and approved at the level of groups of debtors, rather than individual loan clients. The formation of debtor groups is based on self-selection. Self-selection helps mitigate risks typical for micro finance, i.e. adverse selection and moral hazard.

In the MFI context, adverse selection decreases the value of the MFI investment portfolio, as particularly those entrepreneurs in need for help apply for credit. As group lending involves shared liability for all individual debtors for total group borrowings, self-selection mitigates adverse selection and prevents high-risk loan applicants from joining the debtors group. Group lending based on self-selection and shared liability effectively cuts access to MFI loans for those applicants lacking creditworthiness.

Group lending also serves to mitigate moral hazard. In the MFI context, moral hazard results from the inability of MFI loan officers to closely monitor loan clients and introduces the risk of debtors using the loans for inefficient investments or expenses unrelated to their businesses. Group lending reduces moral hazard and encourages group members to monitor each other and use borrowed funds prudentially. Group lending makes individual group members behave more appropriately and truthfully (Aghion, Fally, & Scarpetta, 2007; Armendáriz de Aghion & Morduch, 2000). Groups serve as social networks for exchanging business related information among entrepreneurs. Group lending therefore reduces the risks of loan defaults.

Consistent with the Grameen concept of group lending, the MFI's CSR investments were also aimed at the level of the group, rather than the individual, to further improve social cohesion.

From 2008 onwards, MFI's abandoned the concept of group lending and shifted their focus to the individual debtor. The shift from group lending to individual debtor strategies introduced a lot of problems. First, as opposed to group lending, individual lending does not provide societal cohesion. The provision of social cohesion was the main rationale underlying the establishment of MFI's. Second, training programmes became very expensive as clients needed to be trained individually. Third, the problem of moral hazard which was solved in the group lending model was re-introduced. This was because the loan officers were unable to monitor all loan clients individually. Fourth, the problem of adverse selection which was also solved in the group lending model, was re-introduced as well. The reason was that with individual lending, the loan officer finds it difficult to select loan clients with good character who are likely to repay their loan. This ultimately leads to an increase in loan delinquencies.

Consequently, along with the shift towards the level of the individual loan client, MFI's CSR investments no longer were aimed at communities of debtors. Rather, social investments were tailored to the individual entrepreneur and no longer contributed to social cohesion (Ledgerwood, 1999).

2.3 CSR Investments and Financial Performance

As opposed to financial services investments, CSR investments do not return any profits to the MFI and set the socially responsible MFI at the competitive disadvantage to those that provide loans only. Moskowitz (1972), Parket and Eilbirt (1975) however argue that CSR investments provide the MFI with public support. This helps raise MFI deposits and improves MFI funding opportunities. Elaine and Barton (1998) and Godquin (2004) observe that CSR investments such as primary health care, basic literacy, marketing information and occupational and business skills training positively affect loan repayment rates. Lensink, Mersland, and Nhung (2011) conclude that those MFI's that provide both CSR and financial investments show better returns than do MFI's that specialize in financial services only. Cheng, Ioannou, and Serafeim (2014) and Ribando and Bonne (2010) emphasize that CSR investments should focus on training programmes. Training makes the relationship between loan clients and MFI's more sustainable. This further mitigates the risks for debtors to default on their debts and improves MFI's long term solvency.

2.4 CSR, Information Disclosure and Financial Performance

Many authors (Botosan, 1997; Chen, Chen, & Wei, 2009; El Ghoul, Guedhami, Kwok, & Mishra, 2011; Hail & Leuz, 2006; Khurana & Raman, 2004; Poddi & Vergalli, 2009) have stated that the provision of accurate information for stakeholders is considered as CSR. Availability of truthful and accurate information reduces information asymmetries between the MFI and its investors. Accurate information increases public support and enhances deposit mobilization. Yet most MFI's are not able to provide reliable and truthful information on the interest they pay on clients' fixed deposits. This is because most MFI's use very high interest rates on fixed deposits as baits to attract investment. However, at maturity date the MFI cannot pay the promised interest to clients. This reduces confidence of investors in MFI's. This is critical because MFI's need deposits as an instrument to raise funds for loans to debtors. Reluctance among depositors leads to low investments and therefore low availability of loanable funds. Scarcity of loanable funds further increases interest rates charged to loan clients. According to Hubbard (1997), accurate and adequate information about MFI's leads to lower capital constraints and this ultimately provides the MFI with better investment opportunities.

2.5 Entrepreneurial Climate in Ghana

Entrepreneurship involves taking calculated risks with one's ideas, with the hope of reaping the benefits. An entrepreneur sees opportunities where others do not (Kwasi Buame, 1996). In Ghana today many consider entrepreneurship as a way out of unemployment and poverty (Banson, 1997). Most of the firms in Ghana are locally oriented micro and small-sized businesses. Notwithstanding their small scale, these firms help alleviate poverty. Relative to foreign-owned multinational firms, local firms cannot repatriate their profits and are therefore most valuable to the Ghanaian economy. Local firms' profits are likely to be re-invested in the business or spent locally (Ledgerwood, 1999).

However, small firms cannot easily secure start-up capital from any financial institution or venture capitalist in Ghana. Sjauw-Koen-Faa, Blok, and Omta (2016) indicate that micro farmers are required to hand over price guarantees for their merchandise whereas large scale farmers have easy access to credit. MFI's are more willing to invest in large and already successful ventures than those to be started from scratch (Ledgerwood, 1999). Most MFI'S require loan applicants to show prospective financial statements that prove their business plans are viable. Most of the poorly educated micro clients cannot design these statements and therefore have to rely on their savings and money from family and friends. Access to finance is the most critical issue that hinders the entrepreneurial development in Ghana. In a developing country like Ghana, political instability, paucity of the macro-infrastructures and failing government policies have contributed to the non-development of effective indigenous entrepreneurship. In societies with strong family orientation and social networks, mechanisms for collective survival are used as a countervailing mechanism for developing entrepreneurship (Kwasi Buame, 1996).

3 Challenges Faced by MFI Stakeholders

According to the stakeholder theory, stakeholders are those who are affected or can be affected by the business of the organisation. The six main stakeholders include the employees, customers, investors/shareholders, suppliers, the community and the environment (Papasolomou-Doukakis, Krambia-Kapardis, & Katsioloudes, 2005). The concept of stakeholders is central to CSR, in that it forms the link between the aims and ambitions of the organisations and the expectations of the public (Maon, Lindgreen, & Swaen, 2009). The stakeholder theory also emphasises that the organisations' survival and success hinges on the organisations' ability to generate sufficient wealth, value and satisfaction for its primary stakeholders (Clarkson, 1995).

According to several authors (Galbreath, 2006; Jamali, 2008; Papasolomou-Doukakis et al., 2005; Snider, Hill, & Martin, 2003), the most important stakeholders are the employees, the customers and the owners of the organisations. Therefore in this section we elaborate on challenges faced by MFI management, MFI debtors and MFI depositors. Consistent with stakeholder theory, we take into account the responsibility of MFI management to enable the stakeholders to create value (Freeman, Wicks, & Parmar, 2004). This pushes managers of MFI's to articulate how exactly they intent to do business with their clients and what kind of relationship with their clients they aim for.

MFI management teams most often lack training. The lack of banking knowledge jeopardizes the accuracy of management decisions. Many MFI's violate basic financing principles and fund long term investments with short term deposits.

This easily causes liquidity problems should depositors withdraw their funds (Armendariz & Morduch, 2010; Ledgerwood, 1999).

Loan non-repayments are among the major challenges faced by MFI management. The issue of loan non-repayments is aggravated by high interest rates the MFI's charge their loan clients. MFI's are generally compelled to charge high interest as most MFI's promise high interest on deposits to attract depositors for funding (Armendariz & Morduch, 2010; Ledgerwood, 1999).

MFI management cannot easily retrieve information on loan applicants and pass judgment on creditworthiness. Only recently the Bank of Ghana instituted the register of collaterals. This register details collateral of loan clients and helps MFI's appraise their loan clients. The register does not, however, cancel the need for skilled loan officers to distinguish the viable from the risky loan applicants (Monitor, 2013).

The majority of MFI debtors; lack any business skills and are not able to manage their businesses efficiently. Debtors are therefore likely to fail in their businesses and default on their MFI loans repayments. Among others, several authors (Edgcomb, 2002; Ekpe, Mat, & Razak, 2011; Karlan & Valdivia, 2011; Karnani, 2007; Ployhart & Moliterno, 2011) argue that entrepreneurial training is key for entrepreneurs to perform their businesses better. The literature remains, however, inconclusive as to whether training programmes are supportive for successful entrepreneurship. According to Yunus (1999, 2007), entrepreneurial skills are innate and therefore cannot be trained.

MFI depositors cannot easily distinguish unregistered MFI's from those registered with the Bank of Ghana. In the event of collapse, the Bank of Ghana steps in only for those depositors at registered MFI's. As Ghana MFI failure rates mount up to 20% (Starrfmonline, 2015), many MFI depositors risk losing their savings.

4 Methods

In our discussion on CSR practices we rely on case analysis and interviews. Case analysis allows for close examination of CSR, its embeddedness and benefits for MFI operations. For our case analysis we single out the Ashaiman branch of uniCredit Ghana Limited, for three reasons.

First, immediately after the 2008 credit crunch, both savings deposits and loan repayments dropped significantly. All MFI's were forced to mobilize deposits and to recover their loans that were overdue. This was the only way MFI's would be able to meet depositors' withdrawals, provide new loans to new clients and stay in business. During this period, uniCredit developed a deposit mobilization procedure and loan recovery targets for each of its 12 branches. The branches were also given the mandates to develop new ideas and innovations to meet their deposit and loan recovery targets. The Ashaiman branch chose to develop CSR strategies. These CSR strategies enabled the branch to outperform the other branches in terms of deposit mobilization and loan recovery rates. This branch won the best branch award for 2009 and 2010. The CSR strategy developed by this single branch was later adopted by the entire institution in 2010. This has been the institution's CSR deposit mobilization and loan recovery strategy till date.

Second, uniCredit Ashaiman branch stands out on social empowerment investments and considers CSR key to its success.

Third, the first author has been working for uniCredit as branch manager. For this reason, we have access to data records that allow us to critically examine CSR and its impact on deposit mobilization and loan repayment rates.

Eisenhardt (1989), Gibbs (2002) and Graebner, Martin, and Roundy (2012) suggest that in-depth interviews are useful to gain in-depth understanding and build theory. In this research we interview each of the eleven non-Ashaiman branch managers in order determine the strategies each of the eleven branches adopted in 2009 and 2010. We compare these strategies to those used by the Ashaiman branch during the period in which the latter outperformed the others.

Using T24 Banking application software, we analysed the performance of all 12 branches of uniCredit with regards to deposit mobilization, loan recovery and profitability.

5 Best Practices: CSR in uniCredit Ghana

We briefly introduce the case of uniCredit Ghana Limited (Sect. 5.1), and discuss its best CSR practices (Sect. 5.2).

5.1 uniCredit Ghana

The study site for this research was the Ashiaman branch of uniCredit Ghana Limited. Ashaiman is a densely populated city in Accra, Ghana with a population of approximately 150,312 citizens (City-Population.de, 2010) and a land area of 116.2 km^2 with a yearly growth rate of 4.6%, which is higher than the 2.6% national growth rate. The Census Report estimated that 75,183 of the population are males while 75,129 are females with the majority of the population within the ages of 15 and 34 years. The Ashaiman community has poor sanitation. Apart from warding off possible investors, poor sanitation may trigger the outbreak of contagious diseases. Ashaiman has only one public health centre, 14 private clinics and one private maternity home. According to the Tema Human Development Report (2004), apart from the insufficient number of health facilities, most of the existing ones did not have adequately qualified personnel and operated under poor hygienic conditions. As a result ill health prevents micro clients from achieving their full potential in business. The Ashaiman population predominantly consists of micro clients involved in retail business. The loan request of the clients in this community is most often less than 4000 €, which is an indication of the small size of their businesses. This community witnessed the establishment of over 60 MFI's. Less than 30 have survived the 2008 credit crunch (Monitor, 2013), among these was uniCredit Ghana Limited.

The primary focus of uniCredit is to provide financial services tailored to the needs of micro clients, small and medium size enterprises. The mission of uniCredit is to develop easily accessible financial products. As of 2011, uniCredit had 12 branches in Ghana, with an institutional deposit of USD40 million secured by 70,000 depositors and a loans portfolio worth of USD24 million distributed among 3500 debtors (Mix-Market, 2015).

5.2 uniCredit Ghana's Best Practices

Lack of staff and management training and knowledge of microfinance best practices is one of the major issues responsible for the collapse of most MFI's in Ghana (Monitor, 2013; Starrfmonline, 2015). uniCredit stands out from its competitors in that it systematically invests in staff training. uniCredit is among the first to institutionalize monthly training schedules for loan officers and management teams. uniCredit is in the process of establishing a training school for both customers and staff. This school will serve to develop the cognitive ability of both clients and customers to be able to perform better. MFI staff will be trained to maximize profits, clients will be trained to maximize sales.

The lack of knowledge on business skills is a major reason for the poor business performance of clients. To address this, uniCredit invests in CSR programmes. uniCredit organizes training programmes for its debtors on a quarterly basis. The training is aimed at developing clients' entrepreneurial and business operations skills. uniCredit organizes joint outreach seminars in churches. Free and public educative talks about MFI services not only educate the attendants but also provide uniCredit exposure to the public.

Another problem facing MFI's in Ghana is non-payment of loans. This is due to the high inflation and prime rates, the rate at which the central Bank of Ghana lends to other Banks and MFI's (Bank-of-Ghana, 2014). The inflation and the prime rate, 15% and 19%, respectively, create harsh economic conditions for the loan clients (Joakim & Wismer, 2015). In addressing this, uniCredit relies on CSR strategies to mobilize cheap savings deposits from depositors and translates this into lower interest rates on loans for small businesses. With the lower interest rates clients are able to work more effectively and are able to pay their loans more adequately.

uniCredit organizes quarterly customer appreciation weeks during which staff interact with customers and solve their problems. These customer appreciation weeks include a forum for clients to discuss their challenges with uniCredit management for redress.

All MFI's consider the 5C's as a requirement for credit, i.e., credit history, cash flow or statement of account for client, character reference or a guarantor, capacity of the client to repay the loan and, most important, collateral security to secure the loans. Since most of uniCredit's clients have no collateral, uniCredit provides insurances for the stock of the loan free of charge. At the same time, uniCredit closely monitors its clients. Loan officers regularly visit their debtors' businesses to early intervene and advise clients if necessary. These actions not only improve business performance but also loan repayment probabilities.

Because of health problems many clients are unable to work and pay their loans. uniCredit addresses this problem by inviting officials from the hospitals to register clients who do not have national health insurance cards. This prevents clients from having to spend the income accrued from their business on medical treatments and to default on their loans.

Most MFI's provide inaccurate interest quotes on savings and fixed deposits which has led to a decline in confidence in all MFI's. uniCredit most often pays interest on fixed deposits upfront at date of purchasing the fixed deposit. This is different from other MFI's that pay interest on fixed deposits on maturity date. Even though this strategy reduces the profits for uniCredit, the quantum of deposits that are received as a result of this CSR gesture far and above compensates for the loss.

uniCredit's CSR activities therefore help raise uniCredit's clientele base and deposits significantly (Mix-Market, 2015). Because of the CSR activities, customers have demanded for more branches to be opened within their local communities. Amidst numerous MFI closures, social empowerment investments help uniCredit prosper.

6 Conclusion and Recommendation

We examined our hypothesis that MFI's, in order to improve on their performance, should invest in social empowerment or Corporate Social Responsibility (CSR) programmes. The 2008 credit crunch led many MFI's to abandon the Grameen model and, therefore, CSR investments. The literature review and the case study of uniCredit Ghana shows, however that, those CSR strategies may improve the MFI's competitiveness and financial performance.

Based on the case study of uniCredit, we conclude that CSR investments contribute to public support for the MFI. This helps MFI's raise deposits and improves funding opportunities. CSR investments such as primary health care, basic literacy, marketing and occupational and business skills training positively affect debtors' entrepreneurial performance and therefore loan repayment rates. MFI's that adopt such a CSR strategy are expected to show better returns than do MFI's that specialize in providing financial services only. uniCredit Ghana MFI successfully invests in CSR and considers it key to both its social responsibility and financial objectives. We therefore argue that MFI's should consider CSR investments a business opportunity rather than an expense.

7 Limitations

This research is based on observations and interviews with only eleven managers for a single case study. This limits the generalizability of the results. Further research is needed to systematically compare several MFI's on the implementation of CSR strategies and to assess the significance of CSR investments with regards to public support for MFI's, opportunities for MFI's to raise funds and MFI loan repayment rates.

References

Aghion, P., Fally, T., & Scarpetta, S. (2007). Credit constraints as a barrier to the entry and post-entry growth of firms. *Economic Policy, 22*(52), 732–779.

Armendariz, B., & Morduch, J. (2010). *The economics of microfinance*. Cambridge, MA: MIT Press.

Armendáriz de Aghion, B., & Morduch, J. (2000). Microfinance beyond group lending. *Economics of Transition, 8*(2), 401–420.

Bank-of-Ghana. (2014). Bank of Ghana website. *Generic*.

Banson, M. Y. (1997). *The development of entrepreneurship in Ghana, opportunities for the tertiary graduate:(A case study of UST graduates)*.

Botosan, C. A. (1997). Disclosure level and the cost of equity capital. *Accounting Review*, 323–349.

Chen, K. C., Chen, Z., & Wei, K. J. (2009). Legal protection of investors, corporate governance, and the cost of equity capital. *Journal of Corporate Finance, 15*(3), 273–289.

Cheng, B., Ioannou, I., & Serafeim, G. (2014). Corporate social responsibility and access to finance. *Strategic Management Journal, 35*(1), 1–23.

City-Population.de. (2010). Population census in Ashaiman. *Generic*.

Clarkson, M. E. (1995). A stakeholder framework for analyzing and evaluating corporate social performance. *Academy of Management Review, 20*(1), 92–117.

Edgcomb, E. L. (2002). What makes for effective microenterprise training? *Journal of Microfinance/ESR Review, 4*(1), 99–114.

Eisenhardt, K. M. (1989). Building theories from case study research. *Academy of Management Review, 14*(4), 532–550.

Ekpe, I., Mat, N., & Razak, R. C. (2011). Attributes, environment factors and women entrepreneurial activity: A literature review. *Asian Social Science, 7*(9), p124.

El Ghoul, S., Guedhami, O., Kwok, C. C., & Mishra, D. R. (2011). Does corporate social responsibility affect the cost of capital? *Journal of Banking and Finance, 35*(9), 2388–2406.

Elaine, E., & Barton, L. (1998). *Social intermediation and microfinance programmes: A literature review*. Washington, DC: Micro-Enterprises Best Practices.

Freeman, R. E., Wicks, A. C., & Parmar, B. (2004). Stakeholder theory and "the corporate objective revisited". *Organization Science, 15*(3), 364–369.

Galbreath, J. (2006). Does primary stakeholder management positively affect the bottom line? Some evidence from Australia. *Management Decision, 44*(8), 1106–1121.

Gibbs, G. (2002). *Qualitative data analysis: Explorations with NVivo (Understanding social research)*. Buckingham: Open University Press.

Godquin, M. (2004). Microfinance repayment performance in Bangladesh: How to improve the allocation of loans by MFIs. *World Development, 32*(11), 1909–1926.

Graebner, M. E., Martin, J. A., & Roundy, P. T. (2012). Qualitative data: Cooking without a recipe. *Strategic Organization, 10*(3), 276–284.

Hail, L., & Leuz, C. (2006). International differences in the cost of equity capital: Do legal institutions and securities regulation matter? *Journal of Accounting Research, 44*(3), 485–531.

Hubbard, R. G. (1997). *Capital-market imperfections and investment.* National Bureau of Economic Research.

Jamali, D. (2008). A stakeholder approach to corporate social responsibility: A fresh perspective into theory and practice. *Journal of Business Ethics, 82*(1), 213–231.

Joakim, E. P., & Wismer, S. K. (2015). Livelihood recovery after disaster. *Development in Practice, 25*(3), 401–418.

Joy-online. (2014). Fidelity Bank aquired Procredit. *Generic.*

Karlan, D., & Valdivia, M. (2011). Teaching entrepreneurship: Impact of business training on microfinance clients and institutions. *Review of Economics and Statistics, 93*(2), 510–527. doi:10.1162/Rest_a_00074.

Karnani, A. G. (2007). *Employment, not microcredit, is the solution* (Ross School of Business Paper 1065).

Khurana, I. K., & Raman, K. (2004). Litigation risk and the financial reporting credibility of Big 4 versus non-Big 4 audits: Evidence from Anglo-American countries. *The Accounting Review, 79*(2), 473–495.

Kwasi Buame, S. (1996). *Entrepreneurship: A contextual perspective. Discourses and praxis of entrepreneurial activities within the institutional context of Ghana* (Vol. 28). Lund: Lund University Press.

Ledgerwood, J. (1999). *Microfinance handbook: An institutional and financial perspective.* Washington, DC: World Bank.

Lensink, R., Mersland, R., & Nhung, V. T. H. (2011). *Should microfinance institutions specialize in financial services.* Paper presented at the Second international research conference on microfinance. Groningen: The Netherlands.

Maon, F., Lindgreen, A., & Swaen, V. (2009). Designing and implementing corporate social responsibility: An integrative framework grounded in theory and practice. *Journal of Business Ethics, 87*(1), 71–89.

Mix-Market. (2015). *Transforming financial inclusion data into insight.* Accessed December 5, 2015, from https://www.themix.org/mixmarket/countries-regions/ghana

Monitor, M. C. (2013). *Thirty Microfinance Institutions (MFIs) close in Ghana.* Accessed October 10, 2015, from http://www.microcapital.org/microcapital-brief-thirty-microfinanceinstitutions-mfis-close-in-ghana/

Moskowitz, M. (1972). Choosing socially responsible stocks. *Business and Society Review, 1*(1), 71–75.

Papasolomou-Doukakis, I., Krambia-Kapardis, M., & Katsioloudes, M. (2005). Corporate social responsibility: The way forward? Maybe not! A preliminary study in Cyprus. *European Business Review, 17*(3), 263–279.

Parket, I. R., & Eilbirt, H. (1975). The practice of business social responsibility: The underlying factors. *Business Horizons, 18*(4), 5–10.

Ployhart, R. E., & Moliterno, T. P. (2011). Emergence of the human capital resource: A multilevel model. *Academy of Management Review, 36*(1), 127–150.

Poddi, L., & Vergalli, S. (2009). Does corporate social responsibility affect the performance of firms. *Fondazione Enri Enrico Mattei, 52*(09), 16–21.

Ribando, J. M., & Bonne, G. (2010). *A new quality factor: Finding alpha with ASSET4 ESG data.* Starmine Research Note, Thomson Reuters.

Sjauw-Koen-Faa, A. R., Blok, V., & Omta, S. O. (2016). Critical success factors for smallholder inclusion in high value-adding supply chains by food & agribusiness multinational enterprises. *International Food and Agribusiness Management Review, 19*(1), 83.

Snider, J., Hill, R. P., & Martin, D. (2003). Corporate social responsibility in the 21st century: A view from the world's most successful firms. *Journal of Business Ethics, 48*(2), 175–187.

Starrfmonline. (2015). *BoG has revoked the licenses of 70 MFI's in Ghana* (2015 ed.). Accra: Ghana.

Tang, Z., & Tang, J. (2012). Stakeholder–firm power difference, stakeholders' CSR orientation, and SMEs' environmental performance in China. *Journal of Business Venturing, 27*(4), 436–455.

Wichterich, C. (2012). The other financial crisis: Growth and crash of the microfinance sector in India. *Development, 55*(3), 406–412.

Yunus, M. (1999). The Grameen Bank. *Scientific American, 281*(5), 114–119.

Yunus, M. (2007). *Creating a world without poverty: Social business and the future of capitalism.* New York: PublicAffairs.

Part IV
Corporate Social Responsibility in Times of Crisis: Summary

Corporate Social Responsibility in Times of Crisis: A Summary

Adriana Schiopoiu Burlea, Stephen Vertigans, and Samuel O. Idowu

Throughout this book, Corporate Social Responsibility (CSR) has been looked at and approached from different angles and in different forms. The European aspects of CSR are presented in the framework of the dimensions used in the participating countries: Switzerland, Italy, The Netherlands, Romania and Cyprus. **In Switzerland**, L'EPOQuE is the model of Corporate Social Responsibility (CSR) used by Small and Medium-sized Enterprises (SMEs). During the financial crisis, this model was successfully tested in the Swiss business environment and the results proved that even during financial scarcity the ethical features were important ingredients for the success of small businesses. The Swiss model is culturally embedded and driven by intrinsic and implicit motives, while the CSR agendas of Swiss SMEs stem from corporate cultures that nurture a "raison d'être" beyond profit-maximisation. It appears that the social capital and institutional pressures are powerful drivers for CSR in the case of Swiss firms'. Thereby, the Swiss model is culturally embedded and it is not exposed to often expensive CSR activities and not influenced by difficult economic situations. The other conclusion is related to the "Valais case" that proved that CSR is linked to the geographical presence and the size of companies. Companies with local imprints should develop local CSR practices that are different from other SMEs with regional or national reach.

In **Italy**, the bases for integral development are in the diffusion of authentic anthropological values, characterize companies which "adhere" to the Encyclical Letter principles and render them multicultural companies engaged in the local and

A.S. Burlea (✉)
University of Craiova, Craiova, Romania
e-mail: adriana.burlea@gmail.com

S. Vertigans
Robert Gordon University, Aberdeen, UK

S.O. Idowu
Guildhall School of Business and Law, London Metropolitan University, London, UK

S.O. Idowu et al. (eds.), *Corporate Social Responsibility in Times of Crisis*,
CSR, Sustainability, Ethics & Governance, DOI 10.1007/978-3-319-52839-7_14

global community. The strategic and operational behaviors of the Italian companies are the results of durable CSR orientation and a genuine responsible decision-making process. During the crisis period, the Italian entrepreneurs/managers voluntarily persist in investing in CSR—can promote cultural reorientation, helping others to unlearn the bad habits inspired by the 'turbo-capitalism' and valorize humanity, relationships, and the local/global human community.

In Belgium, Sustainability-Oriented Business Models (SOBMs), be it for generations, integrate societal goals into strategic objectives, organizational architecture, and decision-making processes. These models are based on challenges of CSR that are ranging from poverty-reduction to biodiversity and climate change. The first generation of SOBMs is characterised by the effort of the firms to reduce their negative impact from sustainability through redesigning products and services and applying circular economy principles. The second generation of SOBMs is based on a double impact: an environmental and a social impact, because in order to survive in a period of crisis, firms have to try and increase their market share and profits and taking two competing paradigms into account: creatively trying to synthesize the social and economic purpose of the business. The third generation of SOBMs is designed to generate environmental and social value instead of just financial value. In fact, they aim at creating economic value *through* creating societal values even if some firms consider profitability as a means to an end rather than an end in itself. The fourth generation of SOBMs push the firms to play the role of catalysts for a relatively smooth transition towards a more sustainable system; they need the support of consumers. As a conclusion, both sustainability and native intelligence play an important role on the value creation processes that are characteristic of fourth generation SOBMs.

In Romania with two chapters. The first of the two chapters by Sitnikov and Bocean notes the review of costs and restructuring of CSR activities when times are hard. The consequential effects of such actions are that there is a noticeable reduction in corporate social responsibility programs, both in terms of size and complexity. Interestingly, it was noticed that during the same period some companies did choose to strengthen their social responsibility programs in order to improve their corporate reputations and to gain the trust of the local community where they operate. Whist on other hand, some companies substantially reduced their social responsibility budgets and activities. These two Romanian scholars addressed the perspectives Romanian companies took during the economic crisis both in terms of microeconomic and macroeconomic analyses. They used both qualitative and quantitative analyses of OMV Petrom 2010–2014 social responsibility programs were performed, as OMV Petrom is the largest company in Romanian economy, whose turnover represents around 5% of national GDP. Macroeconomic analysis focused on the solutions that Romanian Government developed as to overcome the threats and challenges faced by financial responsibility due to economic and financial crisis. The financial crisis has a negative impact on corporate social responsibility programs. For this reason, the Romanian Government has developed, since 2011—after crisis period—a National Strategy on Social Responsibility to raise awareness of the importance and benefits of applying social responsibility and increasing involvement of public sector, Romanian companies, multinational corporations, and civil society in implementing social responsibility. This strategy implements ***the European Directives related to CSR****; establish* ***the obstacles*** *in front of a higher CSR activity in the public sector; offers* ***solutions to strengthen the partnership*** *between the public and private partners.*

In Cyprus, the four dimensions of internal CSR practice (e.g. health and safety in the workplace, organisational justice and fairness, employee training and development, and work-life balance) revealed that SMEs use human resource management tools as an integral part of their internal CSR. In Cypriot SMEs the employee training and development is a priority for the managers, and health and safety in the workplace, organizational justice and fairness, and work-life balance are neglected. As a recommendation, the managers have to ensure that all the dimensions of internal CSR are effectively implemented.

In the United States of America (US), the government is an important actor in developing corporate social responsibility policies, environmentally-sustainable behaviours and particular corporate citizenship procedures and expectations. This regulatory position is imposed by the individual approach of CSR by US entities, including bureaus, agencies and non-governmental organizations (NGOs). In this case, the US government as the main driver of CSR policy promotes responsible business practices and provides support and guidance on areas of corporate conduct and sustainable behaviours. Therefore, the U.S. secretary of state's agenda is to ensure effective coordination and partnerships with individual bureaus and offices in order to harness global economic tools that advance U.S. foreign policy goals on responsible initiatives. The weaknesses of U.S. leaders regarding CSR policies consist on the fact that many businesses are characterised by their unsustainable practices such as linear flows of materials, the heavy dependence on fossil fuels, a disregard for renewable resources, and resource use, and the reforms on social and environmental responsibility have not implemented adequate. Therefore, the US CSR policies are based on sound theoretical arguments and tackle the economic, legal, ethical, and discretionary dimensions. As a conclusion, organisations have to be urged to find alternative ways for sustainable **energy** generation, **energy** and water **conservation**, environmental protection and greener transportation systems. Corporate citizenship policies should be promoting socially-responsible investing (SRI), responsible supply chain management and the responsible procurement of sustainable products.

In Zimbabwe, the weak institutional environment and poor governance structure require innovation, integrity and ethical behaviour on the side of businesses to champion their CSR initiatives. The key factors that support or weaken the implementation of CSR in Zimbabwean companies during the challenging times are related to a balanced approach that encompasses the economic, social and legal pillars of business operations. Corruption was one of the sources of the economic mis-management of the country's resources and, as a result, Zimbabwe has one of the most uncompetitive business environments and is ranked among the worst in terms of ease of doing business. The economy is shrinking with many company closures citing liquidity crises and a lot of workers being laid off. Also, the multinational business's lack of connection to the local community and the inability or unwillingness of a central government to assure such a connection reduces accountability and responsibility—core elements of a CSR program—, while many Zimbabwean companies are involved in implementing CSR practices in business and are committed to treating all stakeholders fairly and with dignity. They also recognise the imperatives for corporate social responsibility in the communities of their operations.

In Ghana, Corporate Social Responsibility (CSR) is still a concept that is underdeveloped and it is even so that many businesses seek to propound their involvement in CSR, even though such involvement very often amounts to something purposely altruistic and quite negligible and, at times, clearly done on ad hoc basis. Taking into account that in Ghana the level of poverty is high, local communities have assigned to business a principal role in poverty alleviation and protection of the environment. For example, the mining companies operating in Ghana are involved in implementation of CSR policies during the hard times even after the falling of the gold price, because CSR activities are not necessarily driven by strategy, but by the level of profitability; and uniCredit Limited Ghana MFI used CSR strategies and outperformed the others, it considers CSR investment as a business opportunity rather than an expense. The complicity between companies and government is reflected in political corruption, environmental destruction, labour exploitation and social disruption.

The challenges of engaging in CSR during hard times are different from country to country, depending on levels of economic development, the involvement of the government in the promoting and implementing of CSR policies, and the degrees of corruption. Irrespective of the country, the main characteristic of CSR in hard time is related to a positive perception of the need of integration of CSR policies on national strategies on the purpose to develop a healthy institutional environment.

Index

S.O. Idowu et al. (eds.), *Corporate Social Responsibility in Times of Crisis*, CSR, Sustainability, Ethics & Governance, DOI 10.1007/978-3-319-52839-7

D

E

The manufacturer's authorised representative in the EU is Springer Nature Customer Service Centre GmbH, Europaplatz 3, 69115 Heidelberg, Germany. If you have any concerns regarding our products, please contact ProductSafety@springernature.com

Printed and bound by CPI Group (UK) Ltd, Croydon, CR0 4YY
30/09/2026
02225502-0020